Plant Metabolic Networks

Jörg Schwender
Editor

Plant Metabolic Networks

Foreword by Jacqueline V. Shanks

 Springer

Editor
Jörg Schwender
Department of Biology
Brookhaven National Laboratory
50 Bell Avenue
Upton NY 11973
USA
schwend@bnl.gov

ISBN 978-0-387-78744-2   e-ISBN 978-0-387-78745-9
DOI 10.1007/978-0-387-78745-9
Springer Dordrecht Heidelberg London New York

Library of Congress Control Number: 2009920489

© Springer Science+Business Media, LLC 2009
All rights reserved. This work may not be translated or copied in whole or in part without the written permission of the publisher (Springer Science+Business Media, LLC, 233 Spring Street, New York, NY 10013, USA), except for brief excerpts in connection with reviews or scholarly analysis. Use in connection with any form of information storage and retrieval, electronic adaptation, computer software, or by similar or dissimilar methodology now known or hereafter developed is forbidden.
The use in this publication of trade names, trademarks, service marks, and similar terms, even if they are not identified as such, is not to be taken as an expression of opinion as to whether or not they are subject to proprietary rights.

Printed on acid-free paper

Springer is part of Springer Science+Business Media (www.springer.com)

Foreword

I am delighted to see this pioneering edited book on plant metabolic networks, by Schwender. Now, and increasingly in the future, plant biotechnology will be key element in enabling a renewable and sustainable world. Plants and their products impact several economic sectors of society – food, feed, materials, environmental aesthetics, pharmaceuticals, fuels, and feedstocks for the chemical industry. With the advances in plant genomics, the plant researcher has a wealth of technologies available for enhanced plant productivity, but is left to ponder what the best approaches to link genotype to phenotype are. The linchpin in this link is the plant metabolic networks that govern plant product synthesis. The time is right for training many young scientists and engineers to develop and maximize the knowledge base of plant metabolic networks for the rational design of improved plant varieties.

Unfortunately, there have been relatively few complete and technical books on quantitative analysis of metabolic networks, in plants in particular. Schwender brings his experience in, and enthusiasm for, plant metabolic networks to this book. His personal research experience in flux analysis is clear in guiding the organization and content of this book, and adds a tremendous amount of practical insight and relevance. One of the most impressive aspects of this book is its broad coverage by expert and active researchers in the field of the challenges involved in analyzing plant networks. Two highly divergent approaches are covered – the "omics" view that uses global measurements of network parts (metabolites, transcripts, proteins) and statistics to deduce correlative interactions versus the mechanistic view that uses modeling and is typically not as comprehensive. This book will assist the plant researcher in deciding when to use the "omics" approach versus the mechanistic one, and how information for both approaches can be interpreted and consolidated into working hypotheses of the functioning of plant networks.

Part I sets the stage and discusses the unique complexity of plant metabolism. I am pleased to see that model organisms are described and encouraged as a general framework for describing and understanding plant networks. In addition, two important areas of active research, transport processes and metabolons, are highlighted. The compartmentation within and between cells is one of the distinctive aspects of plant metabolism, which gives plants "plasticity" to adapt to many different environments. Bottlenecks in plant metabolic engineering can often be tied to these two topics. Regarding transport, having the metabolite in the right place at the right time to undergo catalysis may be a rate-limiting step in making the desired product. For metabolons, multienzyme complexes of several enzymes, overexpression of one enzyme in the complex may not affect the overall rate. These complexes will have to be modeled differently than single enzymes – the enzymes will be tightly coupled and coordinated. Hopefully, the material in these chapters will help prevent "irrational" engineering by future plant biotechnologists!

Part II examines the key measuring parts of the metabolic network: i.e., metabolites and enzyme kinetics. The technologies and associated challenges to measure both of these quantities are straightforwardly described. With the tremendous amount and variety of chemistries in the plant cell, and with the number of methodologies for quantification of metabolites, including ^{13}C-labeled ones, this is no small feat! Here, the plant researcher can deduce major strategic decisions that they must grapple with – how to balance breadth versus depth, both in terms of the measurement quality and the statistical analysis required to analyze the information. These questions are ones that must be continuously examined by the researcher in any organization.

Part III discusses the analytical and organizational challenges that all researchers of plant metabolic networks must tackle: how to analyze, interpret, and/or model the data in order to draw relevant conclusions. Virtually, every technologist inevitably wonders whether they need modeling. This is especially true, given the varying mathematical backgrounds of the researchers – leading to a significant fraction with an inherent distrust of models. This section debunks the mystique of the modeling process, to enable a larger audience to comfortably understand the huge potential and limitations of these vital tools for analyzing metabolic networks. Understanding of the model structure, assumptions in building the model, and the uncertainty in data and model parameters will enable the researcher to critically assess the model results, resulting in increased power to generate testable hypotheses that by inspection alone one cannot muster. After reading these chapters, the researcher will become more comfortable with the uncertainty inherent in metabolism.

The quantitative analysis of plant metabolic networks is intrinsically interdisciplinary in nature – hence the challenges in training and utilization of these technologies. In looking through this practical and comprehensive treatment, my first reaction was, "I wish I had available a book like this when I had begun to mentor my students, before they started analyzing plant networks!" A similar reaction from my students might have been, "I wish I had read a book like this when I began my research project so that I knew what she was talking about!" Unfortunately, we had to learn many of the topics covered here in real time.

This book does justice to the current state of the art for analyzing plant metabolic networks. In my experience, the challenges discussed in this book have really proved to be the hurdles to overcome.

With this new book, a new generation will get to share the extensive and deep insights of Schwender and the books' contributors, push the frontiers in quantitative plant network analysis, and become tomorrow's industrial and academic leaders in plant biotechnology.

Ames, Iowa   Jacqueline V. Shanks

Preface

The main motivation behind *Plant Metabolic Networks* was to bring together in one book various, and in part, very diverse approaches that relate to the quantitative analysis of metabolic networks in plants. Although flux analysis and related approaches are not really new, their recent re-emergence in plant science made it worthwhile to summarize recent advances in this area, and to give an overview of the current state of knowledge. Expert authors from leading groups in plant science could be convinced to contribute the 11 chapters to the book. I am very thankful to all the contributors for their time and effort spent in writing the chapters, and I hope that this book might be useful for plant biologists, students, and researchers in related fields in order to stimulate further progress in plant metabolic research and biotechnology.

Plant Metabolic Networks is intended to be in part tutorial, and in part review of recent literature, and hopefully turns out to be more than a collection of review articles. Two basic areas are touched: experimental/analytical techniques needed to produce metabolic data, and the mathematical modeling used to analyze these data. The authors kept the mathematical detail at a minimum to make the book accessible for a broad audience of plant biologists. However, the metabolic research described in this book certainly is of interdisciplinary nature and mathematicians who are interested in analyzing plant metabolic networks should benefit from the book as well.

February 23, 2009
Upton, New York

Jörg Schwender

Contents

1. **Introduction** 1
   Jörg Schwender

2. **Definition of Plant Metabolic Networks** 9
   Andreas P.M. Weber

3. **Metabolite Measurements** 39
   Ute Roessner and Diane M. Beckles

4. **Enzyme Kinetics: Theory and Practice** 71
   Alistair Rogers and Yves Gibon

5. **Quantification of Isotope Label** 105
   D.K. Allen and R.G. Ratcliffe

6. **Data Integration** 151
   Aaron Fait and Alisdair R. Fernie

7. **Topology of Plant Metabolic Networks** 173
   Eva Grafahrend-Belau, Björn H. Junker, Christian Klukas,
   Dirk Koschützki, Falk Schreiber, and Henning Schwöbbermeyer

8. **Network Stoichiometry** 211
   Nanette R. Boyle, Avantika A. Shastri,
   and John A. Morgan

9. **Isotopic Steady-State Flux Analysis** 245
   Jörg Schwender

10. **Application of Dynamic Flux Analysis in Plant
    Metabolic Networks** 285
    Amy J.M. Colón, John A. Morgan, Natalia Dudareva,
    and David Rhodes

11 Kinetic Properties of Metabolic Networks 307
   Jörg Schwender

Index 323

Contributors

D.K. Allen Department of Plant Biology, Michigan State University, East Lansing, MI 48824, USA

Diane M. Beckles Department of Plant Sciences, University of California, Mail Stop 3, 133 Asmundson Hall, One Shields Avenue, Davis, CA 95616, USA

Nanette R. Boyle School of Chemical Engineering, Purdue University, 480 Stadium Mall Dr., West Lafayette, IN 47907, USA

Amy J.M. Colón Department of Horticulture, Purdue University, West Lafayette, IN, USA

Natalia Dudareva Department of Horticulture and Landscape Architecture, Purdue University, West Lafayette, IN 47907, USA

Aaron Fait Ben-Gurion University of the Negev, Jacob Blaustein Insts. for Desert Research, Dept. of Dryland Biotechnology, Midreshet Ben-Gurion, 84990, Israel

Alisdair R. Fernie Max Planck Institute for Molecular Plant Physiology, Department 1, 14059, Golm-Potsdam, Germany

Yves Gibon Max Planck Institute for Molecular Plant Physiology, Department 1, 14059, Golm-Potsdam, Germany

Eva Grafahrend-Belau Department of Molecular Genetics, Leibniz Institute of Plant Genetics and Crop Plant Research (IPK), Corrensstraße 3, 06466 Gatersleben, Germany

Björn H. Junker Department of Molecular Genetics, Leibniz Institute of Plant Genetics and Crop Plant Research (IPK), Corrensstraße 3, 06466 Gatersleben, Germany

Christian Klukas Department of Molecular Genetics, Leibniz Institute of Plant Genetics and Crop Plant Research (IPK), Corrensstraße 3, 06466 Gatersleben, Germany

Dirk Koschützki Department of Molecular Genetics, Leibniz Institute of Plant Genetics and Crop Plant Research (IPK), Corrensstraße 3, 06466 Gatersleben, Germany

A. Marshall-Colon Department of Horticulture and Landscape Architecture, Purdue University, West Lafayette, IN 47907, USA

John A. Morgan Department of Chemical Engineering, Purdue University, 480 Stadium Mall Dr., West Lafayette, IN 47907, USA

R.G. Ratcliffe Department of Plant Sciences, University of Oxford, South Parks Road, Oxford OX1 3RB, UK

David Rhodes Department of Horticulture and Landscape Architecture, Purdue University, West Lafayette, IN 47907, USA

Ute Roessner Australian Centre for Plant Functional Genomics, School of Botany, The University of Melbourne, 3010, Australia

Alistair Rogers Department of Environmental Sciences, Brookhaven National Laboratory, 50 Bell Avenue, Upton, NY 11973, USA

Falk Schreiber Leibniz Institute of Plant Genetics and Crop Plant Research (IPK) Gatersleben, Corrensstr. 3, D-06466 Gatersleben, Germany

Jörg Schwender Biology Department, Brookhaven National Laboratory, 50 Bell Avenue, Upton, NY 11973, USA

Henning Schwöbbermeyer Department of Molecular Genetics, Leibniz Institute of Plant Genetics and Crop Plant Research (IPK), Corrensstraße 3, 06466 Gatersleben, Germany

Avantika A. Shastri School of Chemical Engineering, Purdue University, 480 Stadium Mall Dr., West Lafayette, IN 47907, USA

Andreas P.M. Weber Institute for Plant Biochemistry, Heinrich-Heine-Universität, Bldg. 26.03.01, 40225 Düsseldorf, Germany

Chapter 1
Introduction

Jörg Schwender

1.1 Why a Book on Plant Metabolic Networks?

With several complete plant genomes currently available and with massive amounts of transcriptomic, proteomic, and metabolomic data being generated by plant scientists, one could assume that making sense of all this data is only a minor problem. However, it has been pointed out by many leading plant scientists that this is not the case and that developing appropriate modeling skills and tools may lack behind the technological progress that allows the data generation. In recent years, the plant science community became more and more aware of the importance of different kinds of analysis and modeling approaches, like metabolic flux analysis. Accordingly, in this book, contributions from different expert authors have been assembled to give a current view on plant metabolic networks, from the analysis of the molecular parts to approaches of mathematical modeling of plant metabolic networks at the cellular level. Other processes like gene regulation, cell signaling or models at the whole plant or ecosystem level certainly have their justification [1], but have been mostly excluded here to give cellular metabolism special attention.

1.1.1 What Constitutes Metabolism and Metabolic Networks?

Metabolism can be defined as a highly organized, self-regulated, and continuous process of uptake, transport, chemical modification, and secretion of small molecules. As pointed out by Wiechert et al. [2], metabolism has its manifestation in two quantities: concentration and flux – with analogy to potential and flow in many different physical systems. Flow is causal for potential and vice versa [2]. This means that if the momentary metabolic state of a cell is to be described, both of them have to be determined.

The many specific interactions between the multitude of different metabolites and enzymes constitute a complex network. There is increasing awareness among plant

---

J. Schwender (✉)
Brookhaven National Laboratory, Biology Department, Upton, NY 11973, USA
e-mail: schwend@bnl.gov

scientists that analyzing the systems properties of these networks and the use of predictive models of metabolism will be of major importance for further progress in understanding of plants at the molecular level [3]. Systems properties can describe a system at its whole but are destroyed if the system is dissected into its elements. Systems theory assumes that there are different levels of organization with additional properties arising at each level ("emerging properties") [4]. Metabolic networks are one level at which properties like elementary flux modes, metabolic flux, centrality or robustness emerge.

1.1.2 Why Study Metabolic Networks?

In difference to protein–protein or protein–nucleic acid interaction networks, metabolic networks, in particular, can be modeled in a very exact way. For example, the stoichiometric coefficients in the mass balance of an enzyme-catalyzed reaction are exact integer numbers. This may in part explain the success of the different modeling approaches described and discussed in this book.

In an attempt to categorize current approaches to model plant metabolism, two groups of approaches can be discerned. Global inquiry and discovery (Chapters 3 and 6) is based on global measurement of network compounds (metabolites, transcripts, proteins), typically in response to different kinds of perturbation (e.g., growth condition, transgenic events). Mainly by use of multivariate statistics, this approach typically identifies components that co-respond to a stimulus and therefore should have close functional relationship. In fact, it has been found that the metabolome reacts in a very sensitive and dynamic way to many kinds of perturbation (see Chapter 6). In this way, e.g., metabolites belonging to a metabolic pathway can be identified. This approach can be applied to diverse plant systems and most often captures the effect of perturbation globally. It also can integrate across the organizational levels (metabolome, transcriptome, see Chapter 6). However, the discovered underlying correlative relationship has subsequently to be investigated in detail to reveal the molecular mechanisms behind.

An apparently diametrical approach tries to build mechanistic models based on functions already known with some detail, such as reaction stoichiometry or enzyme mechanisms. Systems properties on the level of metabolism can be analyzed and predicted by these models and then verified on basis of experiments. This approach receives increasing interest in the plant science community during recent years as documented in several chapters of this book. Due to the apparent complexity of metabolism, compromise in the level of detail represented in mechanistic models is unavoidable, i.e., by far not all cellular reactions may be represented. Also, models bridging several organizational levels are still in their infancy, i.e., metabolic models often ignore genetic regulation or the dynamics of the metabolic network during development. Therefore one has always to be aware of the limited validity of this kind of modeling. In addition, the experimental component of this approach can be quite demanding and not easily applicable to whole plants. For example, if metabolic flux is to be measured, issues like metabolic non-steady state or heterogeneous cell types in tissues may result in misleading results.

For many plant scientist, mathematical and computational procedures are a hurdle in applying mechanistic models to plant metabolism – and hopefully this book is helpful to alleviate this problem.

1.2 Contents of the Book

The book can be divided into three parts: First an introductory chapter (Chapter 2), relating to the unique complexity of plant metabolism. The following three chapters describe how to analyze the components that make up the metabolic network, metabolites, and enzymes. Finally, Chapters 6–11 are devoted to network analysis and modeling.

1.2.1 Complexity of Plant Metabolism

Chapter 2 introduces the elucidation of metabolic pathways in plants, pointing out the importance of model organisms. While plants are characterized by a complex organization into different cell types as well as by subcellular compartments with different metabolic activity, big parts of core plant metabolic pathways had been revealed by using simple model organisms such as unicellular green algae. The choice of such a uniform cell type, as an experimental system, was critical to elucidate the structure of key metabolic routes in sufficient detail. Therefore, the author encourages plant scientists to keep using some of these model organisms in research on plant metabolism.

The discovery of pathway/network structure in plants is far from being completed. Due to the complex organization of plant metabolism into different cell types and subcellular compartments, transport processes and transport proteins are critical parts of plant metabolism and by far not all of them have been identified and characterized yet. For some metabolic pathways, the particular steps are distributed across several subcellular compartments and even across different cell types. Here, highly selective metabolite transporters take part in the control of metabolism and cross-talk between compartments. For example, the photorespiratory cycle and C4 photosynthesis are both highly compartmentalized pathways and have been long described in textbooks. However, the transport between the organelles is not yet understood in detail at the molecular level, and many of the transporters involved have yet to be identified.

Chapter 2 goes on to show that another important but not well-studied issue in the organization of plant metabolism is the association of enzymes in metabolons. Metabolons are macromolecular complexes of enzymes in which metabolic intermediates are passed on from the active site of one enzyme to the next. Metabolon organization has been detected in plant glycolysis, in photorespiration, but is also supposed to be present in Calvin cycle and secondary metabolism [5]. Recognizing the importance of macromolecular associations of enzymes, it is clear that more structural studies in enzyme proteins and protein complexes will be necessary.

The definition of metabolic networks is possible based on genomic information. With lower cost and faster throughput in sequencing technology, the number of plant genomes is rapidly increasing. The genome information basically contains all the genes that make up the metabolic network. For the reconstruction of genome scale metabolic networks, it is important to note that such methods do not permit the assignment of functions to unknown genes. Here, the author predicts that that integration of different omics data types will be helpful. The author predicts that integration of multivariate datasets from different "omics" approaches might eventually permit the ab initio computational deduction of complex networks.

1.2.2 Measurement of Network Components

The second part of the book is devoted to the generation of data that quantify and describe the components of metabolic networks – metabolites and the enzymes.

Chapter 3 gives an overview over experimental approaches and technologies that can be used to measure, identify, and quantify the vast amount of metabolites found in plants. While these technologies are constantly evolving, GC/MS with electron impact ionization is probably still the mostly used technology in metabolic profiling and metabolomics, due to instrumentation cost, well-established methods, and readily available mass spectral libraries. To just mention recent technologies not reviewed here: Mass spectrometric imaging approaches (see e.g., Zhang et al. [6]) may represent an important progress in the field. While classical metabolite profiling always uses extraction and separation/analysis of whole tissues (global analysis), these more recent developments may be key to resolve the metabolism of the different cell types.

Chapter 4 introduces the kinetic properties of enzymes and the parameters that describe them and are needed to model metabolic networks dynamically. The chapter then goes on to put emphasis on recent advances in high throughput profiling of total enzyme activities as well as of apparent kinetic constants in extracts of plant tissues. The enzyme analysis platform described here approaches the high throughput scale, which is very desirable and already routine in metabolomics, transcriptomics, or proteomics.

In Chapter 5, the analysis of metabolite labeling in stable isotope labeling experiments is reviewed in much detail. The resulting data are basis for ^{13}C-based metabolic flux analysis. Isotope tracers can be used to follow the metabolic fate of specific atoms through the metabolic network and to determine mass flows. First, procedures commonly used in plant flux analysis for extraction of plant tissue and subsequent fractionation and derivatization of the labeled compounds are described. Quantification of stable isotope label by nuclear magnetic resonance (NMR) is reviewed with respect to one-dimensional and two-dimensional NMR techniques that have been applied. In NMR, fractional enrichment (i.e., position-specific isotope enrichment) as well as isotopomer information can be obtained. Then the often used analysis of label by gas chromatography/mass spectrometry (GC/MS) is discussed. MS has the advantage of high sensitivity and accuracy and is in particular powerful if different fragments of one molecule can be analyzed. Different aspects of chromatographic separation, ionization, ion fragmentation, and detector properties are discussed, which are the basis for accurate quantification of isotope label. Finally, the correction of mass spectrometric isotopomer data for the occurrence of natural isotopes is laid out in detail.

1.2.3 Network Modeling and Analysis

The third part of the book is devoted to the analysis of metabolic networks with different approaches from topological analysis via stoichiometric models to the dynamic simulation of biochemical reactions.

The metabolome, the complete set of small-molecule metabolites, changes its quantitative composition in response to different perturbations and stimuli. Metabolomic data indicate functional relations between metabolites. Accordingly, the processing and analysis of metabolomic data is reviewed in Chapter 6, where also the challenge of integration of different "omics" data is highlighted.

In order to describe topological features of metabolic networks, graph-theoretical mathematical approaches are reviewed in Chapter 7. Different properties like network motifs and diverse centrality measures help to understand network structures and to compare networks. The chapter also introduces various visualization tools that can be used to analyze complex data generated by metabolite profiling or transcriptomics. This includes mapping of such data on network structures. Specific reference is given to dynamic visualization of data.

The subsequent three chapters are all related to constraint-based steady-state flux analysis, which is based on the stoichiometries of biochemical reactions and on reaction directionality/reversibility. In Chapter 8, stoichiometry-based models are discussed, which allow to analyze and predict the theoretical flux capabilities of metabolic networks. The chapter first introduces the fundamentals of stoichiometric modeling. Then flux balance analysis (FBA) is introduced, which allows to predict the usage of pathways under the assumption of a certain cellular objective, such as maximal yield in biomass compounds. Different variants of flux balance analysis like minimization of metabolic adjustment (MOMA) and dynamic FBA are well explained. While these approaches are based on reaction stoichiometry, some information on reaction directionality has to be known as well. This and other thermodynamic considerations that are current limitations to the reliability of the standard FBA approach are given broad space in the chapter. Following this, extreme pathway analysis and elementary mode analysis are introduced, and applications to plant systems discussed. Finally, a section on genome scale models describes model reconstruction based on genomic data. Besides the general problems of incomplete genome annotation, gaps in pathways and dead-end metabolites require manual refinement of the models. A problem specific to eukaryotic cells is the presence of subcellular compartments. This means that for each enzyme, the subcellular localization has to be known as well as the transport proteins that connect the compartments. In addition, to realistically simulate plants, the stoichiometry-based approach has to be extended to specific cell types.

Chapter 9 is focused on the estimation of intracellular flux, based on stable isotope-labeling experiments. The chapter relates in particular to the steady-state labeling approach, i.e., interpretation of labeling pattern in metabolites after the distribution of isotope label has reached steady state. First, application of the methodology to plants and main insights gained from such studies on plant metabolism are reviewed. Then the basic experimental setup of a steady-state labeling experiment is described and important assumptions inherent to the experimental approach, such as the approximation of metabolic and isotopic steady state, are discussed. Some characteristic features of metabolic networks used in ^{13}C-flux analysis are highlighted. Different approaches for the interpretation of labeling data and estimation of fluxes are reviewed with detailed consideration given to the underlying basic relation between labeling signatures and flux. Computational aspects of flux analysis are discussed with regards to the limits to the complexity of networks that can be analyzed, as well as related to the reliability of flux values obtained by flux-parameter fitting. Finally, the software tools available for ^{13}C-MFA are discussed.

In addition to the steady-state isotope labeling described in Chapter 9, Chapter 10 discusses the study of plant metabolic networks by dynamic flux analysis. The mathematical formalism is introduced, and classical examples of dynamic labeling as well as most recent studies on secondary plant metabolism are discussed. Dynamic flux analysis is built on the principle of the classical pulse-chase experiment. After feeding a labeled precursor, labeling pattern in metabolites are monitored in a time course. In difference to steady-state labeling, multiple samples of the time course have to be analyzed, as well as intracellular metabolite concentrations have to be measured. However, as discussed in detail in this chapter, dynamic labeling allows to circumvent some of the shortcomings of steady-state flux analysis. Actually from the famous pioneer work of Calvin Bassham and Benson until today, dynamic labeling has been proved to be a powerful tool in pathway discovery in plants. Careful interpretation of labeling time courses can establish precursor–product relationships.

Chapter 11 gives a basic introduction into kinetic modeling of plant metabolic networks. Based on a literature example, the setup, simulation, validation, and use of a kinetic model is explained. Also, metabolic control analysis (MCA) is addressed as many studies use this approach of sensitivity analysis to characterize kinetic models. This includes experimental MCA, which has regularly been applied to plants.

References

1. Yin X, Struik PC (2008) Applying modelling experiences from the past to shape crop systems biology: the need to converge crop physiology and functional genomics. New Phytol 179: 629–642

2. Wiechert W, Schweissgut O, Takanaga H, Frommer WB (2007) Fluxomics: mass spectrometry versus quantitative imaging. Curr Opin Plant Biol 10: 323–330
3. Sweetlove LJ, Fell D, Fernie AR (2008) Getting to grips with the plant metabolic network. Biochem J. 409: 27–41
4. Bertalanffy L (1976) General System Theory: Foundations, Development, Applications. George Braziller, New York
5. Winkel BSJ (2004) Metabolic channeling in plants. Ann Rev Plant Biol 55: 85–107
6. Zhang H, Cha S, Yeung ES (2007) Colloidal graphite-assisted laser desorption/ionization MS and MS(n) of small molecules. 2. Direct profiling and MS imaging of small metabolites from fruits. Anal Chem 79:6575–6584

Chapter 2
Definition of Plant Metabolic Networks

Andreas P.M. Weber

2.1 Plant Metabolic Networks – More Complex than Anything Else?

To the best of the author's knowledge, the answer to this question is yes. In part, the enormous complexity of plant metabolic networks is due to the high degree of compartmentation of plant cells. That is, plant cells, in contrast to other non-plant eukaryotic cells, display a higher degree of compartmentation, with the chloroplast being the most prominent compartment that is exclusive to plants. In addition to having a higher degree of compartmentation, most higher plants are sessile and thus cannot escape from biotic or abiotic stressors, which promoted the evolution of a host of metabolic adaptations to the sessile lifestyle, including an impressive range of secondary metabolites that are synthesized by plants to serve as defense compounds against pathogen attack, as attractants for pollinators, and as protective agents against abiotic stress.

In this chapter, I will discuss the impact of compartmentation of plant cells on plant metabolic networks. Particular emphasis will be given to the use of unicellular model systems for deciphering metabolic routes, to solute transport across intracellular membranes, to the concept of metabolons and substrate channeling, and to the role of chloroplasts in the plant metabolic network. The role of chloroplasts will also be considered in the context of their evolution from cyanobacterial ancestors through the process of endosymbiosis because this process is a prime example for integration of two separate metabolic entities into one "super-organism" (i.e., the first plant cell).

A.P.M. Weber (✉)
Institute for Plant Biochemistry, Heinrich-Heine-Universität, 40225 Düsseldorf, Germany
e-mail: Andreas.weber@uni-duesseldorf.de

2.2 Keeping It Simple – Why Unicells Are Cool

From above said, it becomes clear that gaining a deeper understanding of the metabolic networks in plant cells is not a trivial pursuit. The high degree of compartmentation, the uncertainty about which metabolites are transported across organellar membranes, and a lack of knowledge about the kinetic constants of many metabolite transporters and of subcellular metabolite concentrations hamper progress. In photosynthetic tissues, rapid randomization of applied label hinders the application of metabolic flux analysis that has been applied very successfully to eukaryotic and prokaryotic unicells and in heterotrophic plant tissues, such as developing oil seeds. As described in more detail later in this chapter, further obstacle comes from the fact that, e.g., leaf tissue of land plants consists of multiple cells types, such as the epidermis, vasculature, and photosynthetic mesophyll tissue, which is frequently multi-layered and differentiated into spongy and palisade parenchyma cells. Hence, tissue extracts always represent a mixture of metabolites and enzymes derived not only from various cellular compartments but also from a variety of tissues, with very different metabolic and enzymatic capacities. Some of these issues can be addressed by subcellular fractionation techniques, such as the non-aqueous fractionation of plant tissues, which provides insights into subcellular metabolite concentrations, although it does not distinguish between the individual cell types that are present in a typical leaf of a land plant. Biological imaging techniques, based on NMR, do provide some information at cellular and subcellular level, but they are restricted to relatively few metabolites, minerals, and water [84, 85, 139]. Laser-micro-dissection of fixed tissues followed by microanalytic methods such as the analysis of secondary metabolites in individual stone cells of Norway spruce [92] and single-cell sampling techniques such as measurements of sugar concentrations of individual cells of castor bean hypocotyls [154] or amino acid analysis in leaf mesophyll cells (MC) [5] is also providing additional insights.

Many of the abovementioned limitations do not apply to unicellular photosynthetic organisms, such as algae, and the "compartmentation issue" can be addressed by investigation of prokaryotic photosynthetic organisms, such as cyanobacteria which are evolutionary connected to plastids and thus might serve as models for plastid metabolism. A prime example for the dissection of a relatively complex metabolic pathway in photosynthetic organisms is the reductive pentose phosphate pathway, also known as the Calvin cycle. In series of milestone papers in the late 1940s and early 1950s, using unicellular algae such as *Scenedesmus* and *Chlorella*, as well as land plant leaves, the path of carbon in photosynthesis was elucidated. In a landmark paper, it was shown that the first labeled organic carbon compound found in *Scenedesmus* cells that were allowed to photosynthesize in the presence of labeled carbon dioxide for five seconds was phosphoglyceric acid and that the first free carbohydrate to appear was sucrose [25]. To cite from the introduction of this classical paper, "The ideal design of an experiment to determine the chemical path of carbon from carbon dioxide to the variety of plant constituents is relatively simple and straightforward. It would consist of feeding a photosynthesizing organism radioactive carbon dioxide for various lengths of time and stopping the reaction by killing the plant. By determining those compounds into which the radioactive carbon has been incorporated for each period of illumination and, further, by determining the distribution of radioactivity within each compound, these data could then be used to construct a family of curves depicting the increase in radioactivity in each compound (and in each carbon atom of each compound) as a function of time. From a complete set of such curves it should be possible to draw a map of the path of carbon as it flows into the plant in the form of carbon dioxide and distributes itself among all the plant constituents." In the opinion of this author, there is no better way of explaining the principle, hence the quotation. This paper goes on to suggest that hexose phosphates (i.e., glucose 1-phosphate and fructose 6-phosphate), not free glucose and fructose, are the precursors of sucrose and that sucrose phosphate might be an intermediate of sucrose biosynthesis. From quantitative analysis of the amount and the kinetics of radiolabel incorporated into the hexose phosphate pool and in the glucose and fructose moieties of sucrose, they estimated the size and turnover of the hexose phosphate pool. In a later publication, the authors also show that the sugar nucleotide UDP-glucose is likely involved in biosynthesis of sucrose [24]. A key benefit of using unicellular algae for these studies was that it was possible to very rapidly stop metabolism after applying the label, thus allowing very rapid kinetic studies of metabolism. Since the algae used in these experiments were kept as cell suspension, label could be injected into the solution and the cells be killed shortly after by dumping the suspension into boiling ethanol. Similar studies would have been very difficult, if not impossible with leaf tissue of land plants. However, other metabolic pathways in plant cells, such as sucrose biosynthesis from oil via the glyoxylate cycle and gluconeogenesis [27, 86, 87], was unraveled by feeding of radioactive tracers to land plant tissues, such as castor bean embryos. Unicellular algae have been used in numerous studies of plant metabolism, and in many cases, metabolic pathways were first worked out using algae as model systems. A specific example is the discovery of the biosynthesis of isoprenoids in plants such as carotenoids, sterols and the prenyl-side chains of chlorophyll and plastochinone by the non-mevalonate pathway [131]. This study took advantage of the fact that *Scenedesmus obliquus* cells can be grown heterotrophically on ^{13}C-labeled glucose or acetate, thus permitting the application of metabolic flux analysis, based on isotopomer distribution obtained by ^{13}C-NMR spectroscopy.

Not only eukaryotic photosynthetic systems have been used to decipher metabolic pathways but also prokaryotic photosynthetic organisms have been used as model systems. For example, the process of nitrogen fixation was for the first time directly demonstrated to be localized in the heterocysts of the N-fixing cyanobacterium *Anabaena cylindrica*, using radioactive ^{13}N that was generated by proton bombardment of ^{13}C powder in combination with micro-autoradiography [174]. This study also permitted some insights into kinetics and flux of fixed nitrogen out of the heterocysts along the filament. In two follow-up papers, the same group used ^{13}N-short-term labeling (1–120s) and pulse-chase experiments to unravel the pathway of nitrogen metabolism after fixation [146, 175]. It was clearly shown that the reaction sequence moved from production of ammonia from nitrogen gas to the formation of glutamine and eventually glutamate. From simultaneous application of specific enzyme inhibitors, it was concluded that fixed nitrogen is metabolized by glutamine synthetase/glutamate synthase pathway [175]. Photosynthetic prokaryotic unicells, such as *Synechocystis* sp. PCC 6803, are currently being rediscovered as model systems to study metabolic fluxes [133, 132]. Cyanobacteria, such as *Synechocystis*, offer the advantage of having only one metabolically active compartment, while the carbon fixation pathway is generally similar to that of higher plants. The genome of *Synechocystis* is fully sequenced, thus permitting the reconstruction of its metabolic network from its genomic sequence. Unfortunately, steady-state metabolic flux analysis is not directly applicable to autotrophic photosynthetic organisms because incorporation of labeled carbon in the form of ^{13}C–CO_2 will eventually lead to uniform labeling of all carbon atoms, thus preventing the deduction of flux information from isotopomer analysis. However, transient metabolic flux analysis, basically following the ideas put forward by Benson and Calvin (see above), in combination with modern analytical techniques, reconstructed stoichiometric metabolic networks from genomic data, and mathematical modeling does allow the deduction of fluxes in photoautotrophic organisms [133].

2.3 Connections Are Everything – Solute Transport and Metabolic Networks

A fundamental property of metabolic networks in plants is the selective partitioning of organic metabolites among different organelles, cells, tissues, and organs. This requires various transport mechanisms to accommodate the directional transport of metabolites. Transporters participate in basic metabolism by partitioning metabolites within and between cells, and they are essential for intermediate and long-distance transport between tissues and organs, respectively. Plants assimilate inorganic carbon and nitrogen into organic compounds required for plant growth; a very large variety of metabolites are produced, and the anabolic and catabolic pathways that they feed into are complex and interconnected. Metabolic pathways are frequently partitioned between organelles, cells, or even tissues and organs. Thus, intracellular and long-distance transport processes are critical for sustaining biosynthesis, catabolism, and growth. Since transport processes potentially affect the availability of substrates or products, they also represent critical sites at which metabolism and growth can be regulated. Hence, transport processes in plants, in particular the location and kinetic properties of transporters, are essential components of metabolic networks since they frequently influence metabolic fluxes, as well as partitioning of nutrients between growth and storage.

The high degree of compartmentation of plant cells and the distribution of many metabolic pathways across several cellular compartments and, in some cases, even different cell types requires massive flux of metabolic intermediates across cellular and organellar membranes. Since most small molecules in plant cells are not membrane permeable, metabolite transporters are required to catalyze the transport of metabolites across membranes. The compartmentation of metabolic pathways provides additional options for regulation, permits the simultaneous operation of pathways that compete for the same substrates within the same cell, and they help avoiding futile cycles. Metabolite transporters thus play critical roles in connecting the parallel and interdependent biosynthetic and catabolic pathways and thus represent the integrating elements in these metabolic networks, similar to interchanges in road networks. In vascular plants, long-distance transport is critical for the allocation of organic carbon and nitrogen compounds from their sites of synthesis to developing or reproductive plant organs that rely on import of the organic compounds for growth and development. Obviously, a plethora of multicompartment pathways and long-distance transport processes could be reviewed here to illustrate the principles; however, due to space constraints, only two specific examples, the photorespiratory C_2 oxidation cycle and the biochemical CO_2 pump of C_4 photosynthesis, will be given.

2.3.1 The Photorespiratory C_2 Oxidation Cycle – A Highly Compartmentalized and Interconnected Metabolic Route

The photosynthetic carbon-assimilating enzyme ribulose bisphosphate carboxylase-oxygenase (Rubisco) is a bifunctional enzyme. That is, it catalyzes both the productive carboxylation and non-productive oxygenation of ribulose 1,5-bisphosphate (RuBP) [20, 21, 111]. Oxygenation of RuBP leads to the production of one molecule of 3-phosphoglycerate (3-PGA) and one molecule of the toxic intermediate 2-phosphoglycolate (2-PG), which enters the photorespiratory carbon cycle (Fig. 2.1).

The detoxification of 2-PG and its recycling to 3-PGA occurs by the complex photosynthetic carbon oxidation cycle [125, 150]. Because this pathway leads to the consumption of oxygen (oxygenation of RuBP) and production of carbon dioxide (during the recycling of 2-PG) in the light, it is also called photorespiration. The specificity factor of the bifunctional enzyme Rubisco for the carboxylation reaction versus the oxygenation reaction is in the range of 80–100 for most land plants [137].

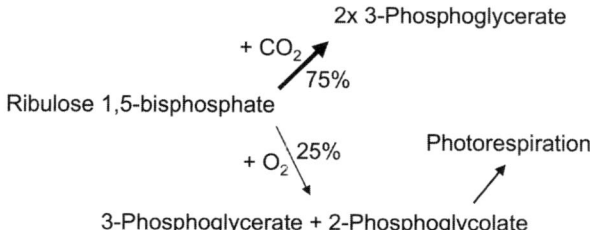

Fig. 2.1 Oxygenation reaction (+ O_2) and carboxylation (+ CO_2) reactions of Rubisco. A ratio of carboxylation to oxygenation reaction of Rubisco of 75:25 is commonly observed in C_3-type land plants

Under current atmospheric conditions, this leads to a carboxylation-to-oxygenation ratio of approximately 3:1. Hence, in C_3 plants, the rate of photorespiration would be 25% of the rate of gross CO_2 assimilation. These rates have been confirmed by short-term labeling studies of the intermediates glycolate, glycine, and serine with $^{18}O_2$ [38], by using $^{13}CO_2$ and mass spectrometry to determine CO_2 fluxes under conditions of steady-state photosynthesis [62], and from the post-illumination photorespiratory CO_2 burst [89].

2-Phosphoglycolate, one of the products of the oxygenation reaction of Rubsico, cannot be further metabolized inside the chloroplast stroma but must undergo a complex and highly compartmentalized reaction pathway in which two molecules of 2-PG are converted into one molecule of 3-PGA. This pathway minimizes the loss of fixed CO_2 and prevents the depletion of intermediates from the Calvin cycle, since 75% of the carbon of 2-PG is recycled in the photorespiratory pathway to yield 3-PGA, while 25% of the carbon is released in the form of CO_2 (Fig. 2.2).

The photorespiratory pathway represents a coordinated network consisting of 14 soluble enzymes that have been compartimentalized between chloroplasts, leaf peroxisomes, mitochondria, and cytoplasm and at least 12 transmembrane transport steps that connect the compartments (Fig. 2.3).

The photorespiratory carbon cycle is initiated in the chloroplast stroma by dephosphorylation of 2-phosphoglycolate, a reaction that is catalyzed by phosphoglycolate phosphatase (PGP). The resulting glycolate leaves the chloroplast by a glycolate/glycerate antiporter [69, 70, 179] and is taken up into the peroxisomes by an unknown transporter. Inside the peroxisomal matrix, glycolate is oxidized to glyoxylate by FMN-dependent glycolate oxidase, which catalyzes the transfer of two electrons from glycolate to O_2, yielding hydrogen peroxide (H_2O_2). Glyoxylate is transaminated to glycine by two aminotransferases – serine:glyoxylate and glutamate:glyoxylate aminotransferase (SGT and GGT, respectively [73, 93, 122]. Glycine then leaves the peroxisomes by an unknown transporter and is taken up into the mitochondria by a glycine/serine transporter [37, 36, 180]. In the mitochondrial matrix, two molecules of glycine are converted to CO_2, NH_3, NADH, and serine by the concerted actions of glycine dehydrogenase [114] and serine hydroxymethyl transferase [156]. Serine leaves the mitochondria, likely by the same transporter that also catalyzes the import of glycine, and is taken up into the peroxisomes, where the amino group of serine is removed by SGT, yielding hydroxypyruvate which is subsequently reduced to glycerate in a reaction consuming NADH by

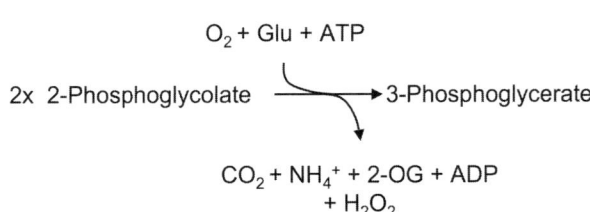

Fig. 2.2 Summary view of the photorespiratory carbon oxidation cycle

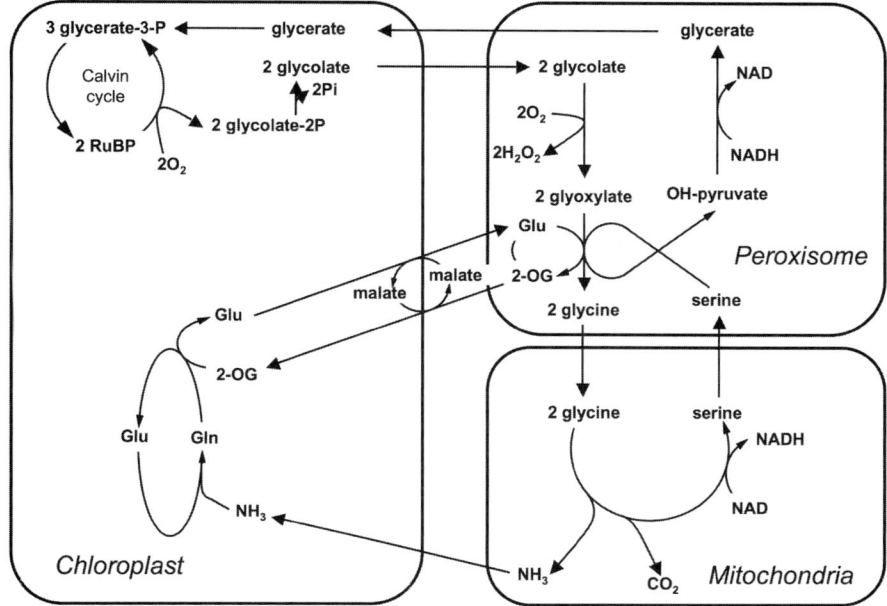

Fig. 2.3 Schematic representation of the photorespiratory carbon oxidation cycle. Please note that the redox shuttles connecting the NADH/NAD pools of mitochondria, cytoplasm, and peroxisomes via the respective isoforms of malate dehydrogenase are not shown for the sake of clarity

hydroxypyruvate reductase (HPR) [151, 181]. It has to be noted, though, that an alternative (NADPH-dependent) pathway for the reduction of hydroxypyruvate to glycerate apparently exists in the cytoplasm [81, 82]. Also, barley mutants deficient in peroxisomal HPR [107] do not display the characteristic conditional lethal phenotype at ambient CO_2 displayed by mutants deficient in other enzymes of the photorespiratory pathway [125], thus supporting the notion that a bypass exists for the reaction catalyzed by peroxisomal NADH-HPR. Glycerate, resulting from the reduction of hydroxypyruvate, is taken up into the chloroplast stroma by abovementioned glycolate/glycerate transporter, where it is phosphorylated by stromal glycerate kinase in an ATP-dependent reaction to yield 3-PGA [14]. 3-PGA enters the Calvin cycle, thus completing the photorespiratory carbon cycle.

Although photorespiration represents one the major carbon fluxes in photosynthetic tissues of C_3 plants, many aspects related to this pathway, such as its regulation and its impact on the interaction between carbon and nitrogen metabolism are not understood [94, 160, 165]. Particularly scarce is our knowledge about the transport of photorespiratory intermediates across the membrane(s) of the three participating organelles. To date, only two metabolite transporters involved in photorespiration have been identified at the molecular level; i.e., the plastidic 2-oxoglutarate/malate and glutamate/malate translocators (DiT1 and DiT2, respectively) [124, 130, 160, 161].

2.3.2 The Biochemical CO_2 Pump of C_4 Photosynthesis – Share of Labor Between Two Cell Types Causing Massive Flux of Metabolic Intermediates

A dramatic rise of atmospheric oxygen levels and a concomitant decline of the CO_2 levels, approximately 300 million years ago, as a consequence of oxygenic photosynthesis and (bio-) geochemical processes [9, 10, 34, 127], increased the selective pressure on photosynthetic organisms to evolve carbon-concentrating mechanisms (CCM), in order to minimize the oxygenase activity of Rubisco and to thus decrease the rate of photorespiration. Whereas many cyanobacteria and algae have evolved highly efficient single-cell carbon-concentrating mechanisms [7, 147], some terrestrial plants and few aquatic plants have evolved the process of C_4 photosynthesis [44, 58, 60, 61, 75, 120]. Because the carbon-concentrating mechanism of C_4 plants alleviates the CO_2 limitation of photosynthesis that occurs in C_3 plants at high temperature due to the relative acceleration of the oxygenation reaction with increasing temperature, C_4 photosynthesis is more efficient in warmer climates or under conditions of drought and salinity. In addition to a higher carbon-use efficiency, many C_4-type plants also have higher nitrogen use efficiency than C_3-type plants [110].

Since the discovery of C_4 photosynthesis in the 1960s, much has been learned about the physiological, biochemical, and anatomical features associated with this mode of photosynthesis. Until recently, it was believed that a distinct anatomical feature, called the Kranz anatomy, is required for C_4 photosynthesis. Kranz is the German word for wreath – a "wreath" of chloroplast-containing mesophyll cells surrounds a layer of chloroplast-containing bundle sheath cells (BSC) that in turn form a second "wreath" around the vascular bundle. The recent discovery of single-cell C_4 photosynthesis in several *Chenopodiaceae* species demonstrated that Kranz anatomy is not essential for C_4 photosynthesis [43, 157, 158]. The unifying principle of single-cell and dual-cell (i.e., Kranz anatomy) C_4 photosynthesis is compartmentation – either within a single cell or between two specialized cell types. In the following, we will focus on dual-cell C_4 plants.

Common to all C_4 (and CAM) plants is that CO_2 (in the form of HCO_3^-) is initially fixed by PEP carboxylase (PEPC):

$$PEP + HCO_3^- \rightarrow OAA + Pi$$

This reaction happens in the cytosol of the mesophyll cells. The further fate of OAA differs between the three biochemical variants ($NADP^+$-malic enzyme type, NAD^+-malic enzyme type, PEP carboxykinase type) of the C_4 pathway. In this review, we will focus on the $NADP^+$-malic enzyme type plant maize; the other two variants will not be discussed.

In maize, OAA is taken up into mesophyll cell chloroplasts, reduced to malate by $NADP^+$-malate dehydrogenase (MDH); malate is subsequently transported back to the mesophyll cell cytosol. Malate is then transported to the bundle sheath cell chloroplasts, where it is decarboxylated and oxidized, yielding pyruvate, NADPH, and CO_2. The liberated CO_2 is refixed by Rubisco. The decarboxylation product, pyruvate, is transported back to the mesophyll cells, taken up into mesophyll cell chloroplasts, and phosphorylated to regenerate the CO_2 acceptor PEP by pyruvate phosphate dikinase (PPDK). Finally, PEP is exported to the mesophyll cytosol to serve as CO_2 acceptor. Simplified, malate generated from PEP can be considered as a bucket for CO_2, and a bucket chain of metabolites shuffles CO_2 from mesophyll cells to bundle sheath cells. In addition, malate also serves as a vehicle for the transport of redox equivalents from mesophyll to bundle sheath cell chloroplasts. The transport processes involved in the C_4-CCM are summarized in Fig. 2.4.

The plastidic electron transfer chain in bundle sheath cell plastids is limited to photosystem I, thus abolishing photosynthetic O_2 production in these cells. The enrichment of CO_2 by malate decarboxylation and minimization of O_2 allows photosynthesis to proceed more efficiently because the oxygenation reaction of Rubisco is minimized due to high CO_2 and low O_2 partial pressures.

CO_2 fixation by Rubisco in bundle sheath cells yields two molecules of phosphoglycerate (3-PGA). The reduction of two 3-PGA to two triose phosphates requires two NADPH. However, the oxidative decarboxylation of one malate yields only one NADPH (and linear electron transport is insignificant in bundle sheath cells), hence one molecule of 3-PGA has to be exported to the MC plastids where it can be reduced to triosephosphate (TP). Two thirds of the generated triosephosphate then need to be re-exported to bundle sheath plastids for regeneration of ribulose-1,5-bisphosphate; the remainder can be exported as sucrose to sink tissues.

From above-said, it becomes obvious that C_4 photosynthesis is dependent on the exchange of malate, oxaloacetate, pyruvate, 3-PGA, and triosephosphate between

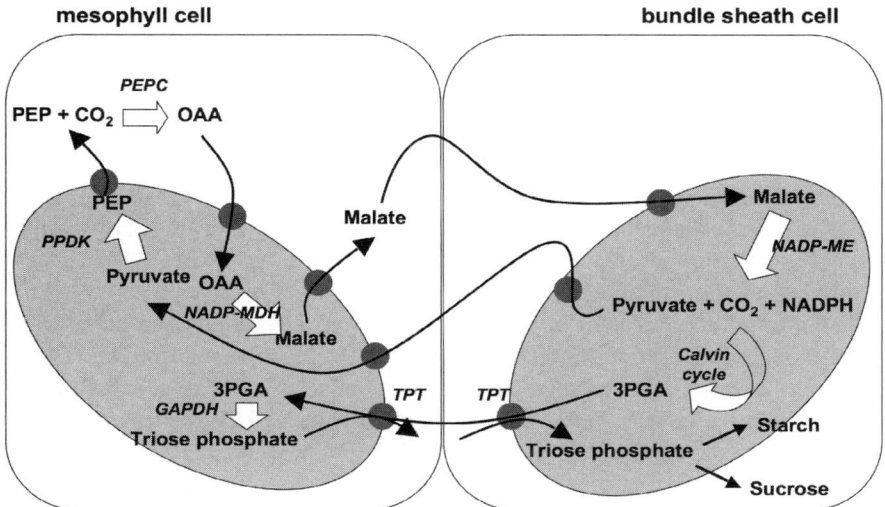

Fig. 2.4 Simplified, schematic representation of transport steps involved in the $NADP^+$-malic enzyme type C_4 carbon-concentrating mechanism

mesophyll and bundle sheath cells and their plastids. It has been hypothesized that symplastic diffusion relying on a concentration gradient accounts for the flux between both cell types. Stitt and Heldt found that gradients for 3-PGA, triosephosphate, and malate were steep enough to drive the required metabolite fluxes between mesophyll and bundle sheath cells [140]. The gradient for pyruvate, however, was minimal and even opposing the direction of transfer. It was proposed that pyruvate could be sequestered to a cellular compartment, most likely the plastids, thereby decreasing the cytosolic concentration and thus generating a concentration gradient between BSC and MC [140]. Weiner and Heldt later determined the subcellular metabolite concentrations and largely corroborated the earlier results [170]. When modeling C_4 photosynthesis, Laisk and Edwards assumed gradients for malate, pyruvate, 3-PGA, and triosephosphate although a gradient for pyruvate was never directly demonstrated [88]. This is likely the reason why the model was not affected by including active pyruvate uptake into plastids. The Laisk–Edwards model succeeds in reproducing data and trends generated earlier by Leegood and von Caemmerer [90]. The maximum carbon (C) fixation rate in their model is 55 μmol C per m^2 leaf area and second, well within the range of experimentally determined values. Maximal carbon fixation capacity may vary between leaf samples as do malate concentrations and transport rates. The enrichment of CO_2 in bundle sheath cells is dependent on overcycling, a higher fixation rate in the MC as compared to the BSC. Overcycling has been calculated and estimated experimentally to be about 10%. A fixation rate of 55 μmol/m^{2}s and overcycling of 10% thus requires a metabolite flow of $\sim$60 μmol/m^{2}s: (1) OAA, pyruvate, and 3-PGA into mesophyll plastids; malate, PEP, and triosephosphate out of mesophyll plastids; (2) malate and triosephosphate into bundle sheath plastids; pyruvate and 3-PGA out of bundle sheath plastids. These fluxes are well above those needed to sustain C_3 photosynthesis. As a specific example, the phosphate/triosephosphate translocator (TPT), which is the most abundant protein in C_3 plant plastid envelopes accounting for as much as 10–15% of total envelope protein, catalyzes a maximum flux of 5 μmol substrate/m^{2}s [48, 49, 54] and TPT limits maximal photosynthetic capacity in C_3 plants [63, 64]. Hence, it is reasonable to hypothesize that transporters involved in C_4 photosynthesis are likely very abundant proteins.

The transporters catalyzing the flux of C_4 photosynthetic intermediates are mostly unknown. Intercellular transport is likely driven by diffusion [140, 170]; thus, there are no apparent requirements for specific transporters at the plasma membrane. In addition, mesophyll and bundle sheath cells in C_4 species are connected by an unusually high number of plasmodesmata.

The intracellular transport of triose phosphate and 3-PGA into and out of plastids is thought to occur via the triose phosphate/phosphate translocator (TPT). This transporter has initially been characterized from spinach [50] and was subsequently found in all plant species that fix carbon, including maize [113]. TPT catalyzes the strict counter-exchange of triose phosphate with 3-PGA or inorganic phosphate.

A PEP/phosphate translocator (PPT) from maize has been demonstrated in isolated maize chloroplasts by Huber and Edwards [72]. A gene encoding a PPT from maize was characterized later [47]. This gene is highly expressed in maize endosperm but transcript abundance was very low in leaves [47], making it unlikely that this particular transporter is involved in the high flux of PEP in the C_4 pathway.

After PEP is exported, OAA is formed by PEPC. OAA levels in maize are too low to be measured, indicating a very rapid conversion of OAA to malate, which can accumulate in considerable amounts. As MDH is localized in plastids, OAA needs to be imported efficiently. Hatch and colleagues have described a high-affinity OAA transporter (OAT) with a K_m for OAA in the range of 0.05–0.07 mM and a corresponding K_i for malate of about 7 mM [59]. The V_{max} was very high in C_4 plastids of maize compared to C_3 plastids from spinach, although K_m and K_i values were similar, indicating that the transporter protein is much more abundant in C_4 plants. We and others have recently demonstrated that the plastidic 2-oxoglutarate/malate translocator DiT1 is able to characterize the specific counter-exchange of OAA with malate [124, 141, 142, 160]. However, the K_i values for malate determined with recombinant DiT1 from maize and spinach are one order of magnitude higher than those determined by Hatch et al. for the OAT, hence DiT1 might not be identical with the OAT that was characterized by Hatch et al. [124, 142].

The export of malate from mesophyll plastids occurs in counter-exchange with oxaloacetate by OAT [59]. However, the malate importer of bundle sheath plastids has not been characterized biochemically or at molecular level to date. For mesophyll cells, a 1:1 stoichiometry for the exchange of malate with OAA fits with the biochemistry of the pathway; however, in bundle sheath cells, for one malate that goes in, one pyruvate goes out. It is thus reasonable to hypothesize the presence of a malate/pyruvate exchanger in bundle sheath chloroplasts of maize. Alternatively, two distinct uniporters could be posited – one that catalyzes the uptake of malate into chloroplasts, and a second one that exports pyruvate.

Pyruvate transport has been characterized in isolated maize mesophyll cell plastids by Huber and Edwards [71]. Later, Flügge et al. analyzed pyruvate uptake into maize leaf plastids and found that the uptake of pyruvate is protein dependent and also dependent on the proton gradient across the inner envelope membrane generated by light [55]. Kanai's group, however, has reported that pyruvate transport, in some C_4 species, is dependent on a sodium gradient [4]. The transport rates of both OAA and pyruvate were calculated and are sufficient to account for the metabolite fluxes during maize photosynthesis [55, 59].

In summary, the transport of 3-PGA and triose phosphates can be accounted for by ZmTPT which is present in large amounts in maize leaf plastids. OAA uptake and pyruvate uptake transporter have been characterized on the biochemical level but their molecular nature is unknown to date. The malate import system of bundle sheath plastids is unknown. A PEP transporter from maize has been reported but its expression pattern argues against a role in C_4 photosynthesis. Overall, most plastidic transporters involved in C_4 photosynthesis are not known at the molecular level.

2.4 Sticking Together – Metabolons and Metabolic Channeling

Compartmentation and distribution of pathways across several cellular organelles is not the only obstacle for constructing networks of plant metabolism. Frequently, several enzymes catalyzing consecutive steps of a multistep pathway are organized as macromolecular complexes in which metabolic intermediates are passed on from the active site of one enzyme to the next in the cascade, without ever reaching the bulk aqueous phase of the cell. This process is called metabolic or substrate channeling, and the macromolecular complexes formed by soluble enzymes and other polypeptides, such as transporters or scaffolding proteins, are commonly termed "metabolons"; the concept metabolic channeling is not restricted to multienzyme complexes, but it can also occur between different cells or between cellular compartments [77, 172]. While the concept of metabolic channeling is discussed in the literature for half a decade now, the term metabolon is a relatively new one and was originally applied to enzymes of the citric acid cycle that form interactions with the inner mitochondrial matrix [138]. In slightly damaged (permeabilized) mitochondria, these particles show, in comparison to solubilized enzymes, kinetic advantage in converting malate to citrate and in fumarate oxidation [138].

Metabolic channeling is particularly interesting from the perspective of cellular compartmentation. The association of enzymes with macromolecular complexes forms "micro-compartments" within cellular compartments, thus allowing substrates and intermediates to be isolated from the surrounding "macro-compartment" [136]. Hence, distinct pools of metabolites can exist in parallel within the same compartment, thus increasing local substrate concentrations above bulk-phase substrate concentrations and thereby permitting high metabolic rates with low overall bulk concentrations of substrates and intermediates [136]. Also, metabolic intermediates are secluded from competing enzymatic reactions; unstable intermediates are protected; and the release of toxic intermediates to the bulk phase is prevented. A further advantage of metabolic channeling is the increased potential for metabolic regulation by (reversible or temporary) association of enzymes with complexes, which is of particular importance in plant secondary metabolism [2, 76, 77, 172]. It also has been proposed that metabolic channeling and the organization of primitive metabolites and catalysts into ordered metabolic complexes predated the evolution of cells and that cellular life originated from those ordered metabolic complexes [45]. From the perspective of kinetic modeling of metabolic networks, the organization of metabolism into metabolons is problematic, since kinetic constants determined in vitro with purified enzymes cannot be applied to the in vivo situation. However, metabolons, such as the glycolytic subcompartment in heart muscle cells, have been included in mathematical multicompartment models of cardiac metabolism, and it has been demonstrated that the model accurately predicted experimental observations, such as rapid activation of glycolysis and lactate production at the onset of ischemia [182].

2.4.1 Substrate Channeling and Membrane Transport

Metabolic channeling is not exclusive to reactions catalyzed by soluble enzymes, such as steps of arginine biosynthesis [1], cysteine biosynthesis [13], or the degradation of branched-chain amino acids [74]. Also, membrane transport steps can be involved in the formation of metabolons and association of membrane transporters with soluble enzymes. For example, it was shown that in red blood cells, the Cl^-/HCO_3^- anion exchanger forms an interaction with carbonic anhydrase via binding of carbonic anhydrase II (CAII) to an acidic motif of the transporter's C-terminus [123]. Carbonic anhydrase is thus positioned in close proximity to the cytosolic domains of the transporter, allowing for efficient hydration of CO_2 to HCO_3^- directly at the site of bicarbonate transport. It was shown that attachment of CAII to the transporter accelerates the transport activity by either producing or consuming bicarbonate, depending on the direction of transport [102]. Association of carbonic anhydrase with the bicarbonate transporter and the proton antiporter NHE1 was also demonstrated in renal tissues, and it was shown that the presence of two distinct carbonic anhydrase isoforms on the *cis* and *trans* sites of the renal membrane, respectively, provides the "push" and "pull" for bicarbonate transport [119]. Also, the activity of the monocarboxylate transporter MCT1 is increased when co-expressed with carbonic anhydrase in *Xenopus* oocytes [8]. Another example for the association of metabolic enzymes with a membrane transporter is the association of hexokinase with the voltage-gated anion channel (VDAC) of the mitochondrial outer membrane. Binding of hexokinase to the mitochondrial VDAC is associated with changes in VDAC structure and in interaction with the adenine nucleotide transporter (ANT) in the inner mitochondrial membrane [159]. Association of hexokinase with VDAC prevents apoptosis by preventing the formation of a mitochondrial transition pore consisting of VDAC and ANT working as a uniporter [6, 11]. Association of hexokinase with mitochondria was also recently shown for Arabidopsis [35]; it was previously shown that hexokinase can be associated with the chloroplast outer envelope membrane [171]. Not only hexokinase but also other glycolytic enzymes such as enolase are tightly associated with the outer mitochondrial membrane in yeast [22] and Arabidopsis [57]. In Arabidopsis, 7 of the 10 glycolytic enzymes were found by proteomics in the mitochondrial outer membrane fraction, and it was shown that the entire glycolytic pathway is associated with mitochondria by enzymatic activity assays [57]. Tracer analysis using ^{13}C-glucose demonstrated that isolated, purified mitochondria were able to convert glucose into intermediates of the tricarboxylic acid cycle, providing further evidence for the association of glycolysis with the mitochondrial surface [57]. In yeast, it was shown that enolase is part of a large macromolecular complex that contains glycolytic enzymes, metabolite transporters of the mitochondrial carrier family, and enzymes of the TCA cycle [57].

2.4.2 Metabolic Channeling in Photorespiratory Metabolism in Peroxisomes

A particular good example for metabolic channeling and organization of multiple enzymes as a metabolon represents the arrangement of enzymes of the photorespiratory C_2 oxidation pathway (photorespiration) in leaf peroxisomes [68, 125].

The membrane surrounding intact isolated peroxisomes can be ruptured by incubation of peroxisomes in hypo-osmotic buffer (osmotic shock). However, the latency of peroxisomal photorespiratory enzymes before and after osmotic shock remains remarkably similar, indicating the organization of these enzymes in a multimeric, macromolecular complex [68]. This assumption was supported by the observation that matrix enzymes of osmotically shocked peroxisomes remained largely associated in the form of matrix particles that could be sedimented at low centrifugal force and that were only slightly smaller in size than intact leaf peroxisomes.

When isolated, peroxisomes were supplemented with intermediates of the photorespiratory pathway, such as glycolate, serine, glutamate, and malate at physiological concentrations [65, 173], and the kinetics of glycolate formation was measured – immediate glycerate formation without any apparent lag phase was observed. Glycerate formation started at a constant rate immediately when using either intact or osmotically shocked leaf peroxisomes [67, 68]. This indicates that the reaction intermediates were efficiently channeled from one peroxisomal enzyme to the next in the reaction sequence without the release of intermediates, even in the absence of an intact peroxisomal membrane. It was indeed demonstrated that photorespiratory intermediates, such as glyoxylate, H_2O_2, and hydroxypyruvate, were not released during in vitro glycolate oxidation by osmotically shocked organelles. Apparently, compartmentalization of the photorespiratory C_2 cycle of leaf peroxisomes is not dependent on a membrane surrounding the peroxisomes but on the arrangement of the corresponding enzymes in a multimeric macromolecular complex [67, 68]. The organization of peroxisomal enzymes in a metabolon prevents the release of detrimental membrane-permeable photorespiratory intermediates such as the strong oxidant H_2O_2 and the weak acid glyoxylate, which are strong inhibitors of thioredoxin-activated enzymes, such as Rubisco and stromal fructose bisphosphatase and sedoheptulose bisphosphatase [26, 33, 51] from the peroxisomal matrix. Metabolic channeling further increases the flux of metabolites through the pathway and thus maximizes the rate of glycolate conversion and return of otherwise lost carbon to the Calvin cycle.

2.5 Plastids – Plant-Specific Organelles with a Multitude of Functions

What sets plants (i.e., land plants and algae) apart from all other eukaryotes is the presence of plastids in plant cells, semi-autonomous organelles of endosymbiotic origin that are bounded by two membranes, the inner and outer plastid envelope membranes . Plastids, the defining organelle of all photosynthetic eukaryotes, play essential roles in plants, being the site of photosynthesis and a plethora of other essential metabolic pathways, and they represent one of the major hubs in plant metabolic networks. Plastid function is also closely linked to that of other cellular compartments, with the majority of its proteins being encoded by the nuclear genome and imported into plastids after translation in the cytosol. The metabolism of plastids is heavily intertwined and connected with that of the surrounding cytosol, thus causing massive traffic of metabolic precursors, intermediates, and products between the plastid and the cytosol [162, 164, 168]. Understanding the transport of metabolites between plastid and cytosol is of crucial importance for understanding plant metabolic networks [164, 167]; we will thus discuss the evolutionary context of establishing metabolic connections between plastid and cytosol.

2.5.1 Endosymbiotic Origin of Chloroplasts and Its Relation to Metabolic Networks

When discussing the role of plastids in plant cells, it is important to consider their evolutionary origin – the first plant cell (also called the protoalga) has evolved only after the basic architecture that is common to all eukaryotic cells, including the presence of mitochondria, nucleus, and endomembrane system, had already been established. Approximately 1.5– 2 billion years ago, a primitive eukaryotic unicell engulfed a free-living photosynthetic cyanobacterium, and this engulfed cyanobacterium eventually evolved into a novel organelle, the plastid [12, 97, 98, 103, 126, 129]. This plastid-containing protoalga gave rise to the three photosynthetic eukaryotic lineages containing primary plastids (i.e., the archaeplastida [3]) – the red and green algae and the glaucocystophytes [12]. The two envelope membranes, bounding modern plastids, are relics of this evolutionary history – while the inner envelope membrane is believed to have evolved from the cyanobacterial plasma membrane, the outer envelope membrane is a chimera between bacterial outer membrane and host membrane, likely a food vacuole derived from the endomembrane system [30, 28]. While the inner leaflet of the lipid bilayer of the outer chloroplast envelope membrane consists mostly of lipids with bacterial origin, the outer leaflet has a lipid composition that resembles the ER membrane [39–42]. Whereas the plastids of the archaeplastida are of monophyletic origin, the additional photosynthetic eukaryotes evolved by secondary endosymbiosis – a process that happened multiple times during which the primary-plastid-containing red or green algae were engulfed by other eukaryotic cells and subsequently reduced to secondary plastids that are enveloped by three to four membranes [29, 91, 177, 178].

Why is this important when talking about metabolic networks? It is important because (a) the cyanobacterium introduced an entirely novel metabolic capacity into eukaryotic cells, the process of oxygenic photosynthesis, and (b) establishment of the plastid was accompanied by massive gene transfer from the cyanobacterial genome to the host nucleus by endosymbiotic gene transfer, which contributed approximately 3,000 additional genes to the genome of the newly evolving organism, thus providing a rich source of genetic material for the evolution of novel metabolic functions [99–101, 148]. Amongst many other requirements, such as the evolution of genome–plastome coordination and a protein translocation apparatus, the merger of two free-living organisms into one necessitates the integration and coordination of two previously independent metabolic entities, i.e., a photoautotrophic primary producer and a heterotrophic organism [166].

2.5.2 A Crucial First Step in Establishing Chloroplast–Cytoplasm Metabolic Connection – Evolution of the Triose Phosphate/Phosphate Translocator

Apparently, establishing a metabolic connection between host and endosymbiont is not a trivial process since it requires the establishment of mechanisms for the controlled exchange of metabolic precursors, intermediates, and end products between two previously free-living organisms, i.e., plastid and host. Very likely, of particular importance was the export of photosynthetically assimilated organic carbon compounds from the evolving plastid to the host organism [126, 166]. Controlled export of carbon from the plastid to the host is very important in this context, because continued operation of the reductive pentose phosphate pathway (i.e., the Calvin cycle) in the plastid is only possible if the withdrawal of Calvin cycle intermediates from the plastid occurs at a rate that is equal to or lower than the rate of net CO_2 assimilation, otherwise the cycle would cease operation. Hence, a feedback mechanism is required that ensures controlled release of photoassimilates from the chloroplast. Triose phosphates not required for the regeneration of the CO_2 acceptor Ru-1,5-bP represent the first important branch point in the allocation of recently assimilated carbon dioxide to different metabolic routes. They can either remain inside the chloroplast to enter plastid-localized metabolic pathways or be exported to the cytosol [145, 163]. It is long known that in land plants, the export of the triose phosphates GAP and DAP to the cytosol is mediated by a triose phosphate/phosphate antiporter [49]. This antiporter catalyzes the strict counter-exchange of one molecule of triosephosphate (TP) with one molecule of orthophosphate (Pi); that is, for each molecule of organically bound phosphate that leaves the chloroplast in the form of triosephosphate, one molecule of inorganic phosphate is returned to the chloroplast, thus avoiding phosphate depletion of the chloroplast stroma [48, 53, 66]. The strict stoichiometry of the counter-exchange is essential for maintaining phosphate homeostasis in the stroma, since inorganic phosphate is required for the biosynthesis of ATP from ADP and Pi in the light reaction of photosynthesis. Depletion of the plastidial phosphate pool by unbalanced export of phosphate in the form of triose phosphates would lead to inhibition of photosynthetic electron transport and ultimately to damage of the photosynthetic machinery.

It was recently shown that controlled export of reduced carbon from chloroplasts is catalyzed by a triosephosphate/phosphate antiporter not only in green land plants but in all photosynthetic eukaryotes for which genome or comprehensive expressed sequence tag information is available, including photosynthetic organisms containing secondary plastids, such as diatoms and dinoflagellates [166]. Apparently, the antiport mechanism of the triosephosphate/phosphate translocator is ideally suited to connect metabolic pathways in separate compartments because the withdrawal of a metabolite from one compartment is always strictly coupled with the availability of a suitable counter-exchange substrate in the other compartment. This also allows for flexible adaptation to changing metabolic requirements; that is, the direction of transport is only dependent on substrate concentrations on *cis* and *trans* side of the membrane, and the direction of metabolite flux is thus largely directed by enzymatic activities on both sides of the membrane [166].

Considering the situation of a free-living, coccal cyanobacterium, the presence of a triosephosphate/phosphate antiporter in its plasma membrane would very likely have been detrimental because it would have allowed for the efflux of triose phosphates from the cyanobacterium in the presence of suitable external concentrations of orthophosphate. It is thus reasonable to hypothesize that the triosephosphate/phosphate antiporter was not introduced by endosymbiotic gene transfer from the cyanobacterium, but it was derived from a pre-existing host protein that was directed to the cyanobacterial plasma membrane (now the inner plastid envelope membrane) after the endosymbiont had entered the host cell. This hypothesis was recently tested by phylogenomic and phylogenetic analysis of genomic and EST-sequence data from a broad range of organisms. It was shown that the plastidial triosephosphate/phosphate translocators evolved from transport proteins of the eukaryotic endomembrane system, specifically from sugar nucleotide transporters of the ER and Golgi membranes [166]. These sugar nucleotide transporters are ubiquitous in eukaryotes but absent from prokaryotes and thus likely represent a eukaryote-specific evolutionary innovation. Possibly, the food vacuole that initially engulfed the cyanobacterium was derived from the host endomembrane system and eventually fused with the bacterial outer membrane to become the outer plastid envelope membrane. A metabolite transporter originally residing in the host endomembrane system was then routed on to the bacterial plasma membrane (i.e., the plastid inner envelope membrane) and thus enabled the host to tap into the photosynthetic carbon pool of the endosymbiont. This hypothetic but nevertheless reasonable scenario must have happened very early in plastid evolution because genes encoding members of the plastidic phosphate translocator family have been detected in the genomes of all sequenced photosynthetic eukaryotes, including red algae, green algae, land plants, and photosynthetic organisms containing complex plastids, such as diatoms and dinoflagellates [166]. In other words, plastidic phosphate translocators have already evolved at the stage of the protoalga, clearly before the split of the red and green plant lineages.

In land plants, plastidial phosphate translocators can be classified into four distinct groups with distinctive substrate specificities, namely the triosephosphate/phosphate translocators (TPT), the phosphoenolpyruvate/phosphate translocators (PPT), the glucose 6-phosphate/phosphate translocators (GPT), and the xylose 5-phosphate/phosphate translocators (XPT) (see [52, 83] for detailed reviews of substrate specificities and kinetic constants). At least three of the subgroups have evolved very early on: genes encoding proteins belonging to the TPT and PPT families can be clearly detected in the genomes of ancient red microalgae, as well as in green plants, meaning these proteins have evolved before the split of the red and green lineages. The situation is less clear for the GPT/XPT clade. Proteins belonging to this clade can be detected in red algae, but based on phylogenetic analysis, it is difficult to decide whether they actually represent functional GPTs or XPTs; experimental analysis of their substrate specificity will be required. Nevertheless, clearly three distinct clades of plastidial PTs are already established in basic red algae.

Interestingly, the TPT of red algae, in contrast to that of land plants, has a more narrow substrate specificity. Whereas the protein from land plants is able to exchange orthophosphate (Pi) for triose phosphates (i.e., glyceraldehyde 3-phosphate and dihydroxyacetone phosphate) and 3-phosphoglyceric acid (3-PGA), its red algal paralog does not accept 3-PGA as counter-exchange substrate for either triose phosphates or Pi (unpublished results, M. Linka and A.P.M. Weber). Thus, other than the green plant protein, the red algal protein is not able to serve as a redox shuttle between plastid stroma and cytoplasm by exchanging TPs for 3-PGA. Since red algae, in contrast to green plants, do not store starch in the chloroplast stroma but in the cytosol, red algae do not have the option of depositing recently assimilated carbon inside the chloroplast as transitory starch; hence, most of the assimilated carbon has to be efficiently exported from the chloroplast in the form of triose phosphates to the cytosol where they have to be metabolized to regain the Pi required for continued operation of photosynthesis. Since in red algae all of the assimilated carbon needs to leave the chloroplast via TPT at a rate close to the rate of photosynthetic CO_2 assimilation, a competition of 3-PGA with TPs for transport across the chloroplast envelope membrane would be counter-productive. Hence, evolution of a narrow-specificity TPT in red algae was of advantage.

In the process of secondary endosymbiosis that led to the evolution of secondary plastids in the chromalveolates, a red alga was captured by a non-photosynthetic protist and eventually reduced to a complex plastid that is surrounded by four envelope membranes [91]. The two innermost membranes are thought to be derived from the inner and outer chloroplast membranes of the red algae, the third layer is a remnant of the red algal plasma membrane, and the outermost envelope membrane is derived from the protist endomembrane system. Interestingly, the gene encoding the red algal TPT was transferred from the degenerating red algal nucleus to the nuclear genome of the host, whereas the genes encoding PPT and XPT/GPT apparently were lost in the process. Once arrived at its new location, the TPT gene started to radiate by duplication events and frequently evolved into small gene families [166]. In the case of the apicoplast-containing malaria parasite *Plasmodium falciparum*, the nuclear genome harbors two genes encoding TPTs – one of the gene products is targeted to the inner envelope membrane, whereas the other one is targeted to the outermost envelope membrane [106]. It is interesting that the recruitment of phosphate translocators for the export of reduced carbon from plastids was recapitulated during the evolution of complex plastids in the chromalveolates, thus emphasizing the importance of a phosphate-balanced, controlled export of TPs from the chloroplast.

2.5.3 Integration of Endosymbiont Metabolism with Host Metabolism Was a Host-Driven Process

A comprehensive analysis of the plastid envelope permeome of *Arabidopsis thaliana* showed that at least 50% of the metabolite transporters residing in the plastid envelope evolved from pre-existing host proteins and only a relatively small portion of the permeome was contributed by the endosymbiont [152]. That is, integration of plastid and host metabolism was predominantly a host-driven process, in which transport proteins encoded by the host genome acquired targeting signals for routing to the chloroplast and were inserted into the chloroplast envelope membrane. A surprisingly large share of plastid envelope membrane transporters, such as the adenine nucleotide transporters, the dicarboxylate translocators, and some metal-transporting ATPases have their evolutionary origin in prokaryotic intracellular energy parasites (i.e., *Chlamydia* and/or *Rickettsia*). Most of these genes have already been introduced into the genome of the protoalga by horizontal transfer, and they have been maintained throughout plant evolution, emphasizing their importance for connecting the metabolism of plastid and cytosol [152]. Other plastid transporters belong to the mitochondrial carrier family, such as the folate transporter FOLT1 and the *S*-adenosylmethionine transporter SAMT1 [16, 152, 164]. In this case, proteins that in non-photosynthetic eukaryotes serve as mitochondrial transporters have been recruited to the chloroplast in plants and algae. In most cases, these metabolite transporters function in the antiport mode. Specifically, they catalyze the strict counter-exchange of two substrates that belong to or originate from the same metabolic pathway or enzymatic reaction, such as the substrate pairs ATP and ADP, *S*-adenosylmethionine and *S*-adenosylhomocysteine, oxaloacetate and malate, to mention only few. Apparently, it was of evolutionary advantage to recruit substrate antiporters for the exchange of metabolites across the plastid envelope membrane. More generally, this mode of transport seems to be the preferred one for transport of metabolites across the membranes of mitochondria and plastids, and possibly also of the Golgi apparatus and the ER. This is in contrast to the plasma membrane and the tonoplast of the vacuole, where transport is frequently energized by proton co- or antiport, or, in the case of ABC-type transporters, by hydrolysis of ATP. There are exceptions to this general pattern, though. For example, the transport of pyruvate into chloroplasts was shown to occur by co-transport with protons or sodium, depending on the plant species [4, 55, 71]. To make the situation even more complicated, uniport systems co-exist with antiporters. For example, chloroplasts have in addition to the phosphate translocator family members also a phosphate/proton cotransporter [153]. Although a possible role in cell metabolism was suggested [121], the physiological relevance of this latter transporter is not yet understood, in particular if it serves as a phosphate importer (i.e., import of phosphate into the chloroplast). A phosphate exporter is however required for starch-synthesizing amyloplasts in storage organs, such as potato tubers: starch biosynthesis in amyloplasts of dicotyledonous plants requires glucose 6-phosphate (G6P) and ATP, which are both imported from the cytosol [78, 95, 109, 149]. Whereas the import of G6P is Pi balanced – because for each G6P that is imported, one Pi is exported – ATP import is not. ATP is imported in counter-exchange with ADP; consequently, one Pi is accumulated in the plastid for each ATP that is hydrolyzed to ADP and Pi. It is not known how Pi is exported from starch-synthesizing amyloplasts, although unidirectional export of orthophosphate across the envelope of isolated cauliflower bud plastids has been demonstrated [108].

In summary, it becomes clear that many, if not most, metabolite translocators of the plastid envelope membrane are of host origin, meaning that integration of plastidic with cytosolic metabolism was predominantly a host-driven process [152].

Metabolite transporters of the plastid envelope membrane frequently work as substrate antiporters; as a consequence, rate and direction of transport are commonly dependent on metabolic reactions and substrate concentrations on both sides of the membranes. Put into the evolutionary context, it was apparently of advantage to recruit substrate antiporters for connecting plastid and cytosolic metabolism because this mode of transport, in addition to shuffling metabolites across membranes, also provides a means for crosstalk between compartments and for coordinating metabolic activities on two sides of a membrane. This is a very different situation than substrate uptake across the plasma membrane. In this case, uptake is predominantly dependent on transporter activity.

2.6 Bridging the Divide – Why Genomics, Phylogenetics, Evolution, and Metabolic Networks Belong Together

Genomics is an essential foundation of systems biology, and genome sequences are prerequisite for many other "omics" technologies, such as transcriptomics and proteomics, and for the reconstruction of metabolic and regulatory networks. Recently developed technology has massively increased throughput and decreased cost of genomic sequencing, respectively, thus allowing for many additional genomes to be sequenced, not just from model organisms but also from any source from which DNA can be isolated. In addition, sequencing of transcriptomes will provide additional insights into gene content of genomes that are too large to be sequenced at reasonable cost with current technology [23, 169].

Several novel ultra-low-cost sequencing (ULCS) technologies have recently reduced the cost of DNA sequencing by several orders of magnitude [134]. Using these technologies, a typical microbial genome with a 4 million base genome can be sequenced and assembled within days by a single operator. Also, the novel sequencing technologies do not require cloning. Thus, they can be used to sequence unclonable DNAs with a high GC content and samples that are too degraded (e.g., fossil or ancient DNA) to be reasonably sequenced by traditional approaches. For example, using massively parallel pyrosequencing, sequence information was generated from a wooly mammoth that perished 300,000 years ago [118] and the genome of *Mycoplasma genitalium* (the causal agent of non-gonococcal urethritis) was sequenced in one single instrument run [96]. Sequencing cost will further decrease and throughput will further increase by application of multiplex polony sequencing in the very near future [32, 135].

2.6.1 Orphans, Neighbors, Clusters

Basically, sequence information is just a combination of As, Cs, Gs, and Ts. So how do the recent breakthroughs in genome sequencing technology affect metabolic networks? For this, we need to go beyond annotation, comparative genomics, designing mRNA profiling and whole-genome tiling microarrays, single gene and multigene phylogenies, etc. The objective is to deduce the metabolic network of a recently sequenced organism from its genome, or alternatively, its transcriptome sequence, i.e., ab initio metabolic reconstruction. Usually, a basic reconstruction of the network is founded on functional annotation of gene products that in turn is based on sequence similarities. This strategy is, for example, applied by the KEGG automatic annotation server (KAAS; http://www.genome.jp/kegg/kaas/) [105], which makes heavy use of groups of orthologous genes that are generated from pairwise genome comparisons using bi-directional best-hit (BBH) relationships [143, 144]. While these and other automated procedures for functional annotation and network reconstruction provide a good starting point, a detailed genome-scale reconstruction currently still requires a large amount of manual annotation, literature-based data curation, and significant biochemical and metabolic reaction knowledge. The intense effort required for network reconstruction in terms of time and expertise is thus the reason for the relatively small number of genomes for which detailed and reliable networks are available, such as the unicelluar microbes *Escherichia coli*, *Bacillus subtilis*, and *Saccharomyces cerevisiae* [46, 56, 112], and the multicellular *Homo sapiens* [104]; see also Chapter 8.

Importantly, though, these methods for genome-scale network reconstruction do not permit the assignment of functions to unknown genes and include assigning genes to metabolic functions, for which no gene has been identified yet (orphan metabolic activities) [31, 115]. Possible solutions to this dilemma might come from combining multiple types of associative evidence, such as phylogenetic profiles, gene co-expression patterns, protein interaction data, genetic neighborhood analysis, and the construction of logical relationships [15, 18, 19, 31, 79, 80, 117, 128, 155, 176]. The power of these higher order data analyses increases with the depth of the sequence space because more genomic sequence information permits to generate deeper phylogenetic profiles (co-occurrence profiles across multiple species) [117] and more complex logical relationships [17, 18] and to generate more information from genomic neighborhood analysis [116]. The next challenge will be to overlay and integrate these data with transcript co-expression data, protein–protein interaction and protein fusion data, and with metabolomics data. Eventually, the integration of multiple multivariate datasets might permit the ab initio computational deduction of complex networks and the comprehensive prediction of metabolic capabilities and pathways directly from genome sequence.

References

1. Abadjieva A, Pauwels K, Hilven P, Crabeel M (2001) A new yeast metabolon involving at least the two first enzymes of arginine biosynthesis: Acetylglutamate synthase activity requires complex formation with acetylglutamate kinase. J Biol Chem 276:42869–42880.
2. Achnine L, Blancaflor EB, Rasmussen S, Dixon RA (2004) Colocalization of L-phenylalanine ammonia-lyase and cinnamate 4-hydroxylase for metabolic channeling in phenylpropanoid biosynthesis. Plant Cell 16:3098–3109.
3. Adl SM, Simpson AG, Farmer MA, Andersen RA, Anderson OR, Barta JR, Bowser SS, Brugerolle G, Fensome RA, Fredericq S, James TY, Karpov S, Kugrens P, Krug J, Lane CE,

Lewis LA, Lodge J, Lynn DH, Mann DG, McCourt RM, Mendoza L, Moestrup O, Mozley-Standridge SE, Nerad TA, Shearer CA, Smirnov AV, Spiegel FW, Taylor MF (2005) The new higher level classification of eukaryotes with emphasis on the taxonomy of protists. J Eukaryot Microbiol 52:399–451.
4. Aoki N, Ohnishi J, Kanai R (1992) 2 different mechanisms for transport of pyruvate into mesophyll chloroplasts of C4 plants-a comparative study. Plant Cell Physiol 33:805–809.
5. Arlt K, Brandt S, Kehr J (2001) Amino acid analysis in five pooled single plant cell samples using capillary electrophoresis coupled to laser-induced fluorescence detection. J Chromatogr A 926:319–325.
6. Azoulay-Zohar H, Israelson A, Abu-Hamad S, Shoshan-Barmatz V (2004) In self-defence: Hexokinase promotes voltage-dependent anion channel closure and prevents mitochondria-mediated apoptotic cell death. Biochem J 377:347–355.
7. Badger MR, Price GD (2003) CO_2 concentrating mechanisms in cyanobacteria: Molecular components, their diversity and evolution. J Exp Bot 54:609–622.
8. Becker HM, Broer S, Deitmer JW (2004) Facilitated lactate transport by MCT1 when coexpressed with the sodium bicarbonate cotransporter (NBC) in Xenopus oocytes. Biophys J 86:235–247.
9. Berner RA (1999) Atmospheric oxygen over phanerozoic time. Proc Natl Acad Sci U S A 96:10955–10957.
10. Berner RA (2003) The long-term carbon cycle, fossil fuels and atmospheric composition. Nature 426:323–326.
11. Beutner G, Ruck A, Riede B, Brdiczka D (1998) Complexes between porin, hexokinase, mitochondrial creatine kinase and adenylate translocator display properties of the permeability transition pore. Implication for regulation of permeability transition by the kinases. Biochim Biophys Acta 1368:7–18.
12. Bhattacharya D, Yoon HS, Hackett JD (2004) Photosynthetic eukaryotes unite: Endosymbiosis connects the dots. Bioessays 26:50–60.
13. Bogdanova N, Hell R (1997) Cysteine synthesis in plants: Protein-protein interactions of serine acetyltransferase from *Arabidopsis thaliana*. Plant J 11:251–262.
14. Boldt R, Edner C, Kolukisaoglu U, Hagemann M, Weckwerth W, Wienkoop S, Morgenthal K, Bauwe H (2005) D-Glycerate 3-kinase, the last unknown enzyme in the photorespiratory cycle in Arabidopsis, belongs to a novel kinase family. Plant Cell 17:2413–2420.
15. Bono H, Ogata H, Goto S, Kanehisa M (1998) Reconstruction of amino acid biosynthesis pathways from the complete genome sequence. Genome Res 8:203–210.
16. Bouvier F, Linka N, Isner JC, Mutterer J, Weber APM, Camara B (2006) Arabidopsis SAMT1 defines a plastid transporter regulating plastid biogenesis and plant development. Plant Cell 18:3088–3105.
17. Bowers PM, Cokus SJ, Eisenberg D, Yeates TO (2004) Use of logic relationships to decipher protein network organization. Science 306:2246–2249.
18. Bowers PM, O'Connor BD, Cokus SJ, Sprinzak E, Yeates TO, Eisenberg D (2005) Utilizing logical relationships in genomic data to decipher cellular processes. FEBS J 272:5110–5118.
19. Bowers PM, Pellegrini M, Thompson MJ, Fierro J, Yeates TO, Eisenberg D (2004) Prolinks: A database of protein functional linkages derived from coevolution. Genome Biol 5:R35.
20. Bowes G, Ogren WL (1972) Oxygen inhibition and other properties of soybean ribulose 1,5-diphosphate carboxylase. J Biol Chem 247:2171–2176.
21. Bowes G, Ogren WL, Hageman RH (1971) Phosphoglycolate production catalyzed by ribulose diphosphate carboxylase. Biochem Biophys Res Commun 45:716–722.
22. Brandina I, Graham J, Lemaitre-Guillier C, Entelis N, Krasheninnikov I, Sweetlove L, Tarassov I, Martin RP (2006) Enolase takes part in a macromolecular complex associated to mitochondria in yeast. Biochim Biophys Acta 1757:1217–1228.
23. Bräutigam A, Shrestha RP, Whitten D, Wilkerson CG, Carr KM, Froehlich JE, Weber APM (2008) Massively-parallel pyrosequencing of cDNAs enables proteomics in non-model species: Comparative analysis of a species specific database generated by pyrosequencing

and non-species specific databases for proteome analysis of pea chloroplast envelopes. J Biotechnol 136:44–53.
24. Buchanan JG, Lynch VH, Benson AA, Bradley DF, Calvin M (1953) The path of carbon in photosynthesis. XVIII. The identification of nucleotide coenzymes. J Biol Chem 203: 935–945.
25. Calvin M, Benson AA (1949) The path of carbon in photosynthesis IV: The identity and sequence of the intermediates in sucrose synthesis. Science 109:140–142.
26. Campbell WJ, Ogren WL (1990) Glyoxylate inhibition of ribulosebisphosphate carboxylase oxygenase activation in intact, lysed, and reconstituted chloroplasts. Photosynth Res 23: 257–268.
27. Canvin DT, Beevers H (1961) Sucrose synthesis from acetate in the germinating castor bean: Kinetics and pathway. J Biol Chem 236:988–995.
28. Cavalier-Smith T (2000) Membrane heredity and early chloroplast evolution. Trends Plant Sci 5:174–182.
29. Cavalier-Smith T (2002) Chloroplast evolution: Secondary symbiogenesis and multiple losses. Curr Biol 12:R62–64.
30. Cavalier-Smith T (2003) Genomic reduction and evolution of novel genetic membranes and protein-targeting machinery in eukaryote-eukaryote chimaeras (meta-algae). Philos Trans R Soc Lond B Biol Sci 358:109–134.
31. Chen L, Vitkup D (2006) Predicting genes for orphan metabolic activities using phylogenetic profiles. Genome Biol 7:R17.
32. Church G, Shendure J, Porreca G (2006) Sequencing thoroughbreds. Nat Biotechnol 24: 139.
33. Cook CM, Spellman M, Tolbert NE, Stringer CD, Hartman FC (1985) Characterization of an active-site peptide modified by glyoxylate and pyridoxal-phosphate from spinach ribulosebisphosphate carboxylase-oxygenase. Arch Biochem Biophys 240:402–412.
34. Crowley TJ, Berner RA (2001) Paleoclimate. CO_2 and climate change. Science 292: 870–872.
35. Damari-Weissler H, Ginzburg A, Gidoni D, Mett A, Krassovskaya I, Weber APM, Belausov E, Granot D (2007) Spinach SoHXK1 is a mitochondria-associated hexokinase. Planta 226:1053–1058.
36. Day DA, Wiskich JT (1980) Glycine transport by pea leaf mitochondria. FEBS Lett 112:191–194.
37. Day DA, Wiskich JT (1981) Glycine metabolism and oxaloacetate transport by pea leaf mitochondria. Plant Physiol 68:425–429.
38. de Veau EJ, Burris JE (1988) Photorespiratory rates in wheat and maize as determined by ^{18}O-labeling. Plant Physiol 90:500–511.
39. Dorne A-J, Joyard J, Block MA, Douce R (1985) Localization of phosphatidylcholine in outer envelope membrane of spinach chloroplasts. J Cell Biol 100:1690–1697.
40. Douce R, Block MA, Dorne A-J, Joyard J (1984) The plastid envelope membrane their structure, composition, and role in chloroplast biogenesis. Sub Cell Biochem 10:1–84.
41. Douce R, Joyard J (1980) Chloroplast envelope lipids: Detection and biosynthesis. Methods Enzymology 69:290–301.
42. Douce R, Joyard J (1981) Does the plastid envelope derive from the endoplasmic reticulum? Trends Biochem Sci:237–239.
43. Edwards GE, Franceschi VR, Voznesenskaya EV (2004) Single-cell C4 photosynthesis versus the dual-cell (Kranz) paradigm. Annu Rev Plant Biol 55:173–196.
44. Edwards GE, Furbank RT, Hatch MD, Osmond CB (2001) What does it take to be C_4? lessons from the evolution of C_4 photosynthesis. Plant Physiol 125:46–49.
45. Edwards MR (1996) Metabolite channeling in the origin of life. J Theor Biol 179: 313–322.
46. Feist AM, Henry CS, Reed JL, Krummenacker M, Joyce AR, Karp PD, Broadbelt LJ, Hatzimanikatis V, Palsson BO (2007) A genome-scale metabolic reconstruction for *Escherichia*

coli K-12 MG1655 that accounts for 1260 ORFs and thermodynamic information. Mol Sys Biol 3:121.
47. Fischer K, Kammerer N, Gutensohn M, Arbinger B, Weber A, Häusler RE, Flügge UI (1997) A new class of plastidic phosphate translocators: A putative link between primary and secondary metabolism by the phosphoenolpyruvate/phosphate antiporter. Plant Cell 9:453–462.
48. Flügge UI (1995) Phosphate translocation in the regulation of photosynthesis. J Exp Bot 46:1317–1323.
49. Flügge UI (1999) Phosphate translocators in plastids. Annu Rev Plant Physiol Plant Mol Biol 50:27–45.
50. Flügge UI, Fischer K, Gross A, Sebald W, Lottspeich F, Eckerskorn C (1989) The triose phosphate-3-phosphoglycerate-phosphate translocator from spinach chloroplasts: Nucleotide sequence of a full-length cDNA clone and import of the *in vitro* synthesized precursor protein into chloroplasts. EMBO J 8:39–46.
51. Flügge UI, Freisl M, Heldt HW (1980) The mechanism of the control of carbon fixation by the ph in the chloroplast stroma – studies with acid mediated proton-transfer across the envelope. Planta 149:48–51.
52. Flügge UI, Häusler RE, Ludewig F, Fischer K (2003) Functional genomics of phosphate antiport systems of plastids. Physiol Plantarum 118:475–482.
53. Flügge UI, Heldt HW (1984) The phosphate-triose phosphate-phosphoglycerate translocator of the chloroplast. Trends Biochem Sci 9:530–533.
54. Flügge UI, Heldt HW (1991) Metabolite translocators of the chloroplast envelope. Annu Rev Plant Physiol Plant Mol Biol 42:129–144.
55. Flügge UI, Stitt M, Heldt HW (1985) Light-driven uptake of pyruvate into mesophyll chloroplasts from maize. FEBS Lett 183:335–339.
56. Forster J, Famili I, Fu P, Palsson BO, Nielsen J (2003) Genome-scale reconstruction of the Saccharomyces cerevisiae metabolic network. Genome Res 13:244–253.
57. Giege P, Heazlewood JL, Roessner-Tunali U, Millar AH, Fernie AR, Leaver CJ, Sweetlove LJ (2003) Enzymes of glycolysis are functionally associated with the mitochondrion in Arabidopsis cells. Plant Cell 15:2140–2151.
58. Hatch MD (1971) The C 4 -pathway of photosynthesis. Evidence for an intermediate pool of carbon dioxide and the identity of the donor C 4 -dicarboxylic acid. Biochem J 125:425–432.
59. Hatch MD, Dröscher L, Flügge UI, Heldt HW (1984) A specific translocator for oxaloacetate transport in chloroplasts. FEBS Lett 178:15–19.
60. Hatch MD, Slack CR (1966) Photosynthesis by sugar-cane leaves. A new carboxylation reaction and the pathway of sugar formation. Biochem J 101:103–111.
61. Hatch MD, Slack CR (1967) The participation of phosphoenolpyruvate synthetase in photosynthetic CO_2 fixation of tropical grasses. Arch Biochem Biophys 120:224–225.
62. Haupt-Herting S, Klug K, Fock HP (2001) A new approach to measure gross CO_2 fluxes in leaves. Gross CO_2 assimilation, photorespiration, and mitochondrial respiration in the light in tomato under drought stress. Plant Physiol 126:388–396.
63. Häusler RE, Schlieben NH, Flügge UI (2000) Control of carbon partitioning and photosynthesis by the triose phosphate/phosphate translocator in transgenic tobacco plants (*Nicotiana tabacum*). II. Assessment of control coefficients of the triose phosphate/phosphate translocator. Planta 210:383–390.
64. Häusler RE, Schlieben NH, Nicolay P, Fischer K, Fischer KL, Flügge UI (2000) Control of carbon partitioning and photosynthesis by the triose phosphate/phosphate translocator in transgenic tobacco plants (*Nicotiana tabacum* L.). I. Comparative physiological analysis of tobacco plants with antisense repression and overexpression of the triose phosphate/phosphate translocator. Planta 210:371–382.
65. Heineke D, Riens B, Grosse H, Hoferichter P, Peter U, Flügge UI, Heldt HW (1991) Redox transfer across the inner chloroplast envelope membrane. Plant Physiol 95:1131–1137.
66. Heldt HW, Flügge UI, Borchert S, Brueckner G, Ohnishi J (1990) Phosphate translocators in plastids. Plant Biol 10:39–54.

67. Heupel R, Heldt HW (1994) Protein organization in the matrix of leaf peroxisomes. A multienzyme complex involved in photorespiratory metabolism. Eur J Biochem 220:165–172.
68. Heupel R, Markgraf T, Robinson DG, Heldt HW (1991) Compartmentation studies on spinach leaf peroxisomes. Evidence for channeling of photorespiratory metabolites in peroxisomes devoid of intact boundary membrane. Plant Physiol 96:971–979.
69. Howitz KT, McCarty RE (1985) Kinetic characteristics of the chloroplast envelope glycolate transporter. Biochemistry 24:2645–2652.
70. Howitz KT, McCarty RE (1985) Substrate specificity of the pea chloroplast glycolate transporter. Biochemistry 24:3645–3650.
71. Huber SC, Edwards GE (1977) Transport in C_4 mesophyll chloroplasts: Characterization of the pyruvate carrier. Biochim Biophys Acta 462:583–602.
72. Huber SC, Edwards GE (1977) Transport in C4 mesophyll chloroplasts: Evidence for an exchange of inorganic phosphate and phosphoenolpyruvate. Biochim Biophys Acta 462:603–612.
73. Igarashi D, Miwa T, Seki M, Kobayashi M, Kato T, Tabata S, Shinozaki K, Ohsumi C (2003) Identification of photorespiratory glutamate : Glyoxylate aminotransferase (GGAT) gene in Arabidopsis. Plant J 33:975–987.
74. Islam MM, Wallin R, Wynn RM, Conway M, Fujii H, Mobley JA, Chuang DT, Hutson SM (2007) A novel branched-chain amino acid metabolon. Protein-protein interactions in a supramolecular complex. J Biol Chem 282:11893–11903.
75. Johnson HS, Hatch MD (1969) The C4-dicarboxylic acid pathway of photosynthesis. Identification of intermediates and products and quantitative evidence for the route of carbon flow. Biochem J 114:127–134.
76. Jones P, Vogt T (2001) Glycosyltransferases in secondary plant metabolism: Tranquilizers and stimulant controllers. Planta 213:164–174.
77. Jorgensen K, Rasmussen AV, Morant M, Nielsen AH, Bjarnholt N, Zagrobelny M, Bak S, Moller BL (2005) Metabolon formation and metabolic channeling in the biosynthesis of plant natural products. Curr Opin Plant Biol 8:280–291.
78. Kammerer B, Fischer K, Hilpert B, Schubert S, Gutensohn M, Weber A, Flügge UI (1998) Molecular characterization of a carbon transporter in plastids from heterotrophic tissues: The glucose6-phosphate/phosphate antiporter. Plant Cell 10:105–117.
79. Kharchenko P, Chen L, Freund Y, Vitkup D, Church GM (2006) Identifying metabolic enzymes with multiple types of association evidence. BMC Bioinformatics 7:177.
80. Kharchenko P, Vitkup D, Church GM (2004) Filling gaps in a metabolic network using expression information. Bioinformatics 20 Suppl 1:I178–I185.
81. Kleczkowski LA, Givan CV, Hodgson JM, Randall DD (1988) Subcellular location of NADPH-dependent hydroxypyruvate reductase activity in leaf protoplasts of *Pisum sativum* L. and its role in photorespiratory metabolism. Plant Physiol 88:1182–1185.
82. Kleczkowski LA, Randall DD (1988) Purification and characterization of a novel NADPH(NADH)-dependent hydroxypyruvate reductase from spinach leaves. Comparison of immunological properties of leaf hydroxypyruvate reductases. Biochem J 250:145–152.
83. Knappe S, Flügge UI, Fischer K (2003) Analysis of the plastidic phosphate translocator gene family in Arabidopsis and identification of new phosphate translocator-homologous transporters, classified by their putative substrate-binding site. Plant Physiol 131: 1178–1190.
84. Kockenberger W (2001) Nuclear magnetic resonance micro-imaging in the investigation of plant cell metabolism. J Exp Bot 52:641–652.
85. Kockenberger W, De Panfilis C, Santoro D, Dahiya P, Rawsthorne S (2004) High resolution NMR microscopy of plants and fungi. J Microsc 214:182–189.
86. Kornberg HL, Beevers H (1957) A mechanism of conversion of fat to carbohydrate in castor beans. Nature 180:35–36.
87. Kornberg HL, Beevers H (1957) The glyoxylate cycle as a stage in the conversion of fat to carbohydrate in castor beans. Biochim Biophys Acta 26:531–537.

88. Laisk A, Edwards GE (2000) A mathematical model of C4 photosynthesis: The mechanism of concentrating CO2 in NADP-malic enzyme type species. Photosynth Res 66:199–224.
89. Laisk A, Sumberg A (1994) Partitioning of the leaf CO_2 Exchange into components using CO_2 exchange and fluorescence measurements. Plant Physiol 106:689–695.
90. Leegood RC, von Caemmerer S (1989) Some relationships between contents of photosynthetic metabolites and the rate of photosynthetic carbon assimilation in leaves of *Zea mays*. Planta 178:258–266.
91. Li S, Nosenko T, Hackett JD, Bhattacharya D (2006) Phylogenomic analysis identifies red algal genes of endosymbiotic origin in the chromalveolates. Mol Biol Evol 23:663–674.
92. Li SH, Schneider B, Gershenzon J (2007) Microchemical analysis of laser-microdissected stone cells of Norway spruce by cryogenic nuclear magnetic resonance spectroscopy. Planta 225:771–779.
93. Liepman AH, Olsen LJ (2001) Peroxisomal alanine: Glyoxylate aminotransferase (AGT1) is a photorespiratory enzyme with multiple substrates in *Arabidopsis thaliana*. Plant J 25: 487–498.
94. Linka M, Weber APM (2005) Shuffling ammonia between mitochondria and plastids during photorespiration. Trends Plant Sci 10:461–465.
95. Linke C, Conrath U, Jeblick W, Betsche T, Mahn A, During K, Neuhaus HE (2002) Inhibition of the plastidic ATP/ADP transporter protein primes potato tubers for augmented elicitation of defense responses and enhances their resistance against *Erwinia carotovora*. Plant Physiol 129:1607–1615.
96. Margulies M, Egholm M, Altman WE, Attiya S, Bader JS, Bemben LA, Berka J, Braverman MS, Chen YJ, Chen Z, Dewell SB, Du L, Fierro JM, Gomes XV, Godwin BC, He W, Helgesen S, Ho CH, Irzyk GP, Jando SC, Alenquer ML, Jarvie TP, Jirage KB, Kim JB, Knight JR, Lanza JR, Leamon JH, Lefkowitz SM, Lei M, Li J, Lohman KL, Lu H, Makhijani VB, McDade KE, McKenna MP, Myers EW, Nickerson E, Nobile JR, Plant R, Puc BP, Ronan MT, Roth GT, Sarkis GJ, Simons JF, Simpson JW, Srinivasan M, Tartaro KR, Tomasz A, Vogt KA, Volkmer GA, Wang SH, Wang Y, Weiner MP, Yu P, Begley RF, Rothberg JM (2005) Genome sequencing in microfabricated high-density picolitre reactors. Nature 437:376–380.
97. Margulis L (1971) Symbiosis and evolution. Sci Am 225:48–57.
98. Margulis L (1975) Symbiotic theory of the origin of eukaryotic organelles; criteria for proof. Symp Soc Exp Biol 29:21–38.
99. Martin W (2003) Gene transfer from organelles to the nucleus: Frequent and in big chunks. Proc Natl Acad Sci U S A 100:8612–8614.
100. Martin W, Rujan T, Richly E, Hansen A, Cornelsen S, Lins T, Leister D, Stoebe B, Hasegawa M, Penny D (2002) Evolutionary analysis of Arabidopsis, cyanobacterial, and chloroplast genomes reveals plastid phylogeny and thousands of cyanobacterial genes in the nucleus. Proc Natl Acad Sci U S A 99:12246–12251.
101. Martin W, Stoebe B, Goremykin V, Hansmann S, Hasegawa M, Kowallik KV (1998) Gene transfer to the nucleus and the evolution of chloroplasts. Nature 393:162–165.
102. McMurtrie HL, Cleary HJ, Alvarez BV, Loiselle FB, Sterling D, Morgan PE, Johnson DE, Casey JR (2004) The bicarbonate transport metabolon. J Enzyme Inhib Med Chem 19: 231–236.
103. Mereschkowsky C (1905) Über natur und ursprung der chromatophoren im pflanzenreiche. Biol Centralbl 25:593–604.
104. Mo ML, Jamshidi N, Palsson BO (2007) A genome-scale, constraint-based approach to systems biology of human metabolism. Mol Biosyst 3:598–603.
105. Moriya Y, Itoh M, Okuda S, Yoshizawa AC, Kanehisa M (2007) KAAS: An automatic genome annotation and pathway reconstruction server. Nucleic Acids Res 35:W182–185.
106. Mullin KA, Lim L, Ralph SA, Spurck TP, Handman E, McFadden GI (2006) Membrane transporters in the relict plastid of malaria parasites. Proc Natl Acad Sci U S A 103: 9572–9577.

107. Murray AJ, Blackwell RD, Lea PJ (1989) Metabolism of hydroxypyruvate in a mutant of barley lacking NADH-dependent hydroxypyruvate reductase, an important photorespiratory enzyme activity. Plant Physiol 91:395–400.
108. Neuhaus HE, Maass U (1996) Unidirectional transport of orthophosphate across the envelope of isolated cauliflower-bud amyloplasts. Planta 198:542–548.
109. Neuhaus HE, Wagner R (2000) Solute pores, ion channels, and metabolite transporters in the outer and inner envelope membranes of higher plant plastids. Biochim Biophys Acta 1465:307–323.
110. Oaks A (1994) Efficiency of nitrogen utilization in C_3 and C_4 cereals. Plant Physiol 106: 407–414.
111. Ogren WL, Bowes G (1971) Ribulose diphosphate carboxylase regulates soybean photorespiration. Nat New Biol 230:159–160.
112. Oh YK, Palsson BO, Park SM, Schilling CH, Mahadevan R (2007) Genome-scale reconstruction of metabolic network in bacillus subtilis based on high-throughput phenotyping and gene essentiality data. J Biol Chem 282:28791–28799.
113. Ohnishi J, Flügge UI, Heldt HW (1989) Phosphate Translocator of Mesophyll and Bundle Sheath Chloroplasts of a C_4 Plant, *Panicum miliaceum* L. Identification and Kinetic Characterization. Plant Physiol 91:1507–1511.
114. Oliver DJ, Raman R (1995) Glycine decarboxylase: Protein chemistry and molecular biology of the major protein in leaf mitochondria. J Bioenerg Biomembr 27:407–414.
115. Osterman A, Overbeek R (2003) Missing genes in metabolic pathways: A comparative genomics approach. Curr Opin Chem Biol 7:238–251.
116. Overbeek R, Fonstein M, D'Souza M, Pusch GD, Maltsev N (1999) The use of gene clusters to infer functional coupling. Proc Natl Acad Sci U S A 96:2896–2901.
117. Pellegrini M, Marcotte EM, Thompson MJ, Eisenberg D, Yeates TO (1999) Assigning protein functions by comparative genome analysis: Protein phylogenetic profiles. Proc Natl Acad Sci U S A 96:4285–4288.
118. Poinar HN, Schwarz C, Qi J, Shapiro B, Macphee RD, Buigues B, Tikhonov A, Huson DH, Tomsho LP, Auch A, Rampp M, Miller W, Schuster SC (2006) Metagenomics to paleogenomics: Large-scale sequencing of mammoth DNA. Science 311:392–394.
119. Purkerson JM, Schwartz GJ (2007) The role of carbonic anhydrases in renal physiology. Kidney Int 71:103–115.
120. Rao SK, Magnin NC, Reiskind JB, Bowes G (2002) Photosynthetic and other phosphoenolpyruvate carboxylase isoforms in the single-cell, facultative C(4) system of hydrilla verticillata. Plant Physiol 130:876–886.
121. Rausch C, Zimmermann P, Amrhein N, Bucher M (2004) Expression analysis suggests novel roles for the plastidic phosphate transporter Pht2;1 in auto- and heterotrophic tissues in potato and Arabidopsis. Plant J 39:13–28.
122. Rehfeld DW, Tolbert NE (1972) Aminotransferases in peroxisomes from spinach leaves. J Biol Chem 247:4803–4811.
123. Reithmeier RA (2001) A membrane metabolon linking carbonic anhydrase with chloride/bicarbonate anion exchangers. Blood Cells Mol Dis 27:85–89.
124. Renné P, Dreßen U, Hebbeker U, Hille D, Flügge UI, Westhoff P, Weber APM (2003) The *Arabidopsis* mutant *dct* is deficient in the plastidic glutamate/malate translocator DiT2. Plant J 35:316–331.
125. Reumann S, Weber APM (2006) Plant peroxisomes respire in the light: Some gaps of the photorespiratory C_2 cycle have become filled – Others remain. Biochim Biophys Acta 1763:1496–1510.
126. Reyes-Prieto A, Weber APM, Bhattacharya D (2007) The Origin and establishment of the plastid in algae and plants. Annu Rev Genet 41:147–168.
127. Royer DL, Wing SL, Beerling DJ, Jolley DW, Koch PL, Hickey LJ, Berner RA (2001) Paleobotanical evidence for near present-day levels of atmospheric CO_2 during part of the tertiary. Science 292:2310–2313.

128. Sato T, Yamanishi Y, Kanehisa M, Toh H (2005) The inference of protein-protein interactions by co-evolutionary analysis is improved by excluding the information about the phylogenetic relationships. Bioinformatics 21:3482–3489.
129. Schimper AFW (1885) Untersuchungen über die Chlorophyllkörner und die ihnen homologen Gebilde. Jahrb Wiss Bot 16:1–247.
130. Schneidereit J, Häusler RE, Fiene G, Kaiser WM, Weber APM (2006) Antisense repression reveals a crucial role of the plastidic 2-oxoglutarate/malate translocator DiT1 at the interface between carbon and nitrogen metabolism. Plant J 45:206–224.
131. Schwender J, Seemann M, Lichtenthaler HK, Rohmer M (1996) Biosynthesis of isoprenoids (carotenoids, sterols, prenyl side-chain of chlorophylls and plastochinone) via a novel pyruvate/glyceraldehyde 3-phosphate non-mevalonate pathway in the green algae *Scenedesmus obliquus*. Biochem J 316:73–80.
132. Shastri AA, Morgan JA (2005) Flux balance analysis of photoautotrophic metabolism. Biotechnology progress 21:1617–1626.
133. Shastri AA, Morgan JA (2007) A transient isotopic labeling methodology for ^{13}C metabolic flux analysis of photoautotrophic microorganisms. Phytochemistry 68:2302–2312.
134. Shendure J, Mitra RD, Varma C, Church GM (2004) Advanced sequencing technologies: Methods and goals. Nat Rev Genet 5:335–344.
135. Shendure J, Porreca GJ, Reppas NB, Lin X, McCutcheon JP, Rosenbaum AM, Wang MD, Zhang K, Mitra RD, Church GM (2005) Accurate multiplex polony sequencing of an evolved bacterial genome. Science 309:1728–1732.
136. Spivey HO, Merz JM (1989) Metabolic compartmentation. Bioessays 10:127–130.
137. Spreitzer RJ, Salvucci ME (2002) Rubisco: Structure, regulatory interactions, and possibilities for a better enzyme. Annu Rev Plant Biol 53:449–475.
138. Srere PA, Sumegi B, Sherry AD (1987) Organizational aspects of the citric acid cycle. Biochem Soc Symp 54:173–178.
139. Stark M, Manz B, Ehlers A, Kuppers M, Riemann I, Volke F, Siebert U, Weschke W, Konig K (2007) Multiparametric high-resolution imaging of barley embryos by multiphoton microscopy and magnetic resonance micro-imaging. Microsc Res Tech 70:426–432.
140. Stitt M, Heldt HW (1985) Generation and maintenance of concentration gradients between mesophyll and bundle sheath in maize leaves. Biochim Biophys Acta 808:400–414.
141. Taniguchi M, Taniguchi Y, Kawasaki M, Takeda S, Kato T, Sato S, Tabata S, Miyake H, Sugiyama T (2002) Identifying and Characterizing Plastidic 2-Oxoglutarate/Malate and Dicarboxylate Transporters in *Arabidopsis thaliana*. Plant Cell Physiol 43: 706–717.
142. Taniguchi Y, Nagasaki J, Kawasaki M, Miyake H, Sugiyama T, Taniguchi M (2004) Differentiation of dicarboxylate transporters in mesophyll and bundle sheath chloroplasts of maize. Plant Cell Physiol 45:187–200.
143. Tatusov RL, Koonin EV, Lipman DJ (1997) A genomic perspective on protein families. Science (New York, NY) 278:631–637.
144. Tatusov RL, Natale DA, Garkavtsev IV, Tatusova TA, Shankavaram UT, Rao BS, Kiryutin B, Galperin MY, Fedorova ND, Koonin EV (2001) The COG database: New developments in phylogenetic classification of proteins from complete genomes. Nucleic Acids Res 29: 22–28.
145. Tegeder M, Weber APM (2006) Metabolite transporters in the control of plant primary metabolism. In: Plaxton WC, McManus MT (eds) Control of Primary Metabolism in Plants. Blackwell Publishing Ltd, Oxford, UK, pp. 85–120.
146. Thomas J, Wolk CP, Shaffer PW, Austin SM, Galonsky A (1975) The initial organic products of fixation of 13 N-labeled nitrogen gas by the blue-green alga *Anabaena cylindrica*. Biochem Biophys Res Commun 67:501–507.
147. Thyssen C, Hermes M, Sultemeyer D (2003) Isolation and characterisation of *Chlamydomonas reinhardtii* mutants with an impaired CO_2-concentrating mechanism. Planta 217:102–112.

148. Timmis JN, Ayliffe MA, Huang CY, Martin W (2004) Endosymbiotic gene transfer: Organelle genomes forge eukaryotic chromosomes. Nat Rev Genet 5:123–135.
149. Tjaden J, Möhlmann T, Kampfenkel K, Henrich G, Neuhaus HE (1998) Altered plastidic ATP/ADP-transporter activity influences potato (*Solanum tuberosum* L.) tuber morphology, yield and composition of tuber starch. Plant J 16:531–540.
150. Tolbert NE (1997) The C2 Oxidative Photosynthetic Carbon Cycle. Annu Rev Plant Physiol Plant Mol Biol 48:1–25.
151. Tolbert NE, Yamazaki RK, Oeser A (1970) Localization and properties of hydroxypyruvate and glyoxylate reductases in spinach leaf particles. J Biol Chem 245:5129–5136.
152. Tyra H, Linka M, Weber APM, Bhattacharya D (2007) Host origin of plastid solute transporters in the first photosynthetic eukaryotes. Genome Biol 8:R212.
153. Versaw WK, Harrison MJ (2002) A chloroplast phosphate transporter, PHT2;1, influences allocation of phosphate within the plant and phosphate-starvation responses. Plant Cell 14:1751–1766.
154. Verscht J, Tomos D, Komor E (2006) Sugar concentrations along and across the *Ricinus communis* L. hypocotyl measured by single cell sampling analysis. Planta 224:1303–1314.
155. Vitkup D (2004) Biological networks: From physical principles to biological insights. Genome Biol 5:313.
156. Voll LM, Jamai A, Renne P, Voll H, McClung CR, Weber APM (2006) The photorespiratory Arabidopsis *shm1* mutant is deficient in *SHM1*. Plant Physiol 140:59–66.
157. Voznesenskaya EV, Franceschi VR, Kiirats O, Artyusheva EG, Freitag H, Edwards GE (2002) Proof of C4 photosynthesis without Kranz anatomy in *Bienertia cycloptera* (Chenopodiaceae). Plant J 31:649–662.
158. Voznesenskaya EV, Franceschi VR, Kiirats O, Freitag H, Edwards GE (2001) Kranz anatomy is not essential for terrestrial C4 plant photosynthesis. Nature 414:543–546.
159. Vyssokikh M, Brdiczka D (2004) VDAC and peripheral channelling complexes in health and disease. Mol Cell Biochem 256-257:117–126.
160. Weber A, Flügge UI (2002) Interaction of cytosolic and plastidic nitrogen metabolism in plants. J Exp Bot 53:865–874.
161. Weber A, Menzlaff E, Arbinger B, Gutensohn M, Eckerskorn C, Flügge UI (1995) The 2-oxoglutarate/malate translocator of chloroplast envelope membranes: Molecular cloning of a transporter containing a 12-helix motif and expression of the functional protein in yeast cells. Biochemistry 34:2621–2627.
162. Weber APM (2004) Solute transporters as connecting elements between cytosol and plastid stroma. Curr Opin Plant Biol 7:247–253.
163. Weber APM (2006) Synthesis, export, and partitioning of the end products of photosynthesis. In: Wise RR, Hoober JK (eds) The Structure and Function of Plastids. Springer, Dordrecht, NL, pp. 273–292.
164. Weber APM, Fischer K (2007) Making the connections – the crucial role of metabolite transporters at the interface between chloroplast and cytosol. FEBS Lett 581:2215–2222.
165. Weber APM, Kaiser WM (2005) Rapid modulation of nitrate reduction in leaves by redox-coupling of plastidic and cytosolic metabolism. In: van der Est A, Bruce D (eds) Photosynthesis: Fundamental Aspects to Global Perspectives. Allen Press, Inc., Lawrence, KS, pp. 810–812.
166. Weber APM, Linka M, Bhattacharya D (2006) Single, ancient origin of a plastid metabolite translocator family in Plantae from an endomembrane-derived ancestor. Eukaryot Cell 5:609–612.
167. Weber APM, Schneidereit J, Voll LM (2004) Using mutants to probe the in vivo function of plastid envelope membrane metabolite transporters. J Exp Bot 55:1231–1244.
168. Weber APM, Schwacke R, Flügge UI (2005) Solute transporters of the plastid envelope membrane. Annu Rev Plant Biol 56:133–164.
169. Weber APM, Weber KL, Carr K, Wilkerson C, Ohlrogge JB (2007) Sampling the Arabidopsis transcriptome with massively parallel pyrosequencing. Plant Physiol 144:32–42.

170. Weiner H, Heldt HW (1992) Inter- and intracellular distribution of amino acids and other metabolites in maize (*Zea mays*) leaves. Planta 187:242–246.
171. Wiese A, Gröner F, Sonnewald U, Deppner H, Lerchl J, Hebbeker U, Flügge UI, Weber A (1999) Spinach hexokinase I is located in the outer envelope membrane of plastids. FEBS Lett 461:13–18.
172. Winkel BS (2004) Metabolic channeling in plants. Annu Rev Plant Biol 55:85–107.
173. Winter H, Robinson DG, Heldt HW (1994) Subcellular volumes and metabolite concentrations in spinach leaves. Planta 193:530–535.
174. Wolk CP, Austin SM, Bortins J, Galonsky A (1974) Autoradiographic localization of 13 N after fixation of 13 N-labeled nitrogen gas by a heterocyst-forming blue-green alga. J Cell Biol 61:440–453.
175. Wolk CP, Thomas J, Shaffer PW, Austin SM, Galonsky A (1976) Pathway of nitrogen metabolism after fixation of ^{13}N-labeled nitrogen gas by the cyanobacterium, *Anabaena cylindrica*. J Biol Chem 251:5027–5034.
176. Yamada T, Kanehisa M, Goto S (2006) Extraction of phylogenetic network modules from the metabolic network. BMC Bioinformatics 7:130.
177. Yoon HS, Hackett JD, Pinto G, Bhattacharya D (2002) The single, ancient origin of chromist plastids. Proc Natl Acad Sci U S A 99:15507–15512.
178. Yoon HS, Hackett JD, Van Dolah FM, Nosenko T, Lidie KL, Bhattacharya D (2005) Tertiary endosymbiosis driven genome evolution in dinoflagellate algae. Mol Biol Evol 22:1299–1308.
179. Young XK, McCarty RE (1993) Assay of proton-coupled glycolate and D-glycerate transport into chloroplast inner envelope membrane vesicles by stopped-flow fluorescence. Plant Physiol 101:793–799.
180. Yu C, Claybrook DL, Huang AH (1983) Transport of glycine, serine, and proline into spinach leaf mitochondria. Arch Biochem Biophys 227:180–187.
181. Yu C, Huang AH (1986) Conversion of serine to glycerate in intact spinach leaf peroxisomes: Role of malate dehydrogenase. Arch Biochem Biophys 245:125–133.
182. Zhou L, Yu X, Cabrera ME, Stanley WC (2006) Role of cellular compartmentation in the metabolic response to stress: Mechanistic insights from computational models. Ann N Y Acad Sci 1080:120–139.

Chapter 3
Metabolite Measurements

Ute Roessner and Diane M. Beckles

3.1 Introduction

Metabolites are the products of enzyme-catalyzed reactions that occur naturally within living cells. Metabolites are synthesized by the cell for the purpose of performing a useful, if not indispensable, function in the maintenance and survival of the cells by, for example, contributing to its infrastructure or energy requirements. To do so, they have to be recognized and acted upon by enzymes, which will change the properties of the metabolites by means of a chemical reaction. Therefore, the properties of metabolites and their functionality as they interact within their natural environment determine the chemistry of life. Thus, it can be argued that the metabolome in a biological system represents the final result of the expression of multiple genes in a cell. The analysis of metabolites has been an important part of any biological sciences. A large number of technologies have been developed for the analysis of metabolites in order to study metabolism in great detail. Today, the accumulation and combination of knowledge on analytical biochemistry from the last 50 years is commonly called metabolomics, and large investments are made to its application toward developments of new technologies with greater sensitivity, comprehensiveness, robustness, and higher throughput.

Oliver et al. introduced the term metabolomics in 1998 to describe the change of relative metabolite concentrations due to the alteration of gene expression [62]. Later, the definition of metabolomics describing the "comprehensive and quantitative analysis of all small molecules in a biological system" was introduced [26]. Today, metabolomics is commonly considered as the combination of analytics for metabolite determination with appropriate informatics for data extraction, mining, and interpretation.

U. Roessner (✉)
School of Botany, Australian Centre for Plant Functional Genomics, The University of Melbourne, 3010 Australia
e-mail: u.roessner@unimelb.edu.au

The range of chemical compounds synthesized by plants is astonishing. There may be more than 200,000 chemicals produced within the plant kingdom, representing a diverse array of structures, functional groups, and chemicals with different solubilities and reactivities [90]. This presents an enormous challenge for the researcher trying to assay these compounds in a massively parallel fashion. Thus far, only 50,000 compounds have been identified from a large number of species [20], highlighting the difficult nature of the task and the existence of thousands of undiscovered compound with different physical and chemical properties. Approximately 5,000–25,000 compounds may be produced in a single plant at a given developmental age and environmental state [61, 95, 99]. In contrast, about 500 are produced in bacteria, about 730 in yeast, and approximately 3,000 in human beings [96]. Apart from the sheer number and diversity of chemical structures, there are unique problems to cataloging the steady-state level of the plant metabolome. Reliable sampling and capture of the subset of compounds in a specific organ and subcellular location associated with the biological process of interest is, while not impossible, technically challenging and adds yet another layer of complexity to our ability to accurately record or view the metabolic state of the plant cell. The overall picture that emerges is that current extraction, analytical methods, and instrumentation are inadequate for truly assessing the plant metabolome, which may well remain a lofty vision to aspire toward in the future. These issues explain in part why the field of metabolomics lags behind genomics, transcriptomics, proteomics, and even glycomics. Identifying, reproducing, and curating the chemical species that make up the genome and transcriptome is comparatively facile, and even with the more difficult proteome and glycome, this goal is still feasible. This is easy to understand when the chemical structures of these compounds are considered. Nucleic acids, proteins, and glycoproteins have a common chemical structure(s) [12, 37, 40, 84]. Both DNA and RNA have a phosphodiester bond; all proteins have the peptide bond; glycoproteins have peptide bonds and glycoconjugates. In comparison, there is an astounding array of chemical groups that make up even the simplest of plant compounds. For example, several primary metabolites have inorganic and organic phosphates, amines, esters, hydroxides, carboxyl groups, and this does not even scratch the surface when one considers the exotic functional groups of secondary metabolites and that there are geometric and stereoisomers of many of these chemicals.

In the following, a short overview of the technologies and methodologies most commonly applied in metabolomics are described, with an emphasis on plant-based applications.

3.2 Technologies for Metabolite Analyses

The most commonly used platforms for the detection and measurement of metabolites involve their separation by gas chromatography (GC), liquid chromatography (LC), or capillary electrophoresis (CE) coupled with subsequent mass spectrometry

3 Metabolite Measurements

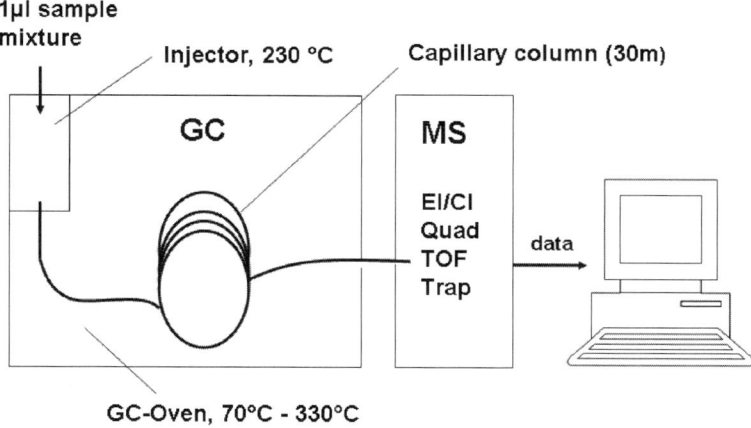

Fig. 3.1 Schematics of a typical GC-MS setup in metabolomics approaches

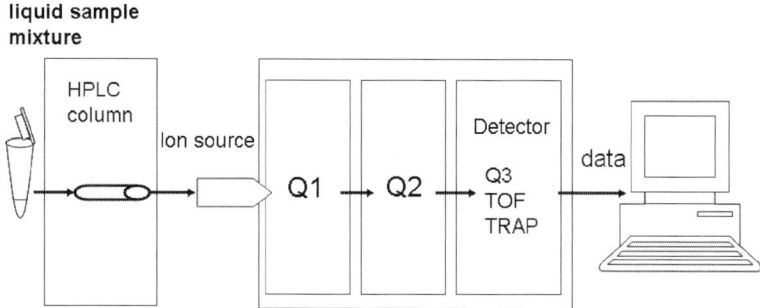

Fig. 3.2 Schematics of possible LC-tandem MS setups in metabolomics approaches

(MS) of the separated molecules (Figs 3.1 and 3.2). Compounds may also be measured directly without chromatographic separation. Fourier transform ion cyclotron resonance mass spectrometry (FT-ICR-MS) and nuclear magnetic resonance spectroscopy (NMR) are two such examples (for review see [66]). The advantages and disadvantages of these technologies will be discussed in turn.

3.2.1 Mass Spectrometry

Different chemicals have different masses, and this fact is used in a mass spectrometer (MS) to determine which chemicals are present in a sample. The underlying principle of all MS is that the paths of gas phase ions in electric and magnetic fields are dependent on their mass-to-charge ratios which are then used by the mass analyzer to distinguish the ions from one another. The most important requirement of mass spectrometry is that compounds have to be vaporized and ionized (in an ion source). Techniques for ionization have been key to determining what types of samples can be analyzed by mass spectrometry. Electron impact ionization (EI, Fig. 3.3) and chemical ionization (CI, Fig. 3.4) are mainly used for volatile compounds, e.g., in combination with gas chromatography (GC; see below, Fig. 3.1). In chemical ionization sources, the analyte is ionized by chemical ion–molecule reactions during collisions in the source.

Three ionization techniques often used with liquid and solid biological samples include electrospray ionization (ESI) (Fig. 3.5, also see Fig. 3.2), atmospheric pressure chemical or photon ionization (APCI/APPI), and matrix-assisted laser desorption/ionization (MALDI). Inductively coupled plasma sources are used primarily for metal analysis on a wide array of sample types.

Others include glow discharge, fast atom bombardment (FAB), thermospray, desorption/ionization on silicon (DIOS), secondary ion mass spectrometry (SIMS), and thermal ionization. The resulting ions are represented by their specific mass and charge, which means that ions from different chemical compounds will have different speed and directions within an electric or magnetic field. The ions are accelerated to a high speed and are thus separated in an electric and/or magnetic field. This process happens in the mass analyzer. There are many types of mass analyzers, using

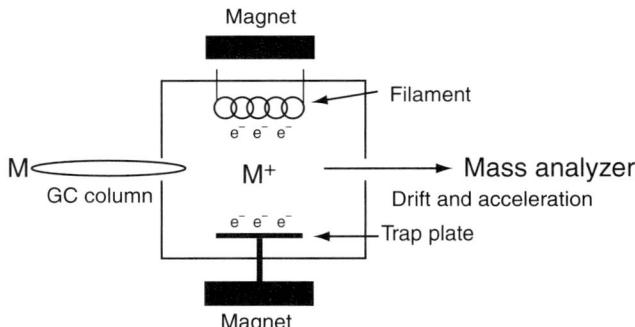

Fig. 3.3 Schematics of electron impact ionization. "M" is the molecule to be analyzed

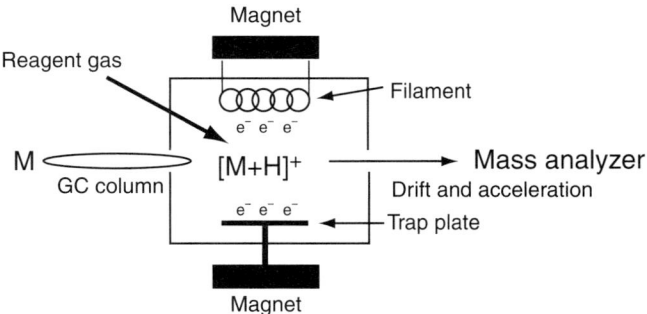

Fig. 3.4 Schematics of chemical ionization. "M" is the molecule to be analyzed

Fig. 3.5 Schematics of electrospray ionization

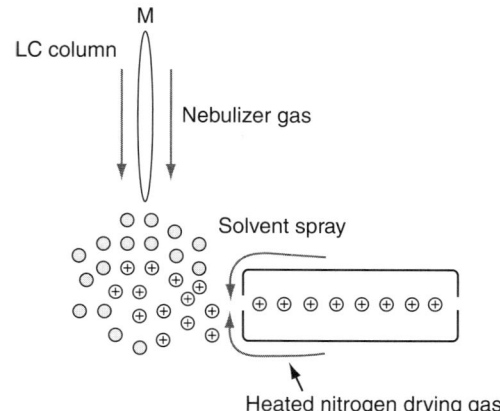

either static or dynamic fields and magnetic or electric fields, but all operate according to this same law. Most commonly used mass analyzers in biological applications include time-of-flight analyzer (TOF), ion trap analyzer (TRAP), quadrupole (Q), or Fourier transform ion cyclotron resonance MS (FT-ICR-MS).

Perhaps the easiest to understand is the time-of-flight (TOF) analyzer (Fig. 3.6A). It uses an electric field to accelerate the ions through the same potential, and then measures the time taken to reach the detector. If all the particles have the same charge, then their kinetic energies will be identical and their velocities will depend only on their masses. Lighter ions will reach the detector first and the heavier slower ions will be last.

Quadrupole mass analyzers use oscillating electrical fields to selectively stabilize or destabilize ions passing through a radio frequency (RF) quadrupole field (Fig. 3.6B). A quadrupole mass analyzer acts as a mass selective filter and is closely related to the quadrupole ion trap. An ion trap may use an electric or an electric and magnetic field to capture a cloud of ions in a region of a vacuum system or tube (Fig. 3.6C). The two most common types of ion traps are the Penning trap and the Paul trap. The Penning trap uses electric and magnetic fields while the

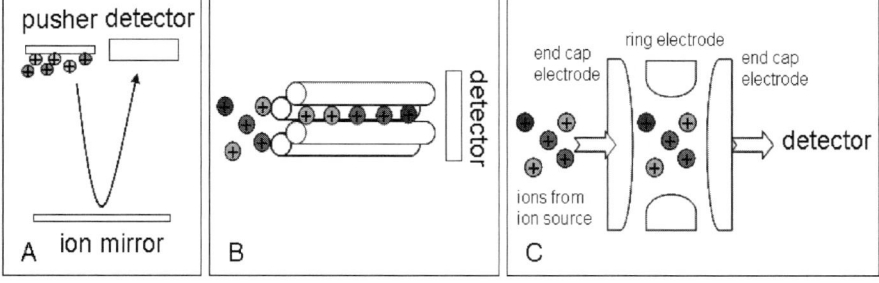

Fig. 3.6 Schematics of (**A**) time-of-flight, (**B**) quadrupole, and (**C**) ion trap mass analyzers (*see also* Color Insert)

Paul trap uses an electric field only. An ion trap MS may incorporate a Paul trap or the Orbitrap (which like the Paul trap, uses only electric field), and other types of mass spectrometers may also use a linear quadrupole ion trap as a selective mass filter.

The Orbitrap is the most recently introduced mass analyzer (commercially available since 2005, ThermoFisher). In the Orbitrap, ions are electrostatically trapped in an orbit around a central, spindle-shaped electrode. The electrode confines the ions so that they both orbit around the central electrode and oscillate back and forth along the central electrode's long axis. This oscillation generates an image current in the detector plates, which is recorded by the instrument. The frequencies of these image currents depend on the mass-to-charge ratios of the ions in the Orbitrap. Mass spectra are obtained by Fourier transformation of the recorded image currents. Similar to FT-ICR-MS (see below), Orbitraps have a high mass accuracy, high sensitivity, and a good dynamic range, but it is cheaper and less complex.

Fourier transform ion cyclotron resonance MS or FT-ICR-MS is regarded as the most complex type of mass analyzer. It determines the mass-to-charge ratio of ions based on the frequency of rotation of the ion inside a homogeneous magnetic field. The ions are first trapped in a magnetic field with electric trapping plates (Penning trap). The magnetic field causes the ions to adopt a circular motion perpendicular to the field. When an RF pulse is applied across the electric trapping plates, the ions are excited into a larger circular motion called the cyclotron frequency. The frequency of rotation is determined by the mass-to-charge ratio of each individual ion. Each ion will produce a current that is almost equivalent to the cyclotron frequency which is then recorded on a pair of detector plates. The resulting signal is called a free induction decay (FID), transient, or interferogram. The useful signal is extracted from this data by performing a Fourier transformation to give the mass spectrum. FT-ICR-MS is characterized by extremely high resolution in that masses can be determined with very high accuracy. Many applications of FT-ICR-MS use this mass accuracy to help determine the composition of molecules. This is possible due to the mass defect of the elements. Another place that FT-ICR-MS is useful is in dealing with complex mixtures since the resolution (narrow peak width) allows the signals of two ions of similar mass to charge ratio (m/z) to be detected as distinct ions. This high resolution is also useful in studying large macromolecules such as proteins with multiple charges, which can be produced by electrospray ionization. These large molecules contain a distribution of isotopes that produce a series of isotopic peaks. Because the isotopic peaks are close to each other on the m/z axis, due to the multiple charges, the high resolving power of the FT-ICR is extremely useful. FT-ICR-MS differs significantly from other MS techniques in that the ions are not detected by hitting a detector such as an electron multiplier but only by passing near detection plates. Additionally, the masses are not resolved in space or time as with other techniques but only in frequency. Thus, the different ions are not detected in different places as with sector instruments or at different times as with TOF instruments, but all ions are detected simultaneously over some given period of time.

Each analyzer type has its strengths and weaknesses. Many mass spectrometers use two or more mass analyzers for tandem mass spectrometry (MS/MS) (Fig. 3.7).

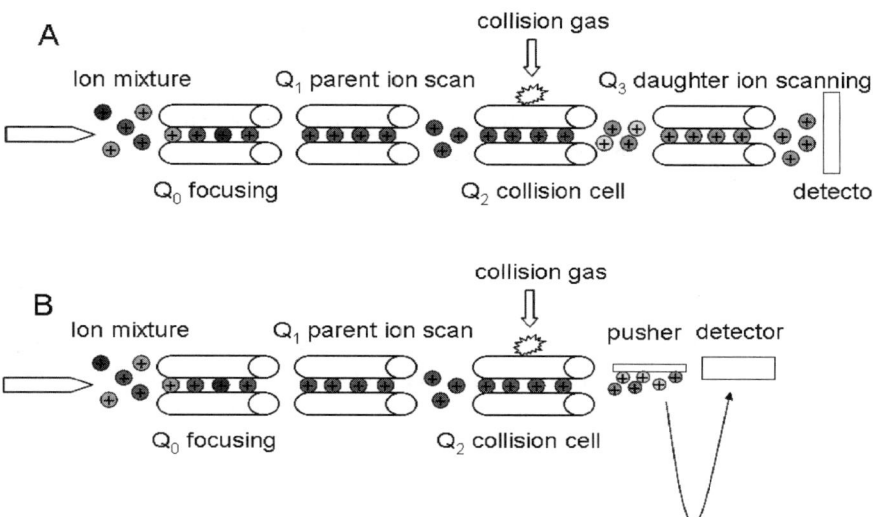

Fig. 3.7 Schemata of tandem MS/MS. (**A**) Triple quadrupole mass spectrometer (QqQ) and (**B**) quadrupole time-of-flight mass spectrometer (QqTOF) (*see* also Color Insert)

Tandem mass spectrometry involves multiple steps of mass selection or analysis, usually separated by some form of fragmentation. A tandem mass spectrometer is one capable of multiple rounds of mass spectrometry. For example, one mass analyzer can isolate one compound and determine its molecular weight. A second mass analyzer then stabilizes the molecular ion while it collides with a gas, causing them to fragment by collision-induced dissociation (CID). A third mass analyzer then catalogs the fragments produced from the original compound. Tandem MS can also be done in a single mass analyzer over time as in an ion trap. There are various methods for fragmenting molecules for tandem MS, including collision-induced dissociation (CID), electron capture dissociation (ECD), electron transfer dissociation (ETD), infrared multiphoton dissociation (IRMPD), and blackbody infrared radiative dissociation (BIRD). An important application, using tandem mass spectrometry, is in the structural elucidation of compounds.

The final element of the MS is the detector. The detector records the charge induced or current produced when an ion passes by or hits a surface. In a scanning instrument, the signal produced in the detector during the course of the scan vs. where the instrument is in the scan (at what m/q) will produce a mass spectrum, a record of ions as a function of m/q. Typically, some type of electron multiplier is used, though other detectors including Faraday cups and ion-to-photon detectors are also used. Because the number of ions leaving the mass analyzer at a particular instant is typically quite small, significant amplification is often necessary to get a signal. Microchannel Plate Detectors are commonly used in modern commercial instruments. In FT-ICR-MS and Orbitraps, the detector consists of a pair of metal surfaces within the mass analyzer/ion trap region which the ions only pass near as they oscillate. No DC current is produced, only a weak AC image current is generated in a circuit between the electrodes.

In metabolomics, mass spectrometry is mostly used as a detector system following a specific separation procedure. Metabolite extracts from a biological source are characterized by great complexity, which has to be reduced before the compounds enter the mass spectrometer. Gas or liquid chromatography-based separation techniques allow the separation of complex mixtures with great precision. It involves passing a mixture dissolved in a "mobile phase" through a stationary phase, which separates the analyte to be measured from other molecules in the mixture and allows it to be isolated. Even very similar components, such as oligosaccharides that may only vary by a single sugar monomer or bond structure, can be separated with chromatography. In fact, chromatography can purify basically any soluble or volatile substance if the right adsorbent material, carrier fluid, and operating conditions are employed. Second, chromatography can be used to separate delicate compounds since the conditions under which it is performed are not typically severe. For these reasons, chromatography is quite well suited to a variety of uses in the field of biotechnology.

3.2.2 GC-MS

Gas chromatography linked to mass spectrometry is an effective and longstanding method for chemical analysis (Fig. 3.1). GC-MS presents several advantages; it is relatively easy to use, low in cost, gives reproducible results and excellent resolution of separated compounds. Hence GC-MS has been widely used in several separation applications since its first demonstrated use in 1964 for the quantitative and qualitative determination of polar/organic constituents of a sample [24].

The basis on which components are separated is on differential partitioning between a mobile gas phase and a solid stationary phase. Samples for GC-MS must first be converted from solid or liquid phase to a gas. This process called volatilization is accomplished by exposing the sample to high temperatures (up to 250°C). Once in the gas or mobile phase, the components are forced along a series of columns containing the solid or stationary phase. The volatilized compounds are partitioned between the two phases and the extent to which this occurs is determined by their chemical properties. Compounds that partition primarily in the mobile phase are eluted from the column faster than those with a greater affinity for the stationary matrix. The chemical behavior of each constituent, as it is eluted from the column, is recorded by its retention time.

Each separated compound eluting from the column must be subjected to ionization before entering the MS. Electron impact (EI) ionization produces electrons using a standardized filament voltage of 70 eV that effectively ionizes compounds (Fig. 3.3). These electrons are of high energy, and when they collide with separated compounds in the MS, they cause the compounds to fragment. The fragmentation patterns are curated into mass spectral libraries for peak identification [51, 97]. Those produced from EI are highly reproducible which makes construction of spectral libraries using this technique consistent across experiments.

Chemical ionization (CI)-MS in contrast to EI uses gases, such as methane or ammonia, to provide the collision energy for fragmentation (Fig. 3.4). It is often described as a "softer" form of ionization, especially when compared to EI. The applications for CI, however, are limited. The fragmentation patterns produced by this method are less reproducible than for EI because it is difficult to control temperature and pressure of the ion source conditions. However, even with this drawback, CI-MS is an advantage for some types of applications, such as identifying compounds based on both the mass of the parent ion and its isotopic pattern. For example, most metabolic flux experiments use stable isotope labeling to monitor the fate of the target compound when metabolized in the fed tissues [25, 77]. Here, CI-MS is a better option for quantifying isotopic label in individual compounds produced from the labeled precursor, because the ligand–label complex likely remains intact [69].

After ionization, the fragmented compounds are analyzed by mass detection by which quantification of ions can be achieved. Mass detection may be performed using quadrupole, ion-trap technology, or TOF detectors. The low-resolution quadrupole-type instruments are most commonly used, although fast scanning time-of-flight (TOF) MS detectors are becoming more popular.

The components eluted from the GC can be identified with a high degree of accuracy by comparing mass spectral and specific retention time indices of the eluted compounds to that of a reference database [51, 98]. Still, the identity of each compound should be verified by co-elution of authentic standards when available, and recovery assays should be performed to assess accuracy of measurements. Accomplishing these seemingly minor objectives is not as straightforward as they should be, especially when analyzing plant extracts. Purified standards of most plant-derived compounds are not commercially available. There are more than 350,000 curated chemical structures in GC-EI-MS libraries (e.g., NIST: http://www.nist.gov/srd/nist1a.htm and MSRI: http://csbdb.mpimp-golm.mpg.de/csbdb/gmd/msri/gmd˙msri.html); yet a significant fraction of the detected compounds, in a typical GC-MS chromatogram, are not yet chemically identifiable. This is due, in part, to our ignorance of the diversity of chemicals manufactured in plants; it may also be due to the artifacts produced during extraction, GC ionization, and analysis. Recovery assays, although useful, add a significant time and cost factor to the analysis; however, it should be used in instances when assaying novel, low abundant, or unstable compounds.

For the researcher wishing to perform a basic characterization of polar compounds in plant extracts, GC-MS presents several advantages; it is reasonably inexpensive and easy to perform. Capital outlay is substantially lower when compared with other techniques such as LC-MS or NMR, and bench-top GC-MS instrumentation is readily available. From a quality control standpoint, GC-MS is exceptional in that there are only minimal matrix effects, i.e., the ionization efficiency of an analyte is not confounded by the presence of co-eluting substances [43]. Finally, there are a growing number of easily adaptable methods available to survey the catalog of low molecular weight compounds in different plant systems (for review see [40, 96]).

There are some limitations to the application of GC-MS for metabolite profiling of any complex mixture, especially plant extracts. GC-MS is only suitable for analyzing compounds that can be volatilized either at high temperatures or by chemical modification. The result is that only small compounds (~1kD) can be effectively evaluated and thermolabile compounds are excluded. Further, samples must be derivatized, i.e., chemically modified to make them adaptable for GC-MS analysis. Derivatization simultaneously increases the volatility, thermostability, and detection limits of low-abundant compounds, and it is often achieved by silylation, alkylation, and acylation of extracts. Trimethylsilylation (TMS) is the most common method used in GC-MS-based metabolomics because it can derivatize a broad spectrum of compounds simultaneously, including sugars, amines, alcohols, amino, and organic acids [49], thus minimizing the probability that a chemical bias is introduced due to selective modification of one class of compounds. Derivatization is achieved when acidic protons in the target compound are effectively exchanged with TMS via a nucleophilic attack. It is imperative to use rigorous analytical procedures and standardized conditions when using TMS derivatives, as the derivatized compounds, especially amino acids, are sensitive to water and oxygen and are thermodynamically unstable [5]. An alternative derivatization method based on tert-butyldimethylsilylation (TBS) increases the chemical stability of amino acids and a range of other compounds thus widening the coverage of analytes potentially detected and measured by GC-MS [44]. A typical approach of GC-MS analysis using both types of derivatization is presented in Fig. 3.8.

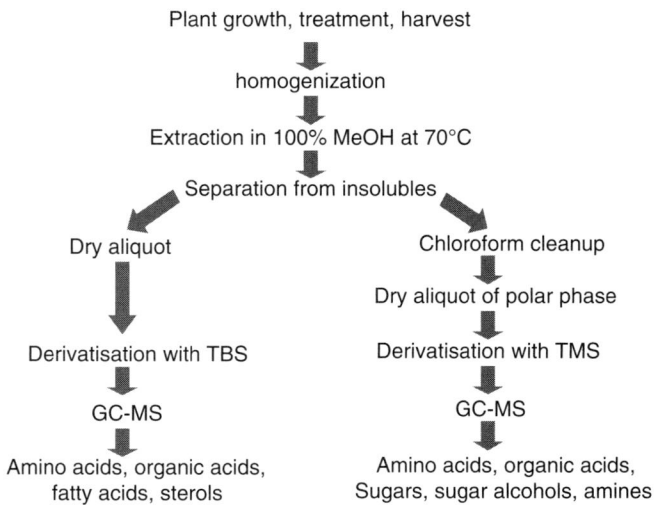

Fig. 3.8 Simplified workflow for GC-MS analysis of metabolites using TMS and TBS derivatization

3.2.3 LC-MS

LC-MS is gaining wider adoption as new technologies have allowed the high pressure liquid chromatography (HPLC) module to be interfaced with the low pressure (vacuum) of the MS (Fig. 3.2). Normal HPLC separation is accomplished by coupling LC with UV/VIS or diode-array detectors; however, the use of mass spectrometry enhances the specificity and selectivity of the system, thus improving the likelihood that determining the structural properties of compounds in a complex matrix can be identified. A requisite step in LC-MS is to convert the eluting compounds from a solute to a gas phase ion. This involves solute vaporization with concomitant ionization at the interface with the MS. Electrospray ionization (ESI, Fig. 3.5) [101], and atmospheric pressure ionization (API) [13] are good options for solute transformation. However, if the sample has a high concentration of salts or other ionizable compounds, these methods produce matrix effects such as ion suppression and ion enhancement [43], requiring additional downstream validation of the results [8].

LC-MS however presents several advantages over GC-MS. Samples do not have to be volatilized, which maintains the compounds in their native state before separation. Classes of compounds incompatible for analysis by GC such as those of higher molecular mass, greater polarity, and lower thermostability can be detected. Processing time may also be reduced in LC-MS because direct infusion methods can be used for analysis. The downside of direct infusion is that matrix effects are often a problem because of the larger number of compounds present in crude plant samples that can simultaneously enter the MS. Therefore, in most applications, LC separation prior to MS is advisable in order to reduce the complexity of ions to be scanned, even though this is less expedient.

A wide selection of column matrices is now available that support LC separations based on ion exchange, reversed phase, and hydrophobic interaction chromatography. Further, LC protocols that optimize the elution of constituents in complex compound mixtures have been developed. Very recently, a new sophisticated LC technology for nanoelectrospray application has been introduced. Nanospray LC-MS provides much higher sensitivity than normal flow electrospray LC-MS, however it uses several small capillary tubing connections which lead to frequent clogging and/or leaking at the column and spray needle. This feature has made nanospray-based LC-MS applications challenging. Agilent Technologies, Inc. has developed an HPLC-Chip interface for mass spectrometry (Fig. 3.9). All components required for LC and spray into the mass spec are integrated directly onto a reusable biocompatible polymer chip. This supports the delivery of solvent and sample, high pressure switching of flows, automated chip loading, and spray performance into the MS. An illustrating video of how the chip technology works can be retrieved from Agilent Technologies, Inc. website (www.agilent.com). Chips are being manufactured with different column packings, and users are even able to provide their own packing for custom-built chips. The HPLC-Chip has been already extensively applied in proteomics and peptidomics applications showing very high reproducibility, long shelf-life, and ease of use. Currently, its potential

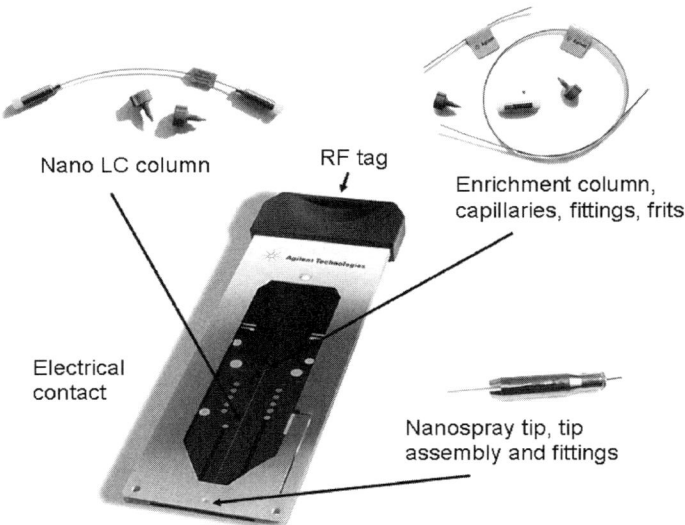

Fig. 3.9 HPLC-Chip and its integrated components related to the components of conventional LC (picture provided in courtesy of Agilent Technologies, Inc.) (*see* also Color Insert)

and applicability is being explored in the metabolomics field. LC, used in tandem with MS, produces a spectrum of separated compounds which can be detected with great selectivity. The fragmentation patterns produced provide information on the chemical structure of the compounds, and the system is also able to detect low-abundance metabolites. LC-MS has found utility in the separation and analysis of both primary and secondary metabolites in plant extracts [42, 89]. Mass detection may be dramatically increased to 5,000 signals from a single plant extract if Fourier transform ion cyclotron resonance mass spectrometry (FT-ICR-MS) is used with LC [2].

The major obstacle that hinders the widespread use of LC-MS for many biological applications is that it is difficult to establish robust mass spectral libraries for peak identification [58]. The type of mass spectra produced by LC-MS is largely dictated by the instrument used (i.e., QqQ, QqTOF, Ion Trap, etc.), and the reference LC-MS spectral libraries constructed are of limited use because they are instrument specific. Each research group has to develop its own "in-house" LC-MS reference library infrastructure, which is often difficult and beyond the capabilities of the average plant biology lab.

3.2.4 CE-MS

Capillary electrophoresis (CE), either coupled to MS or to laser-induced fluorescence (LIF) detection, is a highly efficient and sensitive method for both targeted and unbiased profiling in plant extracts. It has not been widely adopted in high-throughput metabolomic approaches, but has proven useful in particular applications because of its high sensitivity, especially compared to LC-MS. Derivatization is not necessary; solvent consumption is less; separation of low-molecular weight compounds with minimal pre-treatment is easily achieved; small sample sizes can be used; separation runs are faster than with LC-MS [65]; and unlike LC-MS, the separated species do not need to be ionized because they are already charged [80]. However, there are some complications to the use of CE for plant metabolomics. Only charged compounds or those that can be charged by changing the pH of the solution can be analyzed, and because there are limits to the volume of sample that can be injected onto the capillary, this can make detecting some compounds difficult. Still, some promising results have been already achieved. CE coupled with UV analysis has supported targeted profiling of some plant compound classes, e.g., organic acids [98], flavonoids [23] and amino acids [100]. When coupled to MS, more than 80 primary metabolites belonging to glycolytic, photorespiratory, and oxidative pentose phosphate pathways can be analyzed in rice leaf extracts [75]. However, the resolution power of CE coupled to LIF holds even greater promise for analyzing micro-amount of biological fluids. It has the sensitivity to separate and quantify a large number of amino acids and sugars in only ~50 picoliters of phloem sap, or the pooled sap of five leaf mesophyll cells from *Cucurbita maxima* [3]. Once the technology becomes more routinely used, it is likely to become very important for the development of cell-type-specific metabolite analyses.

3.2.5 NMR

NMR is potentially one of the most potent but underused methods available for plant metabolomics. NMR is a non-destructive, non-targeted fingerprinting technique that can detect a multitude of different metabolite classes irrespective of size, charge, volatility, or stability [21]. It is a powerful tool for comparative high-throughput profiling of plant extract, for determining metabolite structure and for elucidating metabolic fluxes. Although NMR use in the plant field lags that in others, e.g., medical research, there is renewed interest and increasing applications of this technology to addressing basic biological questions related to plant processes (67).

NMR spectroscopy uses the magnetic properties of atoms that make up the chemical structure of compounds. A strong magnetic field is combined with radio frequency pulses to produce high-energy spin states in nuclei with odd atomic or mass numbers (e.g., ^{1}H or ^{13}C). The radiation emitted when these nuclei return to the lower energy spin state is detected and used to eventually construct the chemical structure of the analyte [21]. One of the most attractive features of NMR is that metabolites can be measured non-destructively; permitting in vivo measurement of metabolites in intact tissues [54]. Most importantly, NMR provides high-resolution structural information about the metabolites for unambiguous identification. When combined with stable-isotopic labeling, NMR becomes very informative. Real-time in vivo flux of compounds can be monitored and even resolved between subcellular compartments [54, 66, 73]. For example, the location of ^{31}P in different plastids can be determined based on the chemical shift caused by the different pH of the subcellular compartments [73].

There are a number of factors that need to be considered when NMR is applied to plant metabolomics. NMR is in general less sensitive and less discriminating than other established separation techniques [21]. The signal-to-noise ratio and resolution of in vivo samples are much lower when compared to those in vitro extracts; metabolites below 5 nmol will not be detected by NMR [52], and although volumes of around $2\mu l$ may be used [38], NMR generally requires larger sample input. In addition, NMR involves a considerably greater capital investment in instrumentation, and when the number of compounds actually detected is considered vs. cost, it is not always the best option. NMR is best suited for specialized applications involving in vivo real-time imaging of flux, structural elucidation of compounds, and for metabolite fingerprinting. There are emerging technologies that may supersede NMR in some applications because of their resolving power. Once such example is imaging matrix assisted laser desorption ionization mass spectrometry (I-MALDI) coupled with TOF-MS. This was used to detect and measure a range of metabolites even at micromolar concentrations in different subcellular fractions of the plant cell [14].

3.2.6 Other Novel, Highly Valuable Approaches for Metabolite Analysis

As mentioned, the described technologies are currently the most commonly applied analytical instrumentations for metabolite analyses. A range of well-established and routinely applicable methodologies are available for the analysis of a number of different compound classes, ranging from primary metabolites, such as sugars, amino, organic or fatty acids to highly complex secondary metabolites, such as alkaloids and flavonoids [40]. Although claimed as being high-throughput methods, they often are limited by time per sample due to time-consuming chromatographic separation upfront MS analysis. A recently introduced separation technique that shows great promise is based on ion mobility [94]. Ion mobility spectrometry coupled to mass spectrometry (IMS-MS) is a technique where ions are first separated by drift time through a neutral gas which was given an electrical potential gradient before being introduced into a mass spectrometer. The IMS technique separates and detects ions that have been sorted according to how fast they travel through an electrical field in a tube. Small ions travel very fast, and they reach the detector first, with successively larger ions following along. Unlike the mass spectrometry technique, which relies on very low pressures to keep the ions from colliding with each other, IMS operates at normal atmospheric pressure, and the ions collide with each other repeatedly. The ionized gas moves through an electrical field inside a drift tube. Smaller ions collide less frequently than large ions because they present a smaller target and are harder to hit. Thus, they move through the tube relatively unimpeded and reach the detector first. The largest ions take several seconds longer to travel to the detector because they collide more frequently with other ions along the way. Because IMS only sorts molecules by size, and not by chemical properties or other identifying features, it is not a particularly good technique for making a positive identification of unknown compounds. However, the duty cycle of IMS is short relative to LC or GC separations and can thus be coupled to such techniques producing triply hyphenated techniques such as LC-IMS-MS. Perhaps IMS's greatest strength is the speed at which the separations occur – typically on the order of 10s of milliseconds. As a research tool, ion mobility has already shown great strides toward the analysis of biological materials, specifically in proteomics and metabolomics [16, 56, 85, 93].

A powerful way of assaying low-abundant metabolites has recently been reintroduced by Gibon and colleagues [33]. A large number of metabolites are present in very small quantities, especially metabolic intermediates and cofactors in plant cells. Also, in many cases, only small samples sizes can be harvested. To overcome these obstacles, Gibon et al. developed robotized microplate-based activity assays using enzymes involved in central carbon and nitrogen metabolism as tools to assay the metabolites involved in the reaction mechanism. The distinct advantage of this high-sensitivity assay is that limited sample size or metabolites in low abundance are not insurmountable problems for analysis [31, 32].

3.3 Data Analysis

Analyzing the large volumes of data produced by metabolite profiling technologies is still a major challenge facing researchers. The best approach for data analysis will depend on the aim and application of the specific experiment or process investigated. Although accurate detection of compounds is necessary regardless of research objectives, slightly different emphases on the downstream analyses will be required depending on if the primary aim is to identify hitherto unknown compounds, determine differences between samples, or evaluate broad changes in entire pathways. The first step is to ensure adequate quality control of the raw data, i.e., accurate peak identification, assignment, and quantification. Next, robust statistical validation of raw data is critical as it affects the interpretation of the data if comparisons are made between different systems. Finally, transformation and presentation of the data in a manner that allows efficient and maximal extraction of biological information from the system to be studied in an intuitive user-friendly interface are desirable.

Once validated, the data needs to be mined and presented in a manner to make hypothesis testing facile, to highlight pattern or relationships among variables, or to drive the generation of new biological questions. Several methods are currently used including cluster analysis, pathway mapping, and comparative overlays, as well as heatmaps. They each provide slightly different insight into multidimensional datasets and may even be complementary.

3.3.1 Data Extraction from Analytical Instrumentation

This is the most important, but often under-estimated process of data analysis in metabolomics applications both based on mass spectrometry or NMR. The process includes raw data format transformation (if necessary), chromatogram deconvolution, peak detection, peak alignments, baseline corrections, noise reduction, peak assignment (identification), peak quantification, and data validation. Most importantly, the procedure has to be applicable for each batch measurement with the desired or necessary level of correctness. A vast number of software packages are available, both commercially and as free open-public packages. The description of those software packages is beyond the scope of this chapter. Table 3.1 lists a few of the most common software packages available for mass spectral data analysis. Although much progress has been made in developing software application, manual inspection and validation of raw data are still essential and represent a limiting step in a process that could be automated. Therefore, constant evolution and improvements of sophisticated programs capable of evaluating raw data are critical. This would have the dual advantage of increased confidence that the effects of false positives are minimized and that operator time on data analysis is reduced. However, the usefulness of such programs will be determined largely by how it meets the needs of each individual application. Each metabolomics user should invest the time and effort to test some of those packages to decide best suitability for their analytical method as well as comfort in using the package.

3.3.2 Statistics

Data analysis is the process of transforming data with the aim of extracting useful information from which to draw conclusions and develop new working hypotheses. Depending on the type of data and the question asked, this includes the application

Table 3.1 Some examples of software available for mass spectral data analysis

Name	URL
MSFACTs	http://www.noble.org/PlantBio/MS/MSFACTs/MSFACTs.html [22]
MET-IDEA	http://www.noble.org/plantbio/ms/MET-IDEA/index.html [11]
HiRes	http://hatch.cpmc.columbia.edu/highresmrs.html [103]
MZmine	http://mzmine.sourceforge.net/ [48]
XCMS	http://metlin.scripps.edu/download/ [79]
SpectConnect	http://spectconnect.mit.edu [83]
AMDIS	http://chemdata.nist.gov/mass-spc/amdis/ [81]
AnalyzerPro	www.spectralworks.com
metAlign	www.metalign.nl
SIEVE	http://www.thermo.com/
GeneSpringMS	http://www.agilent.com/

of statistical methods, selecting or discarding certain subsets based on specific criteria, or other mathematical methods.

In applying statistics to a scientific problem, one begins with a biological process or sets of organisms to be studied. This might be a population of transgenic plants compared to untransformed wild type, plants treated with a specific stress elicitor compared to unstressed plants, or plants growing in one environment vs. another. It may instead be a comparison observed at various times; data collected about this kind of "population" constitute what is called a time series. For practical reasons, a chosen subset of the population – a sample – is studied, rather than compiling data about an entire population. Once data are collected about the sample, they can be subjected to statistical analysis which serves two related purposes: description and inference. Descriptive statistics can be used to summarize the data, either numerically or graphically, to describe the sample. Basic examples of numerical descriptors include the mean and standard deviation which can be presented as tables or various kinds of graphs and charts. Inferential statistics is used to model patterns in the data, accounting for randomness and drawing inferences about the larger population. These inferences are then presenting a form of hypothesis testing, estimation of numerical characteristics, correlation (description of associations), or regression (modeling of relationships).

If a sample is representative of the population, then inferences and conclusions made from the sample can be extended to the population as a whole. A major problem lies in determining the extent to which the chosen sample is representative. Statistics offers methods to estimate and correct for randomness in the sample and in the data collection procedure, as well as methods for designing robust experiments in the first place. The fundamental mathematical concept employed in understanding such randomness is probability.

The use of any statistical method is valid only when the system or population under consideration satisfies the basic mathematical assumptions of the method. Inappropriate use can produce subtle but serious errors in description and interpretation – subtle in that even experienced professionals sometimes make such errors, and serious is that they may affect data interpretation and decisions. Even when statistics is correctly applied, the results can be difficult to interpret for a non-expert – for example, the statistical significance of a trend in the data, which measures the extent to which the trend could be caused by random variation in the sample and may not agree with one's intuitive sense of its significance. Therefore, a set of basic statistical skills (and skepticism) is needed in the process of metabolomics data analysis and interpretation to deal with the vast amount of information metabolomics is providing.

The most common statistical methods applied for metabolomics data are Student's t test and analysis of variance (ANOVA). A t test is any statistical hypothesis test in which the test statistic has a Student's t distribution if the null hypothesis is true. A requirement is that the means of two normally distributed populations are equal. Given two datasets, which are characterized by their means, standard deviations, and a number of data points, the t test is used to determine whether the means are distinct, provided that the underlying distributions can be assumed to be normal. Two versions of the *t* test exist; the two samples are either independent or dependent of each other. When samples are dependent of each other, or paired, it means that each member of one sample has a unique relationship with a particular member of the other sample. Once a *t* value is determined, a *p* value can be determined using lookup tables or integral calculus (http://www.danielsoper.com/statcalc/calc08.aspx). A threshold chosen for statistical significance (usually below 0.05 Student's *t* distribution for *P* or with 95% confidence level) indicates that the two sample groups differ from each other.

ANOVA is a collection of statistical models, in which the observed variance is partitioned into components due to different variables. There are several types of ANOVA depending on the number of comparing pairs (treatment, genotype, time) under analysis. One-way ANOVA is used to test for differences among three or more independent groups. Factorial or two-way ANOVA is used when the effects of two or more treatment variables are under investigation. The most commonly used two-way ANOVA is the "two-by-two" design, where there are two independent variables and each variable has two levels or distinct values – for example, a control and a mutant plant at two developmental stages.

Three problems occur when both *t* test and ANOVA are applied to metabolomics data. First, both methods assume that the data under analysis are normal distributed which may or may not be true for metabolite data. Secondly, inadequate sample sizes for the large number of variables (metabolites, signals) measured that are required to prove a metabolite are discriminant, and thirdly, multiple hypothesis testing using only univariate statistical tests across all metabolites in parallel [9]. It is worth to note here that the same issues are of concern in other "omics" or high-throughput biological sciences, where a high number of variables are determined in a comparably small sample collection. The issues will increase the so-called false discovery rate (FDR), leading to the identification of apparently significant metabolites/biomarkers which are in fact incorrect. The false discovery rate (FDR) can be described as the expected percent of false prediction from the whole number of prediction and increases with the number of tests performed on a given dataset.

The first problem is difficult to tackle, in theory for each metabolite measured the distribution has to be determined and the appropriate statistical analysis chosen. This is not always feasible and commonly it has been agreed to treat metabolomics data as not normally distributed datasets. Therefore, before applying either *t* test or ANOVA data should be transformed (e.g., by either $\log_2$ or $\log_{10}$ transformation) to achieve close to normal distribution. The second problem is only solvable if the number of biological replications per sample would be increased dramatically, which again is not feasible for time, cost, and other logistical factors. Therefore, additional more stringent statistical methods have to be applied in order to deal with the inadequate sample sizes. The third issue of multiple hypothesis testing using univariate methods can be in fact only excluded by applying multivariate data analysis methods, which will be described below. Broadhurst and Kell [9] summarized those issues arising in metabolomics experiments very clearly and provided a framework of tackling the issues in an easy way to implement. This method is based on the Bonferroni correction [1] in which a new, more stringent *p* value is calculated by dividing the originally chosen p value threshold by the number of variables under t test or ANOVA test ($p\prime = p/n$). For example, if 100 metabolites are determined and the p value threshold was set to 0.05, the new $p\prime$ value threshold would be 0.0005, leading to a substantially smaller number of statistically significant metabolites (with now p value below 0.0005) giving much higher confidence for biological interpretation. The Bonferroni correction is often criticized as being too stringent. Metabolites that fail to meet the new p threshold may still be involved in the biological process studied. Therefore, the validity of using this parameter should be examined on a case-by-case basis.

3.3.3 Data Mining, Classification, and Visualization

Data mining is the extraction of potentially useful information from large datasets by multidimensional analysis. Dataset are sorted, analyzed, and sifted through to extract non-intuitive, but potentially important information. The broad aims are to discover new patterns between the variables examined, to test existing scientific models, to refine the theoretical understanding of a system to the point where it is fact based, and to enhance the ability to predict behavior or trends. Data mining is distinct from data analysis in that it is the platform itself and the dynamic interaction between the user and that platform that eventually provides the user with insight into the data not readily obvious by a precursory examination.

Common data mining methods in metabolomics applications include clustering, such as hierarchical clustering (HCA), multivariate data analysis, principle or independent component analysis, and co-response or correlation analysis. It is outside the scope of this chapter to describe the methods in detail and the ways in which they are applied in metabolomics applications; therefore, the reader is referred to Chapter 6 of this book. In addition, a number of summarizing reviews are available for more detailed understanding (e.g., [96] and [45] and references therein).

A number of web-based tools for the visualization and interpretation of omic data are becoming rapidly available. Ideally, these tools should allow the user to easily extract meaningful biological data from large datasets in a highly interactive, user-driven manner. With metabolomics, the ultimate goal is to understand the regulation of biochemical pathways. This perhaps can be best achieved by integrating gene, protein, and metabolite profiling data and examining the interrelationships of each level of the system. Here we give a brief overview of some of the more popular resources available for analyzing omic data generated from plant species. Each has advantages and disadvantages, and their usefulness will depend on the specific needs of the individual researcher.

The software module of the Pathway Tools database called the "Omics Viewer" (Pathway Tools Omics Viewer – PTOV) is one of the most extensive available. Gene and protein expression data, metabolite levels, and metabolic flux analysis data can be integrated onto a single metabolic map or interface [63]. The full utility of this tool is limited to species which have near-complete genomic sequence, i.e.,

Arabidopsis (dicot; AraCyc) (http://www.arabidopsis.org:1555/expression.html), rice (monocot; RiceCyc) (http://pathway.gramene.org/expression.html), and *Medicago* (legumes; MedicCyc) [92]. The output shows an overview of grouped metabolic pathways where measured metabolites, transcript, etc. can be color-coordinated based on their relative amount in the two samples compared. It is anticipated that as resources grow for other model species, they will be interfaced with PTOV. For example, preliminary data from the *Solanaceace* genomic projects are already being integrated at RiceCyc. MetNet (*Met*abolic *Net*work Exchange) is another plant-specific tool for functional genomics, which permits the user to visualize and overlay pathway components from comparative metabolite and transcriptomic studies. What distinguishes MetNet from similar programs is that it incorporates regulatory pathway information and contains applications for network modeling. In addition, MetNet contains rigorous statistical capabilities and it dovetails with AraCyc at PTOV. One disadvantage is that it is developed solely for *Arabidopsis*. MapMan [87] is a popular interface that is being increasingly used because it is a standalone, is relatively easy to navigate, and can be easily adapted to other species, e.g., legumes [34] and tomato [91]. It, however, depicts pathway data that are categorized by function, and only two experiments can be compared on the same map. Other publicly available tools include kaPPA-View and VANTED [46]. kaPPA-View was developed using *Arabidopsis* genomic data; it boasts an extensive secondary metabolic pathway depiction (added from other plant species), and gene expression data from multiple enzyme isoforms can be shown on the pathway. The user also has the option of showing transcript and metabolite data side by side, which is not possible with any other programs. VANTED (*V*isualization and *A*nalysis of *N*etworks with related *E*xperimental *D*ata) has been more recently introduced and allows the user to show detailed information for individual metabolites, which is advancement over most other tools. It is also possible to compare more than two experiments on the same map, but the trade-off is that this limits the number of pathway steps shown [46].

3.4 Metabolomics Approaches

After some confusion over the correct nomenclature to accurately describe various metabolite analytical approaches, Fiehn has provided clear definitions for each [27] which are now commonly accepted and used in the metabolomics community.

Target analysis describes the determination and quantification of a small set of known metabolites using one particular analytical technique, such as HPLC with UV detection. This type of approach has been done for many decades, before even the word metabolomics was invented. A number of different columns and elution protocols for many different compounds or compound classes have been developed and successfully applied for metabolic studies. Methods exist for small sugar analysis [15], amino acids [41], nucleotides [57, 68], flavonoids [59], volatiles [6], or alkaloids [30].

The second type of approach is called metabolite profiling, which includes the analysis of a larger set of both, identified and unknown metabolites, in a more unbiased manner, for instance using GC-MS. A number of exciting applications of GC-MS in plant metabolomics have been already described. To date, GC-MS is considered as the "workhorse" of separation and profiling technologies because the methodologies are well established and routinely applicable. Off the shelf, comparably affordable instrumentation is available from a range of vendors. Existing methods allow the detection and quantification of a large number of small metabolites, such as amino, organic and fatty acids, sugars, sugar alcohols, amines, or sterols. The development of a GC-MS mass spectral library (MSRI) in the public domain [51] is a great help to the researcher planning to use GC-MS technology. This library contains not only the electron impact originated mass spectra of metabolites but also their corresponding retention time indices which together, provides a highly valuable tool for peak identification. In the past decade, the GC-MS technology has been successfully applied for comparative metabolomics approaches in a range of different plant species, for example in *Arabidopsis thaliana* [28], *Solanum tuberosum* [71], *Medicago truncatula* [10], *Lotus japonicus* [17] or *Hordeum vulgare* [72], *Oryza sativa* [86], and *Solanum esculentum* [68].

Metabolomics itself would represent the determination and quantification of as many metabolites as possible, again both identified and unidentified, using complementary analytical methodologies to ensure maximal comprehensiveness, e.g., LC-MS/MS, GC-MS, and CE-MS. A schematic approach for metabolomics starting from sample harvest to data analysis is presented in Fig. 3.10.

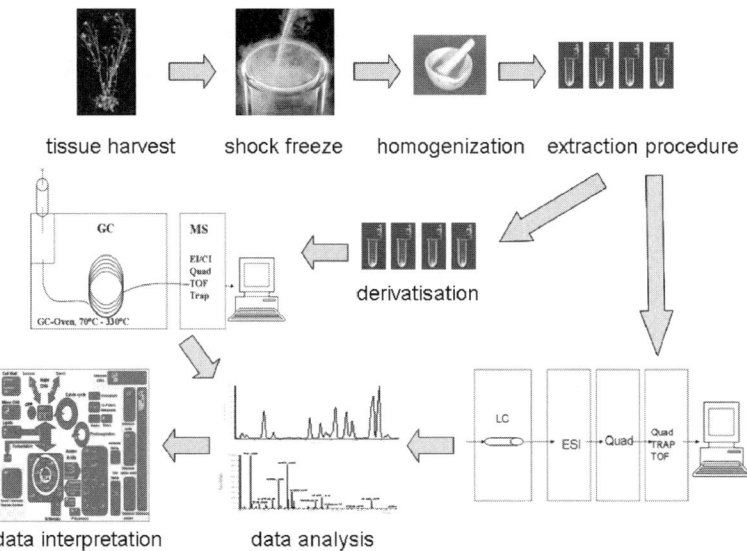

Fig. 3.10 Schematic workflow of a metabolomics approach from tissue harvest to data interpretation using complementary analytical instrumentation for greater comprehensiveness of metabolite detection and quantification (*see* also Color Insert)

The study of Tohge et al. [88] illustrates the utility of using multiple technologies platforms to investigate gene functions in rice. The aim of the study was to determine the effects of over-expression of the *PAP1* gene encoding an MYB transcription factor on *Arabidopsis* plants. This T-DNA activation-tagged line is known to produce higher levels of anthocyanins compared to untransformed plants. In order to unravel the mode of action of this transcription factor, the transgenic plants were analyzed by targeted profiling confirming the higher levels of flavonoids using liquid chromatography with photodiode array detection followed by mass spectral analysis (LC-PAD-MS). In addition, amino acid levels were determined by HPLC and anions and sugars by CE-MS, and a non-targeted analysis was performed using FT-ICR-MS. One outcome of the study was that the sample origin (plant organ) and the growth conditions have greater influences on the metabolite composition than the transgenic event. When different analytical platforms are employed for comprehensive metabolite analysis, attention must be paid to sample harvest (sufficient tissue for the different methods) and data integration. Most importantly, when data are integrated for data mining and interpretation, each type of analysis should be performed using the same sample, or more ideally, the same homogenate. The aliquots of this homogenate may then undergo different extraction procedures to prepare the sample to be amendable for the chosen analytical techniques. If possible, a single extraction method for all measurements would be preferable to reduce influences of variability of extraction onto the data generation.

The last more general approach is called metabolic fingerprinting, in which a metabolic "signature" of the sample is generated, for instance, using direct infusion ESI-MS (DIMS) or FT-ICR-MS. This approach is often used for high-throughput screening of large mutant collection or mapping populations. In the first instance, no attempt is made to assign detection signals to a particular metabolite but once a significantly discriminating signal is obtained, identification becomes crucial for interpretation. A recent example is provided by Oikawa and colleagues who described a strategic approach of using FT-ICR-MS for high-throughput metabolomics in plants [60]. The authors suggest a scheme for plant sample analysis starting from the data generation in a reproducible manner using FT-ICR-MS without any further chromatographic separation of the complex extracts. Obtained mass fingerprints (m/z values with respective ion intensity) were then mass-error corrected and directly submitted to multivariate analysis using a newly developed software tool (DMASS). Marker metabolites were identified by searching the open-source metabolite relationship database KNApSAcK, and putative identifications were confirmed by structural analysis using MS/MS mode of the FT-ICR-MS. The described scheme will be extremely helpful for high-throughput metabolic phenotyping studies not only in plants but also in other biological systems.

3.5 Application Examples in Plant Sciences

Even with the challenges and limitation of truly producing a comprehensive metabolite profile of plant extracts, metabolomics has already been successfully applied to many fields in plant science. This discipline is rapidly becoming useful to address fundamental and longstanding questions in biochemistry and physiology, which in turn has, and will continue to broaden our understanding of plant biology. Metabolomics has found applications for the comprehensive phenotyping of genetic varieties or genetically modified plants (GMOs), to determine gene function, to monitor plant behavior, responses to biotic and abiotic stress, and to make determinations of substantial equivalence. Because metabolites are the end products of gene expression, metabolic profiling also has the potential to bridge the gap between genotype and phenotype [36] and thus provides a more comprehensive and integrated view of how cells function in multicellular organisms. In addition, metabolic profiling has the potential to uncover new or dramatic changes in specific metabolites that can point to new hitherto undiscovered regulatory mechanisms or to a prediction of gene or protein function. Academically intriguing questions, testable hypotheses, and potentially new biotechnological targets can be generated and identified from analysis and data mining of metabolic metadata sets. Here we summarize a selection of metabolomics applications in plant research. A vast amount of primary research literature and excellent books and reviews demonstrating and summarizing the fast-growing field of plant metabolomics are available, and the reader is referred to [40, 74, 96] or [78]. In this chapter, we present only a small selection of examples for potential applications of metabolomics.

Metabolomics can be used for comprehensive phenotyping of genetic varieties or genetically modified plants (GMOs). This was capably demonstrated in pioneering work by Fiehn et al. [29] in which GC-MS-based metabolomics was applied to compare the metabolite profiles of two *Arabidopsis* ecotypes (parentals) and two respective, well-characterized mutants produced from each of the two parental lines. This study showed that the differences in the metabolomes of the ecotypes were far greater than those observed in the respective mutants. Roessner et al. [71] used GC-MS-based metabolite profiling to investigate the influence of transgenic modifications of the sucrose degradation and starch synthetic pathway in potato tubers. Large differences in the metabolite profile as a result of the genetic modification were detected. However, when the wildtype tubers were incubated under different in vitro culture conditions, the metabolite profiles of wildtype tubers could be drastically altered, such that even greater modulations in metabolite levels were introduced than that in the transgenics [71]. These two works established the discriminatory power of metabolomics to resolve and molecularly separate organisms based on genotype and culture conditions. Thereafter, metabolomics has been used to test substantial equivalence of GMO crops to examine potential "unintended" effects on the chemical constitution of the cell caused by transgenic manipulation [4, 16, 53, 104]. Although limited in scope and thus requiring caution in interpretation, the results thus far tend to support the view that traditional breeding, different ecotypes/cultivars, or growth in different environments may produce inherently more genetic variation in plants than that from some transgenic manipulations.

Metabolic profiling has also helped to redirect our knowledge of plant primary metabolism. For example, Roessner-Tunali et al. established that the de novo synthesis of amino acids occurs in non-photosynthetic organs (potato tuber) and was independent on import from leaves which was a departure from classical textbook description of this pathway [70]. Another long-held view was that import of sugars into tomato fruit was symplastic early in development and then switched to an apoplastic route during ripening [64]. This was challenged after analysis of metabolite profiles in transgenic tomatoes with reduced expression of a sucrose importer, as the results implied that sugar import was apoplast [39]. These two examples show the power of the metabolomics approach in putting new knowledge within reach by taking an unbiased approach to studying metabolism.

Metabolomics is set to become important as a high-throughput method to screen and accurately quantify the phenotypes of large genetic mapping populations. An important tool for identifying novel genetic variation and new genes determining plant performance and fitness is quantitative trait locus (QTL) mapping. Almost by necessity, traditional QTL analysis was done on easily scorable phenotypes such as fruit color, yield, or stress tolerance. However, the availability of novel technologies for high-throughput and simultaneous analysis of transcripts or metabolites has proved to be efficient for the rapid and efficient dissection of multiple traits at the molecular level, offering unprecedented access to QTLs. Large recombinant populations can therefore be characterized with greater precision for desired features, e.g., carotenoids, vitamins, acid, and/or sugar content in fruits, thus connecting DNA content to the measurable change in phenotype.

Novel QTLs that control the level of a single or a network of metabolites can be rapidly identified by direct comparison of the metabolite profile of progeny with the parents from whom they were derived. For example, Schauer et al. [76] used GC-MS technology to profile fruit from a well-characterized mapping population made from a cross of the cultivated tomato species (*Solanum lycopersicum*) and a wild, non-ripening tomato species (*Solanum pennellii*). Genomic regions from *S. lycopersicon*, which were populated with markers, were introgressed into homologous regions of *S. pennellii* thus allowing mapping of the genes responsible for contrasting traits, e.g., yield, fruit color, fruit sugars between the parentals [76]. A large number of single metabolite QTLs were identified; for example, four lines with overlapping genomic regions correlated with an increase in malate, suggesting that the QTL may map to potential metabolic enzyme(s) or regulatory gene(s) that controls or regulates malate levels. Many QTLs that affected entire pathways and/or metabolic networks in tomato fruit were also discovered, and due to careful measurement of physiological parameters, the study also showed that events in source or photosynthetic tissues has a large control over traits in fruit, even though there is large degree of spatio-temporal separation of biological activities of the two organs. This work will undoubtedly provide tomato breeders with an arsenal of new gene targets that could drive future tomato genetic improvement strategies. Further, it has heightened our awareness of the inseparable and indivisible nature of the biochemical and physiological behavior of crop plants in determining quality traits.

In another example, Keurentjes et al. [50] used a non-targeted LC-qTOF-MS method to produce metabolic fingerprints of 14 *Arabidopsis thaliana* accessions which could serve as parentals in a subsequent mapping experiment. The goal of this study is to support QTL detection by comparing the metabolite profiles of these fingerprinted parental accessions and their resulting progeny in a recombinant inbred line (RIL) population. Over 2,000 individual mass peaks were detected by

LC-qTOF-MS, and careful examination of the metabolomes of the two most divergent accessions allowed QTLs to be assigned to about 75% of all mass peaks [50]. If each mass peak can be unequivocally identified and chemically determined, this application has the power to chart new metabolic pathways and simultaneously elucidate their underlying genetic control which is potentially exciting. The two examples we offer here may well be pioneering work for future QTL identification and mapping; regardless, they demonstrate the potential that metabolomics offers to this field dominated by DNA sequencing technologies. As innovations in metabolomics support faster, easier, and more robust measurements, it will be possible to broadly apply metabolomics to study genetic segregation and to identify novel genes that contribute to biotechnologically important phenotypic traits. It can be envisioned that this new "reverse-QTL" approach will identify pathways based on metabolic profiles that underlie a trait of interest thus enabling a new paradigm in QTL screening and mapping.

Metabolomics represents a powerful approach of monitoring response, adaptation, and tolerance mechanisms of plants to challenging environments and is therefore ripe for use as a diagnostic and investigative tool. Plants are often able to survive and protect themselves against an onslaught of environmental stresses such as extreme temperature, aridity, and salinity by modulating intracellular solutes. This is usually achieved by increased synthesis and accumulation of compatible solutes or osmoprotectant such as polyols (including glycerol and sorbitol), amino acids (especially proline), quaternary compound (glycine betaine), or tertiary sulfonium compounds (dimethylsulfoniopropionate) [7, 67, 82, 102]. Integrating metabolomics with physiology is proving to be yet another valuable resource to dissect plant molecular response to abiotic stress and has generated new and exciting results in this emerging field. Kaplan et al. [47] and Cook et al. [19] detailed the plant metabolic adaptations after perturbation of the systems by exposure to variations in temperature. Low temperatures had greater repercussions for metabolite levels than did high temperatures, and several previously unknown adaptive mechanistic responses to cold stress were revealed, including changes in cellular amino acids, intermediates of both the TCA cycle and carbohydrate metabolism [18, 47].

To identify networks responsible for differential adaptation to salinity, Gong et al. [35] examined transcript and metabolite abundances of *Arabidopsis* and a closely related salt-tolerant species, *Thellungiella halophila*. They found that the metabolomes of both species were surprisingly similar except for two key differences – *Arabidopsis* had a greater flux of carbon to protein synthesis and *Thellungiella* appeared to have a pre-adaptation strategy to salinity stress that was not obvious or present in *Arabidopsis* [35]. Identifying novel metabolites or pathways that underlie plant adaptation to abiotic stress could lead to the production of crop species more tolerant to salt stress. It can be envisioned that in the near future, we will be able to compare the metabolite responses of different plant species to a range of different stresses, which will in turn allow the detection of metabolites affected by stress in all or most species (species independent) as well as species/genus-specific metabolite alterations (species specific).

3.6 Summary and Future Outlook

Plant metabolism is dazzlingly complex, and its study is one of the most fascinating and fast-exploding areas of biology. This complexity challenges any approach used to understand the detailed metabolic events occurring at the cellular or subcellular level. However, regardless of the difficulties or limitation of metabolomics, it is central to the development of the systems approach, which argues for and embraces the comprehensiveness, interrelationships, and interconnected nature of all levels of organism organization, i.e., gene, transcript, protein, metabolite, and physiology, in an attempt to complete our understanding of the system. Perhaps the greatest hurdle is the impossibility at the current time, of obtaining an accurate description of the total metabolic composition of a plant cell.

We have presented the current state of the art for identifying, measuring, and fingerprinting metabolites in plant cells. Although these analytical methods for assessing plant metabolites are becoming more efficient, repeatable, and even dynamic, the field of metabolomics is still very much in its infancy and we are just on the cusp of realizing its potential. Several areas of this new and burgeoning field need to be addressed; constant refinement and development of new enabling technologies and statistical and data mining tools to support data interpretation will broaden the realm of what is now possible in the plant metabolomic field. However, we stress that regardless of the metabolomic application used or the desired outcome, the need for quality control and rigorous statistical analysis cannot be compromised.

References

1. Abdi H (2007) Bonferroni and Sidak corrections for multiple comparisons. In: Salkind NJ (ed.) Encyclopedia of Measurement and Statistics. Sage, Thousand Oaks, CA.
2. Aharoni A, de Vos CHR, Verhoeven HA, Maliepaard CA, Kruppa G, Bino RJ, Goodenowe DB (2002) Nontargeted metabolome analysis by use of Fourier transform ion cyclotron mass spectrometry. Omics 6:217–234.
3. Arlt K, Brandt S, Kehr J (2001) Amino acid analysis in five pooled single plant cell samples using capillary electrophoresis coupled to laser-induced fluorescence detection. J Chromatography A 926:319–325.
4. Baker JM, Hawkins ND, Ward JL, Lovegrove A, Napier JA, Shewry PR, Beale MH (2006) A metabolomic study of substantial equivalence of field-grown genetically modified wheat. Plant Biotechnol J 4:381–392.
5. Birkemeyer C, Kopka J (2007) Design of metabolite recovery by variations of the metabolite profiling protocol. In: Nikolau BJ (ed.) Proceedings of the 3rd Congress on Plant Metabolomics. Springer, Dordrecht, The Netherlands.
6. Block E, Naganathan S, Putman D, Zhao SH (1992) *Allium* chemistry – HPLC analysis of thiosulfinates from onion, garlic, wild garlic (Ramsoms), leek, scallion, shallot, elephant (Great-Headed) garlic, chive, and Chinese chive – Uniquely high allyl to methyl ratios in some garlic samples. J Agric Food Chem 40:2418–2430.
7. Bohnert HJ, Jensen RG (1996) Strategies for engineering water-stress tolerance in plants. Trends Biotechnol 14:89–97.

8. Böttcher C, van Roepenack-Lahaye E, Willscher E, Scheel D, Clemens S (2007) Evaluation of matrix effects in metabolite profiling based on capillary liquid chromatography electrospray ionization quadrupole time-of-flight mass spectrometry. Anal Chem 79: 1507–1513.
9. Broadhurst D, Kell DB (2006) Statistical strategies for avoiding false discoveries in metabolomics and related experiments. Metabolomics 2:171–196.
10. Broeckling CD, Huhman DV, Farag MA, Smith JT, May GD, Mendes P, Dixon RA, Sumner LW (2005) Metabolic profiling of *Medicago truncatula* cell cultures reveals the effects of biotic and abiotic elicitors on metabolism. J Exp Bot 56:323–336.
11. Broeckling CD, Reddy IR, Duran AL, Zhao X, Sumner LW (2006) MET-IDEA: Data extraction tool for mass spectrometry-based metabolomics. Anal Chem 78:4334–4341.
12. Brown M, Dunn WB, Ellis DI, Goodacre R, Handl J, Knowles JD, O'Hagan S, Spasic I, Kell DB (2004) A metabolome pipeline: from concept to data to knowledge. Metabolomics 1:39–51.
13. Bruins AP, Covey TR, Henion JD (1987) Ion spray interface for combined liquid chromatography/atmospheric pressure ionization mass spectrometry. Anal Chem 59:2642–2646.
14. Burrell MM, Earnshaw CJ, Clench MR (2006) Imaging matrix assisted laser desorption ionization mass spectrometry: a technique to map plant metabolites within tissues at high spatial resolution. J Exp Bot 58:757–763.
15. Cairns AJ (1992) A reconsideration of fructan biosynthesis in storage roots of *Asparagus-officinalis* L. New Phytol 120:463–473.
16. Catchpole GS, Beckmann M, Enot DP, Mondhe M, Zywicki B, Taylor J, Hardy N, Smith A, King RD, Kell DB, Fiehn O, Draper J (2005) Hierarchical metabolomics demonstrates substantial compositional similarity between genetically modified and conventional potato crops. Proc Natl Acad Sci USA 102:11458–14462.
17. Clark JM, Daum KA, Kalivas JH (2003) Demonstrated potential of ion mobility spectrometry for detection of adulterated perfumes and plant speciation. Anal Lett 36:215–244.
18. Colebatch G, Desbrosses G, Ott T, Krusell L, Montanari O, Kloska S, Kopka J, Udvardi MK (2004) Global changes in transcription orchestrate metabolic differentiation during symbiotic nitrogen fixation in *Lotus japonicus*. Plant J 39:487–512.
19. Cook D, Fowler S, Fiehn O, Thomashow MF (2004) A prominent role for the CBF cold response pathway in configuring the low-temperature metabolome of *Arabidopsis*. Proc Natl Acad Sci USA 101:15243–15248.
20. De Luca V, St Pierre B (2000) The cell and developmental biology of alkaloid biosynthesis. Trends Plant Sci 361:168–173.
21. Dunn WB, Ellis DI (2005) Metabolomics: Current analytical platforms and methodologies. Trends Anal Chem 24:285–294.
22. Duran AL, Yang J, Wang L, Sumner LW (2003) Metabolomics spectral formatting, alignment and conversion tools (MSFACTs). Bioinformatics 19:2283–2293.
23. Edwards EL, Rodrigues JA, Feffeira J, Goodall DM, Rauter AP, Justino J, Thomas-Oates J (2006) Capillary electrophoresis-mass spectrometry characterisation of secondary metabolites from the antihyperglycaemic plant *Genista tenera*. Electrophoresis 27:2164–2170.
24. Eneroth P, Hellström K, Ryhage R (1964) Identification and quantification of neutral fecal steroids by gas-liquid chromatography and mass spectrometry: studies of human excretion during two dietary regimens. J Lipid Res 5:245–262.
25. Fernie AR, Geigenberger P, Stitt M (2005) Flux an important, but neglected, component of functional glenomics. Current Opin Plant Biol 8:174–182.
26. Fiehn O (2001) Combining genomics, metabolome analysis, and biochemical modelling to understand metabolic networks. Comp Funct Genomics 2:155–168.
27. Fiehn O (2002) Metabolomics – the link between genotypes and phenotypes. Plant Mol Biol 48:155–171.
28. Fiehn O (2006) Metabolite Profiling in *Arabidopsis*. In: Salinas J, Sanchez-Serrano JJ (eds) Arabidopsis Protocols 2nd edition. Methods in Molecular Biology (323), Humana Press, Totowa NJ, pp. 439–447.

29. Fiehn O, Kopka J, Dörmann P, Altmann T, Trethewey RN, Willmitzer L (2000) Metabolite profiling for plant functional genomics. Nat Biotechnol 18:1157–1161.
30. Friedman M, Dao L (1992) Distribution of glycoalkaloids in potato plants and commercial potato products. J Agricult Food Chem 40:419–423.
31. Gibon Y, Blaesing OE, Hannemann J, Carillo P, Höhne M, Hendriks JHM, Palacios N, Cross J, Selbig J, Stitt M (2004) A robot-based platform to measure multiple enzyme activities in *Arabidopsis* using a set of cycling assays: Comparison of changes of enzyme activities and transcript levels during diurnal cycles and in prolonged darkness. Plant Cell 16: 3304–3325.
32. Gibon Y, Usadel B, Blaesing OE, Kamlage B, Hoehne M, Trethewey RN, Stitt M (2006) Integration of metabolite with transcript and enzyme activity profiling during diurnal cycles in *Arabidopsis* rosettes. Genome Biol 7:R76.1–R76.23.
33. Gibon Y, Vigeolas H, Tiessen A, Geigenberger P, Stitt M (2002) Sensitive and high throughput metabolite assays for inorganic pyrophosphate, ADPGlc, nucleotide phosphates, and glycolytic intermediates based on a novel enzymic cycling system. Plant J 30:221–235.
34. Goffard N, Weiller G (2006) Extending MapMan: application to legume genome arrays. Bioinformatics 22:2958–2959.
35. Gong Q, Li P, Ma S, Rupassara SI, Bohnert HJ (2005) Salinity stress adaptation competence in the extremophile *Thellungiella halophila* in comparison with its relative *Arabidopsis thaliana*. Plant J 44:826–839.
36. Goodacre R (2005) Metabolomics – the way forward. Metabolomics 1:1–2.
37. Goodacre R, Vaidyanathan S, Dunn WB, Harrigan GG, Kell DB (2004) Metabolomics by numbers: acquiring and understanding global metabolite data. Trends Biotechnol 22: 245–252.
38. Griffin JL, Nicholls AW, Keun HC, Mortishire-Smith RJ, Nicholson JK, Kuehn T (2002) Metabolic profiling of rodent biological fluids *via* ^{1}H NMR spectroscopy using a 1 mm microlitre probe. Analyst 127:582–584.
39. Hackel A, Schauer N, Carrari F, Fernie AR, Grimm B, Kuhn C (2006) Sucrose transporter LeSUT1 and LeSUT2 inhibition affects tomato fruit development in different ways. Plant J 45:180–192.
40. Hall RD (2006) Plant metabolomics: from holistic hope, to hype, to hot topic. New Phytol 169:453–468.
41. Hodisan T, Culea M, Cimpoiu C, Cot A (1998) Separation, identification and quantitative determination of free amino acids from plant extracts. J Pharmaceutical Biomed Anal 18:319–323.
42. Huhman DV, Sumner LW (2002) Metabolic profiling of saponins in *Medicago sativa* and *Medicago truncatula* using HPLC coupled to an electrospray ion-trap mass spectrometer. Phytochemistry 59:347–360.
43. Ikonomou MG, Blades AT, Kebarle P (1990) Investigations of the electrospray interface for liquid chromatography/mass spectrometry. Anal Chem 62:957–967.
44. Jacobs A, Lunde C, Bacic A, Tester M, Roessner U (2007) The impact of constitutive expression of a moss Na$^+$ transporter on the metabolomes of rice and barley. Metabolomics 3: 307–317.
45. Jewett MC, Hansen ME, Nielsen J (2007) *Saccharomyces cerevisiae* metabolomics: a driver for developing integrative analytical tools for discerning metabolic function. In: Jewett MC, Nielsen J (eds) Metabolomics. Springer, Heidelberg, Germany.
46. Junker BH, Klukas C, Schreiber F (2006) VANTED: A system for advanced data analysis and visualization in the context of biological networks. BMC Bioinformatics 7: 109.
47. Kaplan F, Kopka J, Haskell DW, Zhao W, Schiller KC, Gatzke N, Sung DY, Guy CL (2004) Exploring the temperature-stress metabolome of *Arabidopsis*. Plant Physiol 136: 4159–4168.
48. Katajamaa M, Miettinen J, Orešic M. (2006) MZmine: toolbox for processing and visualization of mass spectrometry based molecular profile data. Bioinformatics 22:634–636.

49. Katona ZF, Sass P, Molnár-Perl I (1999) Simultaneous determination of sugars, sugar alcohols, acids and amino acids in apricots by gas chromatography–mass spectrometry. J Chromat A 847:91–102.
50. Keurentjes JJB, Fu J, de Vos CHR, Lommen A, Hall RD, Bino RJ, van der Plas LHW, Jansen RC, Vreugdenhil D, Koornneef M (2006) The genetics of plant metabolism. Nat Gen 38:842–849.
51. Kopka J, Schauer N, Krueger S, Birkemeyer C, Usadel B, Bergmüller E, Dörmann P, Weckwerth W, Gibon Y, Stitt M, Willmitzer L, Fernie AR, Steinhauser D (2005) GMD@CSB.DB: the Golm metabolome database. Bioinformatics 21:1635–1638.
52. Krishnan P, Kruger NJ, Ratcliffe RG (2005) Metabolite fingerprinting and profiling in plants using NMR. J Exp Bot 56:255–265.
53. Kristensen C, Morant M, Olsen CE, Ekstrøm CT, Galbraith DW, Møller BL, Bak S (2006) Metabolic engineering of dhurrin in transgenic *Arabidopsis* plants with marginal inadvertent effects on the metabolome and transcriptome. Proc Natl Acad Sci USA 102: 1779–1784.
54. Lindon JC, Holmes E, Nicholson JK (2003) So what's the deal with metabonomics? Anal Chem 75:384A–391A.
55. Luedemann A, Weicht D, Selbig J, Kopka J (2004) PaVESy: pathway visualization and editing system. Bioinformatics 20:2841–2844.
56. Matz LM, Dion HM, Hill HH (2002) Evaluation of capillary liquid chromatography-electrospray ionization ion mobility spectrometry with mass spectrometry detection. J Chromat A 946:59–68.
57. Meyer R, Wagner KG (1985) Determination of nucleotide pools in plant-tissue by high-performance liquid-chromatography. Anal Biochem 148:269–276.
58. Moco S, Bino RJ, Vorst O, Verhoeven HA, de Groot J, van Beek TA, Vervoort J, de Vos CHR (2006) A liquid chromatography-mass spectrometry-based metabolome database for tomato. Plant Physiol 141:1205–1218.
59. Molnár-Perl I, Füzfai Z (2005) Chromatographic, capillary electrophoretic and capillary electrochromatographic techniques in the analysis of flavonoids. J Chromat A 1073: 201–227.
60. Oikawa A, Nakamura Y, Oqura T, Kimura A, Suzuki H, Sakurai N, Shinbo Y, Shibata D, Kanaya S, Ohta D (2006) Clarification of pathway-specific inhibition by Fourier transform ion cyclotron resonance/mass spectrometry-based metabolic phenotyping studies. Plant Physiol 142:398–413.
61. Oksman-Caldentey K-M, Inzé D (2004) Plant cell factories in the post-genomic era: new ways to produce designer secondary metabolites. Trends Plant Sci 9:433–440.
62. Oliver SG, Winson MK, Kell DB, Baganz F (1998) Systematic functional analysis of the yeast genome. Trends Biotechnol 16:373–378.
63. Paley SM, Karp PD (2006) The pathway tools cellular overview diagram and Omics viewer. Nucleic Acid Res 34:3771–3778.
64. Patrick JW (1997) Phloem unloading: Sieve element unloading and post-sieve element transport. Annual Rev Plant Physiol Plant Mol Biol 48:191–222.
65. Ramautar R, Demirci A, de Jong GJ (2006) Capillary electrophoresis in metabolomics. Trends Anal Chem 25:455–466.
66. Ratcliffe RG, Shachar-Hill Y (2005) Revealing metabolic phenotypes in plants: inputs from NMR analysis. Biol Rev Cambridge Phil Soc 80:27–43.
67. Rathinasabapathi B (2000) Metabolic engineering for stress tolerance: Installing osmoprotectant synthesis pathways. Annals Bot 86:709–716.

68. Roessner-Tunali U, Hegemann B, Lytovchenko A, Carrari F, Bruedigam C, Granot D, Fernie AR (2003a) Metabolic profiling of transgenic tomato plants overexpressing hexokinase reveals that the influence of hexose phosphorylation diminishes during fruit development. Plant Physiol 133:84–99.
69. Roessner-Tunali U, Lui J, Leisse A, Balbo I, Perez-Melis A, Willmitzer L, Fernie AR (2004) Flux analysis of organic and amino acid metabolism in potato tubers by gas chromatography-mass spectrometry following incubation in ^{13}C labelled isotopes. Plant J 39:668–679.
70. Roessner-Tunali U, Urbanczyk-Wochniak E, Czechowski T, Kolbe A, Willmitzer L, Fernie AR (2003b) De novo amino acid biosynthesis in potato tubers is regulated by sucrose levels. Plant Physiol 133:683–692.
71. Roessner U, Luedemann A, Brust D, Fiehn O, Linke T, Willmitzer L, Fernie AR (2001) Metabolic profiling allows comprehensive phenotyping of genetically or environmentally modified plant systems. Plant Cell 13:11–29.
72. Roessner U, Patterson JH, Forbes MG, Fincher GB, Langridge P, Bacic A (2006) An investigation of boron toxicity in barley using metabolomics. Plant Physiol 142:1087–1101.
73. Roscher A, Emsley L, Raymond P, Roby C (1998) Unidirectional steady state rates of central metabolism enzymes measured simultaneously in a living plant tissue. J Biol Chem 273:25053–25061.
74. Saito K, Dixon RA, Willmitzer L (2006) Plant Metabolomics, Berlin Heidelberg, Germany, Springer-Verlag.
75. Sato S, Soga T, Nishioka T, Tomita M (2004) Simultaneous determination of the main metabolites in rice leaves using capillary electrophoresis mass spectrometry and capillary electrophoresis diode array detection. Plant J 40:151–163.
76. Schauer N, Semel Y, Roessner U, Gur A, Balbo I, Carrari F, Pleban T, Perez-Melis A, Bruedigam C, Kopka J, Willmitzer L, Zamir D, Fernie AR (2006) Comprehensive metabolic profiling and phenotyping of interspecific introgression lines for tomato improvement. Nature Biotechnol 24:447–454.
77. Schwender J, Ohlrogge J, Shachar-Hill Y (2004) Understanding flux in plant metabolic networks. Current Opin Plant Biol 7:309–317.
78. Shulaev V, Cortes D, Miller G, Mittler R (2008) Metabolomics for plant stress response. Physiol. Plant 132:199–208.
79. Smith, CA, Want EJ, O'Maille G, Abagyan R, Siuzdak G (2006) XCMS: processing mass spectrometry data for metabolite profiling using nonlinear peak alignment, matching, and identification. Anal Chem 78:779–787.
80. Soga T, Heiger DN (2000) Amino acid analysis by capillary electrophoresis electrospray ionization mass spectrometry. Anal Chem 72:1236–1241.
81. Stein SE (1999) An Integrated Method for Spectrum Extraction and compound identification from GC/MS data. J Amer Soc Mass Spectrom 10:770–781.
82. Stoop JMH, Williamson JD, Pharr DM (1996) Mannitol metabolism in plants: a method for coping with stress. Trends Plant Sci 1:139–144.
83. Styczynski MP, Moxley JF, Tong LV, Walther JL, Jensen KL, Stephanopoulos GN (2007) Systematic identification of conserved metabolites in GC/MS Data for metabolomics and biomarker discovery. Anal Chem 79:966–973.
84. Sumner LW, Mendes P, Dixon RA (2003) Plant metabolomics: large-scale phytochemistry in the functional genomics era. Phytochemistry 62:817–836.
85. Tang KQ, Li FM, Shvartsburg AA, Strittmatter EF, Smith RD (2005) Two-dimensional gas-phase separations coupled to mass spectrometry for analysis of complex mixtures. Anal Chem 77:6381–6388.
86. Tarpley L, Duran AL, Kebrom TH, Sumner LW (2005) Biomarker metabolites capturing the metabolite variance present in a rice plant developmental period. BMC Plant Biol 5:8.
87. Thimm O, Bläsing O, Gibon Y, Nagel A, Meyer S, Krüger P, Selbig J, Müller LA, Rhee SY, Stitt M (2004) MAPMAN: a user-driven tool to display genomics data sets onto diagrams of metabolic pathways and other biological processes. Plant J 37:914–939.

88. Tohge T, Nishiyama Y, Hirai MY, Yano M, Nakajima J-I, Awazuhara M, Inoue E, Takahashi H, Goodenowe DB, Kitayama M, Noji M, Yamazaki M, Saito K (2005) Functional genomics by integrated analysis of metabolome and transcriptome of *Arabidopsis* plants over-expressing an MYB transcription factor. Plant J 42:218–235.
89. Tolstikov VV, Fiehn O (2002) Analysis of highly polar compounds of plant origin: combination of hydrophilic interaction chromatography and elctrospray ion mass trap spectrometry. Anal Biochem 301:298–307.
90. Trethewey RN (2004) Metabolite profiling as an aid to metabolic engineering in plants. Curr Opin Plant Biol 7:196–201.
91. Urbanczyk-Wochniak E, Usadel B, Thimm O, Nunes-Nesi A, Carrari F, Davy M, Bläsing O, Kowalczyk M, Weicht D, Polinceusz A, Meyer S, Stitt M, Fernie AR (2006) Conversion of MapMan to allow the analysis of transcript data from Solanaceous species: Effects of genetic and environmental alterations in energy metabolism in the leaf. Plant Mol Biol 60:773–792.
92. Urbanczyk-Wochniak E, Sumner LW (2007) MedicCyc: a biochemical pathway database for *Medicago truncatula*. Bioinformatics 23:1418–1423.
93. Valentine SJ, Kulchania M, Barnes CAS, Clemmer DE (2001) Multidimensional separations of complex peptide mixtures: a combined high-performance liquid chromatography/ion mobility/time-of-flight mass spectrometry approach. Internat J Mass Spectrom 212:97–109.
94. Valentine SJ, Liu XY, Plasencia MD, Hilderbrand AE, Kurulugama RT, Koeniger SL, Clemmer DE (2005) Developing liquid chromatography ion mobility mass spectometry techniques. Exp Rev Proteomics 2:553–565.
95. Verpoorte R (1998) Exploration of nature's chemodiversity: the role of secondary metabolites as leads in drug development. Drug Discovery Today 3:232–238.
96. Villas-Bôas SG, Roessner U, Hansen M, Smedsgaard J, Nielsen J (2007) Metabolome Analysis: An Introduction, New Jersey, NJ, USA, John Wiley & Sons, Inc.
97. Wagner C, Sefkow M, Kopka J (2003) Construction and application of a mass spectral and retention time index database generated from plant GC/EI-TOF-MS metabolite profiles. Phytochemistry 62:887–900.
98. Wang SY, Kuo YH, Chang HN, Kang PL, Tsay HS, Lin KF, Yang NS, Shyur LF (2002) Profiling and characterization antioxidant activities in *Anoectochilus formosanus* Hayata. J Agricult Food Chem 50:1859–1865.
99. Williams BJ, Cameron CJ, Workman R, Broeckling CD, Sumner LW, Smith JT (2007) Amino acid profiling in plant cell cultures: An inter-laboratory comparison of CE-MS and GC-MS. Electrophoresis 28:1371–1379.
100. Wurtele ES, Li J, Diao L, Zhang H, Foster CM, Fatland B, Dickerson J, Brown A, Cox Z, Cook D, Lee E-K, Hoffman H (2003) MetNet: software to build and model the biogenetic lattice of *Arabidopsis*. Comp Funct Genomics 4:239–245.
101. Yamashita M, Fenn JB (1984) Electrospray ion source. Another variation on the free-jet theme. J Phys Chem 88:4451–4459.
102. Yancey PH (2005) Organic osmolytes as compatible, metabolic and counteracting cytoprotectants in high osmolarity and other stresses. J Exp Biol 208:2819–2830.
103. Zhao Q, Stoyanova R, Du S, Sajda P, Brown TR (2006) *HiRes* – a tool for comprehensive assessment and interpretation of metabolomic data. Bioinformatics 22:2562–2564.
104. Zywicki B, Catchpole G, Draper J, Fiehn O (2005) Comparison of rapid liquid chromatography-electrospray ionization-tandem mass spectrometry methods for determination of glycoalkaloids in transgenic field-grown potatoes. Anal Biochem 336:178–186.

Chapter 4
Enzyme Kinetics: Theory and Practice

Alistair Rogers and Yves Gibon

4.1 Introduction

Enzymes, like all positive catalysts, dramatically increase the rate of a given reaction. Enzyme kinetics is principally concerned with the measurement and mathematical description of this reaction rate and its associated constants. For many steps in metabolism, enzyme kinetic properties have been determined, and this information has been collected and organized in publicly available online databases (www.brenda.uni-koeln.de). In the first section of this chapter, we review the fundamentals of enzyme kinetics and provide an overview of the concepts that will help the metabolic modeler make the best use of this resource. The techniques and methods required to determine kinetic constants from purified enzymes have been covered in detail elsewhere [4, 12] and are not discussed here. In the second section, we will describe recent advances in the high throughput, high sensitivity measurement of enzyme activity, detail the methodology, and discuss the use of high throughput techniques for profiling large numbers of samples and providing a first step in the process of identifying potential regulatory candidates.

4.2 Enzyme Kinetics

In this section, we will review the basics of enzyme kinetics and, using simple examples, mathematically describe enzyme-catalyzed reactions and the derivation of their key constants. However, first we must turn to the mathematical description of chemical reaction kinetics.

A. Rogers (✉)
Department of Environmental Sciences, Brookhaven National Laboratory,
Upton, NY 11973-5000, USA
e-mail: arogers@bnl.gov

4.2.1 Reaction Rates and Reaction Order

4.2.1.1 First-Order Irreversible Reaction

The simplest possible reaction is the irreversible conversion of substance A to product P (e.g., radioactive decay).

$$A \xrightarrow{k_1} P \tag{4.1}$$

The arrow is drawn from A to P to signify that the equilibrium lies far to the right, and the reverse reaction is infinitesimally small. We can define the reaction rate or velocity (v) of the reaction in terms of the time (t)-dependent production of product P. Since formation of P involves the loss of A, we can also define v in terms of the time-dependent consumption of substance A, where [A] and [P] are the concentrations of the substance and product, respectively.

$$v = \frac{\delta[P]}{\delta t} = -\frac{\delta[A]}{\delta t} = k_1[A] \tag{4.2}$$

The transformation of substance A to product P is an independent event and therefore is unaffected by concentration. As substance A is transformed to product P, there is less of substance A to undergo the transformation, and therefore the concentration of substance A will decrease exponentially with time (Fig 4.1A). The rate constant (k_1) of this reaction is proportional to the concentration of A and has the unit s^{-1}. This type of unimolecular reaction is known as a first-order reaction because the rate depends on the first power of the concentration. Integration of Eq. (4.2) from time zero (t_0) to time t gives

$$\ln \frac{[A]}{[A]_0} = -k_1 t \tag{4.3}$$

or

$$\frac{[A]}{[A]_0} = e^{-k_1 t} \tag{4.4}$$

where $[A]_0$ is the starting concentration at t_0. Eq. (4.4) describes how the concentration of A decreases exponentially with time as shown in Fig. 4.1A. When the ln[A] is plotted against time (Fig. 4.1B), a first-order reaction will yield a straight line, where the gradient is equal to $-k_1$.

4.2.1.2 First-Order Reversible Reaction

Few reactions in biochemistry are as simple as the first-order reaction described above. In most cases, reactions are reversible and equilibrium does not lie far to one side.

Fig. 4.1 A first-order reaction showing the decrease of substance A over time expressed as the concentration of A ([A], **Panel A**) and in a semi-logarithmic plot (ln[A], **Panel B**)

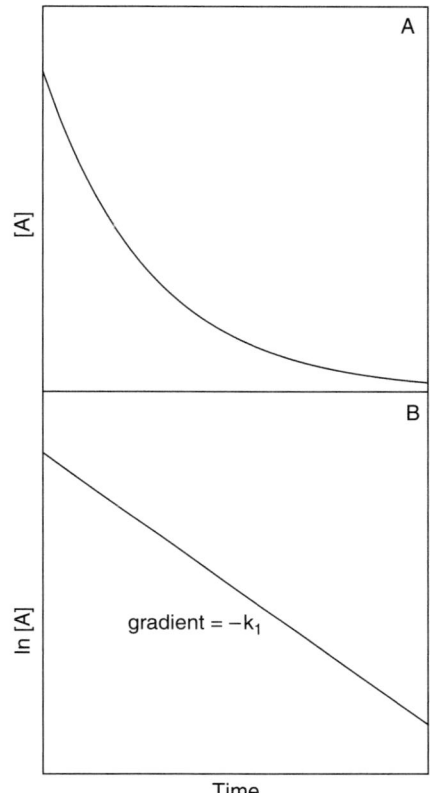

$$[A] \underset{k_{-1}}{\overset{k_1}{\rightleftarrows}} [P] \tag{4.5}$$

Therefore, the corresponding rate equation is

$$v = -\frac{\delta[A]}{\delta t} = k_1[A] - k_{-1}[P] \tag{4.6}$$

where k_1 and k_{-1} are the rate constants for the first-order, forward and reverse, reactions respectively. Consumption of A will stop when the rates of the forward and reverse reactions are equal and the overall reaction rate is zero, i.e., when a state of equilibrium has been attained ($[A]_{eq}$ and $[P]_{eq}$ are the substrate concentrations at equilibrium). Note that in catalyzed reactions, the position of equilibrium is not altered by the presence of an enzyme. The effect of a catalyst is to increase the rate at which equilibrium is attained.

$$0 = -k_1[A]_{eq} + k_{-1}[P]_{eq} \tag{4.7}$$

For this reaction, where the forward and reverse reactions are both first order, the equilibrium constant (K_{eq}) is equal to the ratio of the rate constants for the forward and reverse reactions. For a reaction to precede in the direction of product (P) formation, the equilibrium constant must be large.

$$K_{eq} = \frac{k_1}{k_{-1}} = \frac{[P]_{eq}}{[A]_{eq}} \quad (4.8)$$

4.2.1.3 Second-Order Reaction

In addition to being reversible, most reactions are second order or greater in their complexity. Whenever two reactants come together to form a product, the reaction is considered second order, e.g.,

$$A + B \xrightarrow{k_1} P \quad (4.9)$$

The rate of the above reaction is proportional to the consumption of A and B and to the formation of P. The reaction is described as second order because the rate is proportional to the second power of the concentration; the rate constant k_1 has the unit s^{-1} M^{-1}.

$$v = -\frac{\delta[A]}{\delta t} = -\frac{\delta[B]}{\delta t} = \frac{\delta[P]}{\delta t} = k_1[A][B] \quad (4.10)$$

Integration of Eq. (4.10) yields an equation where t is dependent on two variables, A and B. To solve this equation, either A or B must be assumed to be constant. Experimentally, this can be accomplished by using a concentration of B that is far in excess of requirements such that only a tiny fraction of B is consumed during the reaction and therefore the concentration can be assumed not to change. The reaction is then considered pseudo-first order.

$$v = k_1[A][B]_0 = k'_1[A] \quad (4.11)$$

Alternatively, when the concentration of both A and B at time zero are the same, i.e., $[A_0] = [B_0]$, Eq. (4.10) can be simplified:

$$v = -\frac{\delta[A]}{\delta t} = k_1[A]^2 \quad (4.12)$$

4.2.2 What Does an Enzyme Do?

Transition state theory suggests that as molecules collide and a reaction takes place, they are momentarily in a strained or less stable state than either the reactants or the products. During this transition state, the potential energy of the activated complex increases, effectively creating an energy barrier between the reactants and

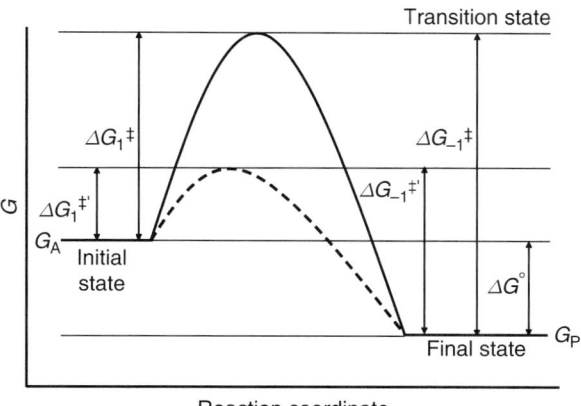

Fig. 4.2 A free energy (G) diagram for a simple reversible exothermic reaction A↔P (*solid and broken lines*). G_A and G_P represent the average free energies per mole for the reactant A and the product P, the initial and final states respectively. The standard state free energy change for the reaction is $\Delta G°$. In order for reactant A to undergo transformation to product P, it must pass through the transition state (indicated at the apex of these plots). The $\Delta G_1^\ddagger$ and $\Delta G_1^{\ddagger\prime}$ indicate the energy of activation necessary to make that transition for the uncatalyzed (*solid*) and catalyzed (*broken*) reactions respectively. The energy of activation for the reverse reaction (P→A) is indicated by $\Delta G_{-1}^\ddagger$ (uncatalyzed) and $\Delta G_{-1}^{\ddagger\prime}$ (catalyzed)

products (solid line, Fig. 4.2). Products can only be formed when colliding reactants have sufficient energy to overcome this energy barrier. The energy barrier is known as the activation energy ($\Delta G^\ddagger$) of a reaction. The greater the activation energy for a given reaction is, the lower the number of effective collisions. The molecular model currently used to explain how an enzyme catalyzes a reaction is the induced-fit hypothesis. In this model, the enzyme binds it's substrate to form an enzyme–substrate complex where the structure of the substrate is distorted and pulled into the transition state conformation. This reduces the energy required for the conversion of a given reactant into a product and increases the rate of a reaction by lowering the energy requirement (broken line, Fig. 4.2) and therefore increasing the number of effective collisions that can result in the formation of the product. In addition, enzymes also promote catalysis by positioning key acidic or basic groups and metal ions in the right position for catalysis. In reality, the free energy diagram for an enzyme-catalyzed reaction is considerably more complicated than the example in Fig. 4.2. Typically an enzyme-catalyzed reaction will involve multiple steps, each with an activation energy that is markedly lower than that for the uncatalyzed reaction.

4.2.3 The Michaelis–Menten Equation

The Michaelis–Menten equation (Eq. 4.26), as presented by Michaelis and Menten and further developed by Briggs and Haldane [6, 34], is fundamentally important to enzyme kinetics. The equation is characterized by two constants: the

Michaelis–Menten constant (K_m) and the indirectly obtained (see Eq. 4.25) catalytic constant, k_{cat}. Although derived from a simple, single-substrate, irreversible reaction, the Michaelis–Menten equation also remains valid for more complex reactions.

The simple conversion of substrate (A) into product (P) catalyzed by the enzyme (E) is described below. As outlined by the induced-fit hypothesis, the first step is substrate binding and the second step is the catalytic step.

$$E + A \underset{k_{-1}}{\overset{k_1}{\rightleftarrows}} EA \xrightarrow{k_2} E + P \tag{4.13}$$

Following Eq. (4.2), we can define the formation of the product in terms of the dissociation rate (k_2) of the enzyme–substrate complex, commonly denoted as k_{cat}, and the concentration of the enzyme–substrate complex ([EA]).

$$v = k_{cat}[EA] \tag{4.14}$$

It is assumed that the dissociation rate (k_{cat} in Eq. 4.14 or k_2 in Eq. 4.13) of the enzyme–substrate complex (EA) is slow compared to association (k_1) and redissociation (k_{-1}) reactions and that the reverse reaction (P→A) is negligible. Figure 4.3 shows how the consumption of substrate, the production of product, and the concentration of the free enzyme and the enzyme–substrate complex change over the course of the reaction. During a very brief initial period, the enzyme–substrate complex is

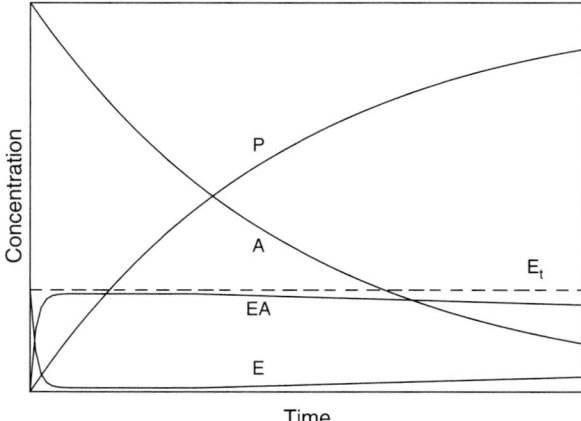

Fig. 4.3 Change in substrate (A), product (P), free enzyme (E), enzyme–substrate complex (EA), and total enzyme (E_t) concentration over time for the simple reaction described in Eq. (4.13). After a very brief initial period, the concentration of the enzyme–substrate complex reaches a steady state in which consumption and formation of the enzyme–substrate complex are balanced. As substrate is consumed, the concentration of the enzyme–substrate complex falls slowly and the concentration of the free enzyme rises. The amounts of enzyme and enzyme–substrate are greatly exaggerated for clarity

4 Enzyme Kinetics

formed and reaches a concentration at which its consumption is matched by its formation. The [EA] then remains almost constant for a considerable time; this period is known as the steady state, and it is this steady-state condition that the Michaelis–Menten equation describes. Eventually, the reaction enters a third phase characterized by substrate depletion in which the [EA] gradually falls.

At steady state, the enzyme–substrate concentration is stable, i.e.,

$$\frac{\delta[EA]}{\delta t} = 0 \tag{4.15}$$

and therefore the formation of the ES complex (association reaction) and the breakdown of the ES complex (the sum of the redissociation and dissociation reactions) are equal.

$$k_1[E][A] = k_{-1}[EA] + k_{cat}[EA] \tag{4.16}$$

Rearrangement of Eq. (4.16) yields

$$\frac{k_1[E][A]}{k_{-1} + k_{cat}} = [EA] \tag{4.17}$$

The three rate constants can now be combined as one term. This new constant, K_m, is known as the Michaelis–Menten constant

$$K_m = \frac{k_{-1} + k_{cat}}{k_1} \tag{4.18}$$

and Eq. (4.17) can be rewritten as

$$\frac{[E][A]}{K_m} = [EA] \tag{4.19}$$

The concentration of enzyme in Eq. (4.19) refers to the unbound enzyme. The amount of free enzyme (E) and enzyme that is bound to the substrate (EA) varies over the course of a reaction, but the total amount of enzyme (E_t) is constant (see Fig. 4.3) such that

$$E = E_t - EA \tag{4.20}$$

Substituting into Eq. (4.19) yields

$$\frac{([E_t] - [EA])[A]}{K_m} = [EA] \tag{4.21}$$

which can be rearranged to yield

$$\frac{[E_t][A]}{K_m + [A]} = [EA] \tag{4.22}$$

Substituting into Eq. (4.14) gives

$$v = \frac{k_{cat}[E_t][A]}{K_m + [A]} \quad (4.23)$$

The maximum possible reaction rate (v_{max}) would be achieved when all the available enzyme is bound to the substrate and involved in catalysis, i.e.,

$$[EA] = [E_t] \quad (4.24)$$

Substituting Eq. (4.24) into Eq. (4.14) under conditions of saturating substrate concentration yields

$$v_{max} = k_{cat}[E_t] \quad (4.25)$$

and substituting Eq. (4.25) into Eq. (4.23) yields what is widely recognized as the Michaelis–Menten equation.

$$v = \frac{v_{max}[A]}{K_m + [A]} \quad (4.26)$$

4.2.4 Key Parameters of the Michaelis–Menten Equation

4.2.4.1 K_m (mol.l^{-1})

Assuming a stable pH, temperature, and redox state, the K_m for a given enzyme is constant, and this parameter provides an indication of the binding strength of that enzyme to its substrate. Michaelis–Menten kinetics assumes that k_{cat} is very low when compared to k_1 and k_{-1}. Therefore, following Eq. (4.18), a high K_m indicates that the redissociation rate (k_{-1}) is markedly greater than the association rate and that the enzyme binds the substrate weakly. Conversely, a low K_m indicates a higher affinity for the substrate (E1 in Fig. 4.4). However, as Eq. (4.18) shows, a large K_m could also be the result of very large k_{cat}. Therefore, care should be taken when using K_m as a proxy for the dissociation equilibrium constant of the enzyme–substrate complex.

4.2.4.2 k_{cat} (s^{-1})

The k_{cat}, also thought of as the turnover number of the enzyme, is a measure of the maximum catalytic production of the product under saturating substrate conditions per unit time per unit enzyme. The larger the value of k_{cat}, the more rapidly catalytic events occur. Values for k_{cat} differ markedly, e.g., 2.5 s^{-1} for rubisco (EC 4.1.1.39) with CO_2 as a substrate to c. 1,150 s^{-1} for fumarase (EC 4.2.1.2) with fumarate as a substrate [13, 49].

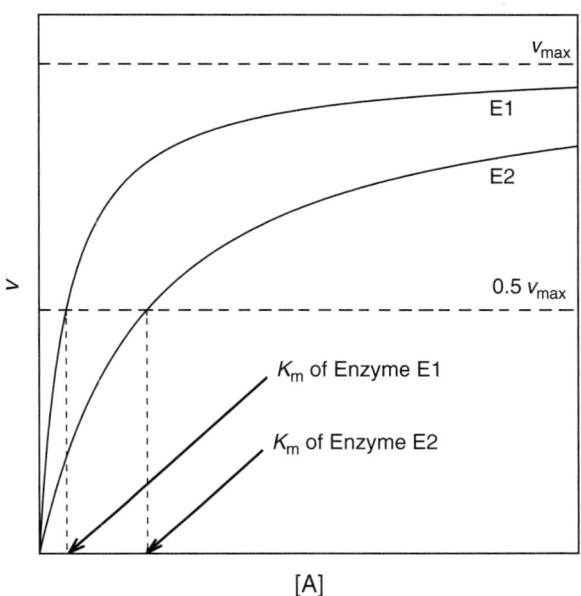

Fig. 4.4 Change in velocity (v) with the concentration of substrate A ([A]) for the reaction shown by Eq. (4.13) catalyzed by two enzymes E1 and E2. The substrate concentration at the point at which the reaction has half its maximum velocity (0.5 v_{max}) is equal to the K_m. Enzyme E2 has a K_m four times greater than enzyme E1 but the same v_{max}

4.2.4.3 Enzyme Efficiency (s^{-1} $(mol.l^{-1})^{-1}$)

The ratio of k_{cat}/K_m is defined as the catalytic efficiency and can be taken as a measure of substrate specificity. When the k_{cat} is markedly greater than k_{-1}, the catalytic process is extremely fast and the efficiency of the enzyme depends on its ability to bind the substrate. Based on the laws of diffusion, the upper limit for such rates, as determined by the frequency of collisions between the substrate and the enzyme, is between 10^8 and 10^9. Some enzymes actually have efficiencies that approach this range, indicating that they have near-perfect efficiency, e.g., fumarase, 2.3×10^8 $s^{-1}(mol.l^{-1})^{-1}$ [13, 50].

4.2.4.4 v_{max}

The v_{max} is the maximum velocity that an enzyme could achieve. The measurement is theoretical because at given time, it would require all enzyme molecules to be tightly bound to their substrates. As shown in Fig. 4.4, v_{max} is approached at high substrate concentration but never reached.

4.2.5 Graphical Determination of Michaelis–Menten Parameters

Since the Michaelis–Menten parameters provide useful information for the network modeler, we need to consider the methods used to estimate K_m, k_{cat}, and v_{max}. There are a number of practical approaches to measuring reaction rates (see Section 4.3). Briefly, we need some way to follow the consumption of substrate or formation

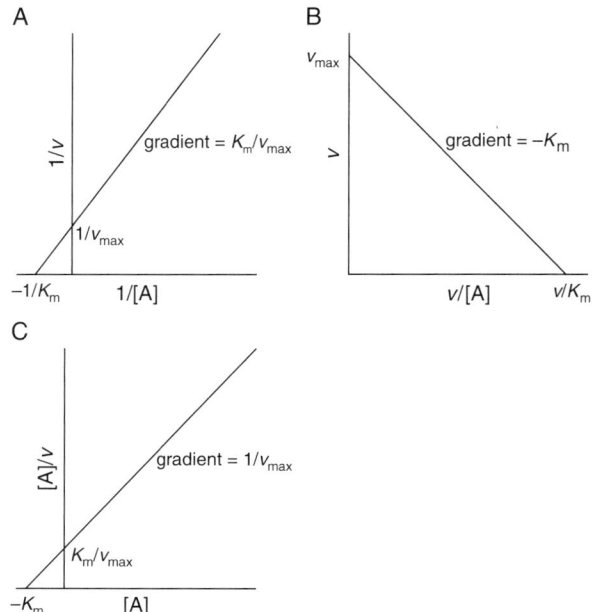

Fig. 4.5 Linear representations of the Michaelis–Menten equation (Eq. 4.26). Lineweaver–Burk (**A**), Eadie–Hofstee (**B**), and Hanes (**C**) plots. The intercepts with the x- and y-axis and the gradient can be used to determine K_m and v_{max}

of product over time. We could simply mix enzyme with substrate and follow the formation of product in a progress reaction (e.g., Fig. 4.3) or conduct several experiments at multiple substrate concentrations and measure initial velocity at each substrate concentration (e.g., Fig. 4.4). However, the graphical evaluation of nonlinear plots to obtain Michaelis–Menten parameters relies on accurate curve fitting. The problems associated with evaluating enzyme kinetics using a nonlinear plot can be avoided by using one of the three common linearization methods to obtain estimates for K_m and v_{max} (Fig. 4.5). However, these methods are not without problems either. Errors in the determination of v at low substrate concentration are greatly magnified in Lineweaver–Burke and Eadie–Hofstee plots and to a lesser extent in Hanes plots. Despite this disadvantage, and in contrast to nonlinear plots, changes in enzyme kinetics, for example, due to the action of an inhibitor, are readily apparent on linear plots (see Fig. 4.6). Clearly, selection of a linear or nonlinear plot should be based on an understanding of the sources of error in the experiment and consistent with the goal of that experiment.

4.2.6 Multisubstrate Reactions

Most biochemical reactions are not simple-, single-substrate reactions, but typically involve two or three substrates that combine to release multiple products. However, the Michaelis–Menten equation is robust and remains valid as reaction complexity increases. When an enzyme binds two or more substrates, the order of the biochemical steps determines the mechanism of the reaction. Below we have detailed the

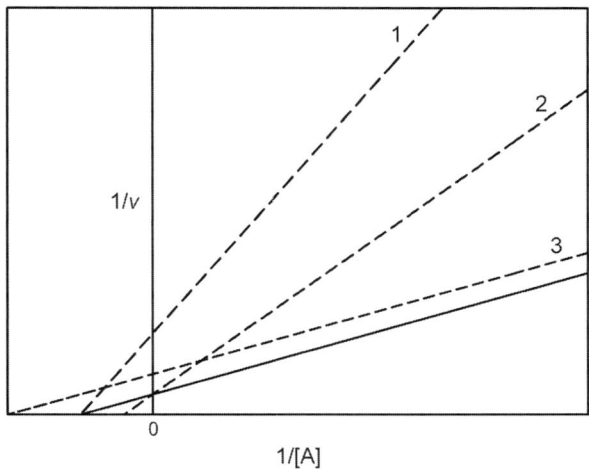

Fig. 4.6 A Lineweaver–Burke plot of an uninhibited enzyme (*solid line*) and the same enzyme in the presence of noncompetitive inhibitor (*plot 1*), competitive inhibitor (*plot 2*), and an uncompetitive inhibitor (*plot 3*). The point where the plots meet the x-axis indicates $-1/K_m$, and the intercept with the y-axis indicates $1/v_{max}$

three major classes of mechanisms for the reaction where two substrates (A and B) react to yield two products (P and Q). Full derivation of the rate equations for these reactions and discussion of more complex mechanisms is covered elsewhere [4, 12] and is beyond the scope of this chapter.

4.2.6.1 Random Substrate Binding

In its simplest form, this mechanism assumes independent binding of substrates and products. Either substrate A or B can be bound first and either product P or Q released first; binding of the first substrate is independent of the second substrate. The catalytic reactions occur in central complexes and are shown here in parentheses to distinguish them from intermediate complexes that are capable of binding substrates. The phosphorylation of glucose by ATP, catalyzed by hexokinase, is an example of a random-ordered mechanism, although there is tendency for glucose to bind first.

$$E+A+B \begin{array}{c} \overset{k_1}{\underset{k_{-1}}{\rightleftarrows}} EA+B \overset{k_2}{\underset{k_{-2}}{\rightleftarrows}} \\ \overset{k_3}{\underset{k_{-3}}{\rightleftarrows}} EB+A \overset{k_4}{\underset{k_{-4}}{\rightleftarrows}} \end{array} \begin{array}{c} (EAB) \\ (EPQ) \end{array} \begin{array}{c} \overset{k_5}{\underset{k_{-5}}{\rightleftarrows}} EP+Q \overset{k_6}{\underset{k_{-6}}{\rightleftarrows}} \\ \overset{k_7}{\underset{k_{-7}}{\rightleftarrows}} EQ+P \overset{k_8}{\underset{k_{-8}}{\rightleftarrows}} \end{array} E+P+Q \quad (4.27)$$

4.2.6.2 Ordered Substrate Binding

In some cases, one substrate must bind first before the second substrate is able to bind effectively. This mechanism is frequently observed in dehydrogenase reactions where NAD^+ acts as a second substrate.

$$E + A + B \underset{k_{-1}}{\overset{k_1}{\rightleftarrows}} EA + B \underset{k_{-2}}{\overset{k_2}{\rightleftarrows}} (EAB - EPQ) \underset{k_{-3}}{\overset{k_3}{\rightleftarrows}}$$
$$EQ + P \underset{k_{-4}}{\overset{k_4}{\rightleftarrows}} E + P + Q$$
(4.28)

4.2.6.3 The Ping-Pong Mechanism

In this mechanism, enzyme E binds substrate A and then releases product P. An intermediate form of enzyme E (E^*), which often carries a fragment of substrate A, then binds substrate B. Finally, product Q is released, and the enzyme is returned to its original form (E). Aminotransferases use this mechanism, e.g., aspartate aminotransferase catalyses the ping-pong transfer of an amino group from aspartate to 2-oxoglutarate to form oxaloacetate and glutamate.

$$E + A + B \underset{k_{-1}}{\overset{k_1}{\rightleftarrows}} (EA - E^*P) + B \underset{k_{-2}}{\overset{k_2}{\rightleftarrows}} E^* + P + B \underset{k_{-3}}{\overset{k_3}{\rightleftarrows}}$$
$$(E^*B - EQ) + P \underset{k_{-4}}{\overset{k_4}{\rightleftarrows}} E + P + Q$$
(4.29)

4.2.7 Regulation

The catalytic capacity for a given process in a cell can be regulated at many levels of biological organization. Coarse control is provided by the regulation of the transcription of genes that encode the enzymatic machinery. Here, we outline the major mechanisms by which the activity of functional enzymes can be altered by fine control mechanisms and show how these mechanisms impact enzyme kinetics.

4.2.7.1 Enzyme Inhibition

Here, we define inhibition as a reduction in enzyme activity through the binding of an inhibitor to a catalytic or regulatory site on the enzyme, or in the case of uncompetitive inhibition, to the enzyme–substrate complex. Inhibition can be reversible or irreversible. Irreversible inhibition nearly always involves the covalent binding of a toxic substance that permanently disables the enzyme. This type of inhibition does not play a role in the fine control of enzyme activity and is not discussed further. In contrast, reversible inhibition involves the noncovalent binding of an inhibitor to the enzyme which results in a temporary reduction in enzyme activity. Inhibitors differ in the mechanism by which they decrease enzyme activity. There are three basic mechanisms of inhibition – competitive, noncompetitive, and uncompetitive inhibition – and these are outlined below using simple examples. The reality is more complex and typically reactions involve mixed and partial mechanisms comprised of these three component mechanisms [4, 12].

4.2.7.2 Competitive Inhibition

A competitive inhibitor is usually a close analogue of the substrate. It binds at the catalytic site but does not undergo catalysis. A competitive inhibitor wastes the enzyme's time by occupying the catalytic site and preventing catalysis. Or put another way, the presence of an inhibitor decreases the ability of the enzyme to bind with its substrate. The reaction scheme below details the mechanism. Here, k_1 and k_{-1} are the rates for the association and redissociation reactions for the enzyme–substrate complex (see Eq. 4.13), and k_2 and k_{-2} are the rates for the association and redissociation reactions between the enzyme and inhibitor (I).

$$\begin{array}{c} E + A \underset{k_{-1}}{\overset{k_1}{\rightleftarrows}} EA \xrightarrow{k_{cat}} E + P \\ + \\ I \\ {\scriptstyle k_{-2}} \updownarrow {\scriptstyle k_2} \\ EI \end{array} \qquad (4.30)$$

At steady state, the enzyme–inhibitor concentration is stable; so following Eqs 4.15–4.18, the association and redissociation rate constants for the enzyme–inhibitor complex can be combined in one term K_i, the dissociation constant for inhibitor, and following Eq. (4.19) can be expressed as follows:

$$K_i = \frac{[E][I]}{[EI]} \qquad (4.31)$$

Since some enzyme is bound to the inhibitor, the equation describing the total amount of enzyme has an extra term and Eq. (4.20) becomes

$$E = E_t - EA - EI. \qquad (4.32)$$

The resulting rate equation becomes

$$v = \frac{k_{cat}[E]_t[A]}{K_m \left(1 + \frac{[I]}{K_i}\right) + [A]} \qquad (4.33)$$

and following Eq. (4.25),

$$v = \frac{v_{max}[A]}{K_m \left(1 + \frac{[I]}{K_i}\right) + [A]} \qquad (4.34)$$

As can be seen from Eq. (4.34), an increase in the concentration of a competitive inhibitor will increase the apparent K_m of the enzyme. However, since an infinite substrate concentration will exclude the competitive inhibitor, there is no effect on v_{max}. The effect of competitive inhibition is readily apparent on a Lineweaver–Burke plot (Fig. 4.6, plot 2).

4.2.7.3 Noncompetitive Inhibition

A noncompetitive inhibitor does not bind to the catalytic site but binds to a second site on the enzyme and acts by reducing the turnover rate of the reaction. The reaction scheme (Eq. 4.35) details the mechanism for a noncompetitive inhibitor. Consider the simplest example of a noncompetitive inhibitor. Here, the binding of the inhibitor and substrate is completely independent, and binding of the inhibitor results in total inhibition of the catalytic step. In this simple case, the association and disassociation rates k_1 and k_{-1} are identical to k_3 and k_{-3} (i.e., K_m), and similarly, k_2 and k_{-2} are equal to k_4 and k_{-4} (i.e., K_i)

$$\begin{array}{c} E + A \underset{k_{-1}}{\overset{k_1}{\rightleftarrows}} EA \xrightarrow{k_{cat}} E + P \\ + \qquad\qquad + \\ I \qquad\qquad I \\ k_{-2} \updownarrow k_2 \qquad k_4 \updownarrow k_{-4} \\ EI + A \underset{k_{-3}}{\overset{k_3}{\rightleftarrows}} EIA \end{array} \qquad (4.35)$$

The apparent k_{cat} for this simple example is given by

$$k_{cat}^{app} = \frac{k_{cat}}{\left(1 + \dfrac{[I]}{K_i}\right)} \qquad (4.36)$$

and the resulting rate equation is

$$v = \frac{k_{cat}^{app}[E]_t[A]}{K_m + [A]} \qquad (4.37)$$

As can be seen from the rate equation, a simple noncompetitive inhibitor will not alter the K_m but will reduce the apparent k_{cat} as inhibitor concentration increases (Fig. 4.6 plot 1).

4.2.7.4 Uncompetitive Inhibition

An uncompetitive inhibitor does not bind to the enzyme but only the enzyme–substrate complex. Consider the simple example where binding of the uncompetitive inhibitor to the enzyme–substrate complex prevents catalysis (Eq. 4.38):

$$E + A \xrightleftharpoons[k_{-1}]{k_1} EA \xrightarrow{k_{cat}} E + P$$
$$+$$
$$I$$
$$k_{-2} \updownarrow k_2$$
$$EI \qquad (4.38)$$

The rate equation is

$$v = \frac{v_{max}[A]}{K_m + [A]\left(1 + \frac{[I]}{K_i}\right)} \qquad (4.39)$$

Sequestration of the enzyme–substrate complex by the inhibitor will reduce the apparent k_{cat} because the inhibited enzyme is less catalytically effective. Apparent v_{max} is reduced (and apparent K_m increased) because binding of the inhibitor cannot be prevented by increasing the substrate concentration (Fig 4.6, plot 3).

4.2.7.5 Substrate and Product Inhibition

The activity of enzymes can also be regulated by their substrates and products. Substrate inhibition, also know as substrate surplus inhibition, occurs when a second substrate molecule acts as an uncompetitive inhibitor binding to the enzyme–substrate complex to form an enzyme–substrate–substrate complex. This mechanism is the same as for uncompetitive inhibition, but here, the inhibitor is replaced by the second substrate molecule. In reversible reactions, the buildup of product can theoretically competitively inhibit the forward reaction by competing with the substrate for the active site. However, since the equilibrium constant is often large, favoring product formation, this type of inhibition is typically negligible. However, the product of a reaction can also behave as a noncompetitive or uncompetitive inhibitor. The mechanisms for these types of inhibition have been described above for the action of inhibitors.

The regulation of enzyme activity by its immediate substrate and or product is not sufficient to allow regulation of complex metabolic pathways with shared substrates. Effective regulation must include the inhibition and activation of enzyme activity by molecules that are distinct from the substrates and products of the regulated rate. These molecules are usually produced by reactions that are multiple biochemical steps away from the regulated enzyme. Allosteric regulation allows this type of control.

4.2.7.6 Allosteric Enzymes

Allosteric enzymes exhibit cooperativity in their substrate binding and regulation of their active site through the binding of a ligand to a second regulatory site. These two traits make allosteric enzymes particularly good at controlling flux through a given metabolic chokepoint when compared to enzymes with classic Michaelis–Menten kinetics. Indeed, classic Michaelis–Menten enzymes require an 81-fold increase in substrate concentration to increase reaction rate from 10% to 90% of the maximal velocity [12].

4.2.7.7 Cooperativity (Homoallostery)

In enzymes with multiple binding sites, cooperative substrate binding describes the phenomenon whereby the binding of the first substrate molecule impacts the ability of the subsequent substrate molecules to bind. In the case of positive cooperativity, binding of the first substrate molecule enhances the ability of the following molecules to bind. An enzyme exhibiting positive cooperativity will appear to have a large K_m at low substrate concentration, but as the substrate concentration rises, the K_m will decrease and the substrate will be bound more readily. Figure 4.7 shows an example of positive cooperativity (plot 3). The physiological advantage of the sigmoidal kinetics is that enzyme activity can be increased more markedly within a narrow range of substrate concentration (gray area Fig. 4.7) when compared to a normal hyperbolic kinetic response (plot 1). The enzyme with positive cooperativity is much more responsive to changes in substrate concentration and can also better maintain a substrate concentration at or below a given threshold. In negative cooperativity, the binding of the first substrate interferes with the occupation of the second site. This can be advantageous when an enzyme needs to respond to a wide

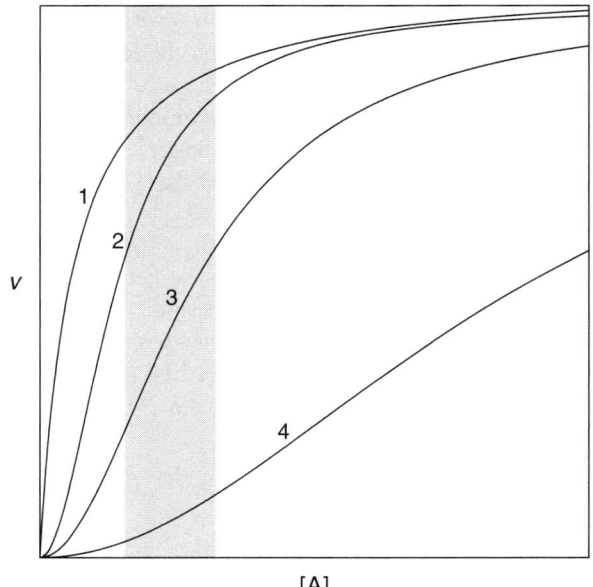

Fig. 4.7 Change in velocity (v) with substrate concentration ([A]) for an enzyme with normal binding (*plot 1*) and positive cooperative binding (*plot 3*) in the presence of an allosteric activator (*plot 2*) and an allosteric inhibitor (*plot 4*). The *gray area* indicates a hypothetical physiological range for this enzyme

range of substrate concentrations. An enzyme with negative cooperativity will be activated by a low concentration of substrate but will not saturate until the substrate concentration is extremely high.

There are two main models that attempt to describe how the enzyme changes its affinity for substrate with cooperative binding [28, 35]. These models share the concept that the subunits of the enzyme can exist in both a tense state (T-state), where substrate binding is weak, and a relaxed state (R-state), where substrate binding is strong, and that the initial binding of the substrate to the T-state enzyme shifts more subunits/enzyme molecules to the R-state, where the substrate can bind more readily.

The Hill equation (Eq. 4.40) describes the fraction of binding sites filled (r) by n molecules of substrate A, where K_d is the dissociation constant for A. The Hill coefficient derived from the gradient of a log-transformed plot of Eq. (4.40) indicates the degree and direction of cooperativity. An enzyme with classic hyperbolic binding behavior has a Hill coefficient of 1; enzymes with positive cooperativity have a Hill coefficient >1; and enzymes with negative cooperativity have a Hill coefficient <1.

$$r = \frac{n[A]^n}{K_d + [A]^n} \quad (4.40)$$

4.2.7.8 Heteroallostery

The second major component of allosteric enzymes is the control of enzyme activity by heteroallosteric effectors (inhibitors or activators). Since allosteric enzymes exhibit cooperativity and can be considered to exist in two states, the T-state and R-state, it follows that an effector that can alter the balance between the T and R states will be able to effect the kinetics. Allosteric inhibitors bind to the subunit and stabilize its T-state. Therefore, a greater substrate concentration is required to compensate for the shift of equilibrium toward the T-state. Activators shift the equilibrium toward the R-state by acting as a cooperative ligand. Figure 4.7 shows the effect of an allosteric inhibitor (plot 4) and activator (plot 2) on the sigmoidal kinetics of an allosteric enzyme.

4.3 Measurement of Enzyme Activity

Enzyme activities have been measured for more than a century in the frame of a wide range of applications ranging from fundamental approaches to industrial applications, from biochemistry to medical diagnostics, and assessment of food quality. The importance of characterizing the catalytic properties of individual enzymes is self-evident to biochemists. Traditionally, enzymes have been purified from individual organisms or tissues and subjected to various in vitro experiments in order to study the corresponding reaction mechanisms (e.g., [3, 5, 22, 33]), and eventually determine their constants (K_m, K_i, k_{cat}, v_{max}); such data can now be found in databases, e.g., BRENDA (www.brenda.uni-koeln.de/).

The next step is to integrate catalytic properties with data describing the structure of proteins, from sequence data to crystal structures. The understanding of structure–function relationships indeed represents one of the major aims in biology [23]. Recent progress in molecular techniques has enabled the design of alterations in the structure of enzymes, via site-directed mutagenesis [43] or tilling [44], and in cases combined with heterologous expression systems [1] which can provide new insight into structure–function relationships. Such approaches are generally focused on a few targets and do not usually involve high throughput techniques.

Another major aim in biology is to link the properties of macromolecules with phenotypes. Variation in the properties of enzymes can indeed have important consequences on metabolism, also on plant form and function. Variation in the sequence of a given structural gene may affect the properties of the corresponding enzyme, and depending or not on growth conditions, affect the phenotype. For example, the introgression of a regulatory subunit of ADP-glucose pyrophosphorylase, from a wild tomato species into a cultivated tomato, has been found to stabilize the active protein and thus maintain a higher activity of this key enzyme in starch synthesis. As a consequence, developing fruits accumulate more starch, which results in the release of more soluble sugars in ripening fruits [38]. Another striking example is the effect of an apoplastic invertase introgressed from another wild tomato species. Its higher affinity for sucrose, due to a single nucleotide polymorphism, results in a higher content of soluble organic compounds in the ripe fruits [15]. Both examples show that variations in properties of enzymes can dramatically affect phenotypes. We can thus predict that many such relationships will be found by screening genotypes (natural populations or mutants) for alterations in the properties of key enzymes.

In addition, environmental parameters like light, temperature, or nutrient availability can influence enzyme activities, via transcriptional, posttranscriptional, posttranslational, or allosteric regulations. Several enzymes, like nitrate reductase [26] and ADP-glucose pyrophosphorylase [1, 25, 51], have been intensively investigated, revealing highly complex regulations, but only a few studies haven been undertaken in a more systematic way. These studies showed, for example, that diurnal changes in transcript levels were not reflected at the level of the activities of the encoded enzymes in leaves of *Arabidopsis* [16], but were generally integrated over time, leading to semi-stable metabolic phenotypes [17, 36, 45, 52].

Finally, the properties and response of enzymes need be understood in their physiological context. In other words, kinetic properties, usually determined in vitro on isolated enzymes, need to be linked to pathway kinetics. Modeling metabolic networks will benefit from the accumulation of data dealing with variations in the levels and catalytic properties of enzymes associated with given genotypes and/or precise growth conditions. Furthermore, high throughput approaches will be crucial to access such data, especially in natural communities where diurnal, seasonal, spatial, and climatic variability requires extensive sampling. We will thus emphasize methodologies dedicated to the determination of activities in complex samples, as they typically represent the first step in the identification of regulatory candidates.

4.3.1 Methodology

Enzyme activity is determined by measuring the amount of product formed or substrate consumed under known conditions of temperature, pH, and substrate concentration. In general, the initial rate, defined as the slope of the tangent to the progress curve at time 0, is determined. It is important to keep in mind that activities express velocities, not concentrations or amounts of molecules. By definition, the determination of a given enzyme activity thus requires *unique* physical and chemical conditions. In consequence and due to the fact that enzymes catalyze very diverse reactions (dehydrogenations, transfers, isomerizations, etc.), the profiling of various enzyme activities implies the application of a wide range of principles. It is therefore at the opposite of "true" profiling approaches, in which a class of molecules (transcripts, metabolites, proteins) sharing similar physical and chemical properties is being analyzed. Below we have detailed some of the key concepts and advances that have made high throughput enzyme analysis possible.

4.3.1.1 Quantification Techniques

Various principles allow the quantification of changes in the concentrations of substrates or products of enzymatic reactions. The most widely used principles are UV-visible spectrophotometry [2], fluorimetry, [19, 20] and, to a lesser extent, luminometry [10, 11, 14]. Spectrophotometric methods benefit from the fact that many reactions involve directly or indirectly the oxidized and reduced forms of NAD(P), the reduced forms absorbing specifically at 340 nm. NAD(P)H can also be determined in a fluorimeter, with a much higher sensitivity [24]. Furthermore, various fluorogenic substrates reacting with a wide range of enzymes are commercially available. The luminometric method involving luciferase and its substrate luciferin is very often used to measure ATP, ADP, and AMP [11], giving access to the quantification of ATP-dependent reactions [10]. Such methodology can benefit from a multiparallel setup (microplates, microfluidic systems) which is well suited for high throughput.

Radioactivity is used when specificity and/or sensitivity cannot be achieved with conventional methods. A typical application is the determination of the incorporation of $^{14}CO_2$ in the carboxylation reaction catalyzed by rubisco (EC 4.1.1.39; [30]). The throughput of such methods is generally low, and their use is increasingly affected by constraints on the use of radioactivity due to increased regulation of environmental health and safety.

Mass spectrometry methods are increasingly developed for the determination of enzyme activities [21]. One major advantage is the possibility to check and quantify almost every type of molecule, given substrates and products can be easily separated and/or have different masses. This technology however requires expensive equipment and considerable expertise.

Electrochemical detection [27] can also be applied to biochemistry, as for example, amperometry which consists of the determination of electrical currents with electrodes eventually coated with enzymes catalyzing ion-producing reactions. The use of such biosensors remains limited and requires sophisticated equipment. This technology is however amenable to high throughput, for example, when combined with microfluidics [41, 54].

In plants, fluorimetric and luminometric methods are difficult to use, due to the presence in extracts of high levels of various compounds interfering at almost every wavelength, e.g., pigments or polyphenols. Without careful fractionation, such compounds will quench the emitted signals, even when present at low concentrations, leading to an underestimation of the actual activities. This is a pity because most recent technological developments in enzymology rely on fluorimetry, in particular, microfluidic chips [40]. In consequence, the microplate[1] format still offers the best compromise between cost and throughput. Microplates can be used manually but are amenable to high throughput applications. Due to the wide range of applications and equipment available nowadays, microplates have almost completely replaced cuvette-based applications. Although microplate readers also exploit the Beer–Lambert law [7], there might be some confusion. In a cuvette photometer, the absorbance (OD for optical density) is defined as follows:

$$OD = \varepsilon.c.l \qquad (4.41)$$

where ε is the extinction coefficient of the substance being measured, c is its concentration, and l, the length of the optical path (generally 1 cm, see Fig. 4.8). In a cuvette, the light path is constant and OD varies with concentration. In a microplate reader, the light path is vertical and dependent on the volume (V) of the solution being measured. Where r is the radius of a well,

$$l = \frac{V}{\pi r^2} \qquad (4.42)$$

since,

$$c = \frac{n}{V} \qquad (4.43)$$

OD will be proportional to the amount (moles) of absorbing molecules (n) and will be independent of the light path, i.e., in Fig. 4.8 the OD for wells B and C will be the same. Thus giving:

---

[1]This format was invented in the early 1950s by the Hungarian G. Takatsky and became popular during the late 1970s with the ELISA application, that's probably the reason why so many researchers call microplates "elisa plates."

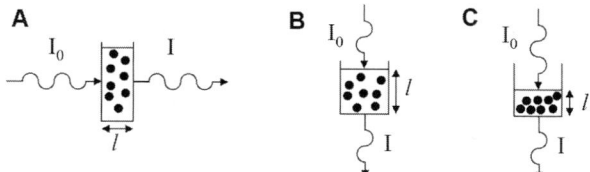

Fig. 4.8 Incident (I_0) and transmitted (I) light in a cuvette (**A**) and two microplate wells (**B** and **C**). The light path (l) is constant in the cuvette (typically 1 cm) but varies in the microplate well. Wells (**B**) and (**C**) have the same amount of analyte (*black dots*) but different concentrations. However, the amount of transmitted light (I) will be the same

$$\text{OD} = \varepsilon.n.\frac{1}{\pi r^2} \qquad (4.44)$$

It is worth noting that the well radius in microplates varies with manufacturer and model and care should be taken to select a plate with a flat bottomed cylindrical well. In addition, the presence of a meniscus in the well can affect this relationship, especially when using low volumes.

4.3.1.2 Continuous and Discontinuous Assays

In a continuous assay, the progress of the reaction is monitored directly in a recorder. This is only possible when changes in either a product or a substrate can be monitored in real time, as is the case with highly active dehydrogenases. In discontinuous assays, the reaction is stopped after fixed time intervals and a product is measured with a second specific reaction. In routine measurements, only two time points may be measured but linearity has to be checked to ensure that the supply of substrates and cofactors has not been depleted.

4.3.1.3 Sensitivity

The sensitivity of an assay can be defined as the smallest quantity that can be determined significantly. When activities are measured in raw extracts, it is also convenient to express it as the smallest amount of biological material that can be assayed. The theoretical detection limit of a standard filter-based photometric microplate reader is 0.001 which represents 0.06 nmol of NADH at 340 nm. In practice, due to experimental noise, the detection limit is much higher, at least 10 times when performing endpoint measurements. Highly sensitive assays allow the determination of activities that are present at very low levels but increased sensitivity also means that interferences can be removed, or significantly reduced by dilution.

A range of highly sensitive methods dedicated to the determination of enzyme activities are available commercially, but as mentioned above, most of them are not suitable for plant extracts, as they rely on fluorimetry or luminometry, an alternative is the use of cycling assays (Fig. 4.9).

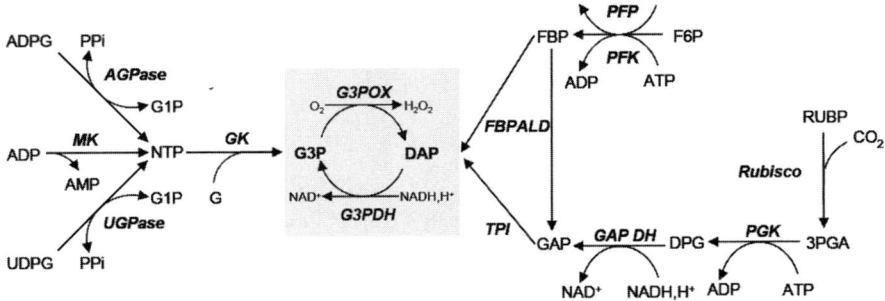

Fig. 4.9 Examples of assay principles based on the glycerol-3-phosphate cycling. Each enzyme activity (represented in *bold italics*) can be determined by adding coupling enzymes and metabolites downstream of its relevant product. After stopping the reaction, the product is determined using the cycling system (*highlighted*), directly or after conversion into either G3P or DAP. The principle is that the net rate of the cycle is a pseudo-zero-order reaction whose rate (δ[analyte measured]/δt = δ[precursor of this analyte]/δt) depends on the initial concentration of G3P and/or DAP being determined. Quantification is achieved by measuring the rate of NADH consumption at 340 nm, and by comparison with a standard curve, in which different concentrations of the G3P and/or DAP are added in the presence of pseudo-extract. Abbreviations: *Metabolites* 3PGA, 3-phosphoglycerate; ADPG, ADP-glucose; DAP, dihydroxyacetone phosphate; DPG, 1,3-diphosphoglycerate; F6P, fructose-6-phosphate; FBP, fructose-1,6-bisphosphate; G, glycerol; G1P, glucose-1-phosphate; G3P, glycerol-3-phosphate; GAP, glyceradehyde-3-phosphate; NTP, nucleotide triphosphate; PPi, pyrophosphate; RUBP, ribulose-1,5-bisphosphate; UDPG, UDP-glucose. *Enzymes* AGPase, ADPG pyrophosphorylase; FBPALD, FBP aldolase; GAPDH, NAD-GAP dehydrogenase; GK, glycerokinase; G3PDH, G3P dehydrogenase; G3POX, G3P oxidase; MK, myokinase; PFK, ATP-phosphofructokinase; PGK, phosphoglycerokinase; PFP, PPi-phosphofructokinase; rubisco, ribulose-1,5-bisphosphate carboxylase/oxygenase; TPI, triosephosphate isomerase; UGPase, UDPG pyrophosphorylase

Cycling assays are less prone to interferences coming from raw extracts as they can be used with standard microplate photometers and provide a 100–10,000 increase in sensitivity compared to direct or endpoint spectrophotometric methods. Cycling assays were developed by Warburg et al. [53] and made popular by the efforts of Lowry et al. [31, 32]. However, these assays are time consuming and tedious when used in cuvettes. Cycling assays prove to be much easier in microplates [18] and can be adapted for the determination of a number of enzyme activities via discontinuous assays [16]. A wide range of reactions can be measured provided they can be coupled to the production or consumption of NAD(H), NADP(H), glycerol-3-phosphate, dihydroxyacetone phosphate, or nucleotide triphosphates (Fig. 4.9).

4.3.1.4 Coupling Reactions

The majority of enzyme activities cannot be monitored directly. One or more coupled reactions are needed to convert a product of the enzyme reaction being measured into a quantifiable product. For example, phosphoglucose isomerase can be assayed by coupling the production of glucose-6-phosphate to NADPH production, using glucose-6-phosphate dehydrogenase, as shown below:

$$F6P \xrightarrow{PGI} G6P \xrightarrow{G6PDH} NADPH \tag{4.45}$$

Abbreviations: F6P, fructose-6-phosphate; G6P, glucose-6-phosphate; NADPH, reduced nicotinamide adenine dinucleotide phosphate; PGI, phosphoglucose isomerase (EC5.3.1.9); G6PDH, glucose-6-phosphate dehydrogenase (EC1.1.1.49)

Coupling reactions may also be used when the primary reaction has an unfavorable equilibrium constant, e.g., malate dehydrogenase in its forward direction

$$Malate + NAD^+ \xleftrightarrow{MDH} Oxaloacetate + NADH + H^+$$
$$K_{eq} = 5.94 \times 10^{-13} \tag{4.46}$$

Abbreviation: MDH, malate dehydrogenase (EC 1.1.1.37). The value for K_{eq} is from Outlaw and Manchester [37].

The addition of citrate synthase and acetyl coenzyme A will consume oxaloacetate and thus displace the equilibrium of the primary reaction:

$$Oxaloacetate + Acetyl\ coenzyme\ A \xrightarrow{CS}$$
$$Citrate + Coenzyme\ A \tag{4.47}$$

Abbreviation: CS, citrate synthase (EC 2.3.3.1).

A number of theoretical studies have been undertaken to model and optimize coupled enzyme assays [2, 46]. Coupled assays are valid if the velocity of the coupling system equals the velocity of the reaction of interest. Thus, an efficient coupling is only possible if steady-state concentrations of the product of the primary reaction are much smaller than the corresponding K_m [46]. The fact that a coupling reaction may increase the time lag required to reach steady state has also to be taken into account. Equations can be used to optimize the concentrations of the coupling enzymes, generally in order to reduce costs or to avoid interfering reactions. It is however recommended to test a range of concentrations of coupling enzymes and to check the duration of the lag phase for each of them.

4.3.1.5 Interferences with Other Components of the Extract

The use of raw extracts to determine kinetic properties of enzymes is subject to various interferences. Undesired substrates of reactions under investigation, including coupling reactions, can lead to underestimations or overestimations of actual activities, especially when non-saturating conditions are being used. Specific or non-specific inhibitors or activators may also interact with the reactions under study. Another possible source of error is the presence of numerous enzymes in the extract, as some of them may react with constituents of the assay.

Running blanks is a way to retrieve interferences; it is particularly useful when an enzyme yielding a common product is present. A typical example is given by the assay for glutamine synthetase, an enzyme involved in nitrogen assimilation and in photorespiration [39]:

$$\text{Glu} + NH_4^+ + \text{ATP} \xrightarrow{GS} \text{Gln} + \text{ADP} + \text{Pi} \qquad (4.48)$$

Abbreviations: Glu, glutamate; ATP, adenosine triphosphate; Gln, glutamine; ADP, adenosine diphosphate; Pi, orthophosphate; GS, glutamine synthetase (EC 6.3.1.2).

Pyruvate kinase and lactate dehydrogenase are then used as coupling enzymes, to convert the ADP into NAD^+:

$$\text{PEP} + \text{ADP} \xrightarrow{PK} \text{Pyruvate} + \text{ATP}$$
$$\text{Pyruvate} + \text{NADH, H}^+ \xrightarrow{LDH} \text{Lactate} + \text{NAD}^+ \qquad (4.49)$$

Abbreviations: PEP, phosphoenolpyruvate; PK, pyruvate kinase (EC 2.7.1.40); $NADH,H^+$, reduced nicotinamide adenine dinucleotide; NAD+, oxidized nicotinamide adenine dinucleotide; LDH, lactate dehydrogenase (EC 1.1.1.27).

In the presence of AMP (generally present as a contaminant of commercial preparations of ATP), adenylate kinase from the extracts will also yield ADP:

$$\text{ATP} + \text{AMP} \xrightarrow{AK} \text{ADP} + \text{ADP} \qquad (4.50)$$

Abbreviations: AMP, adenosine monophosphate; AK, adenylate kinase (EC 2.7.4.3).

Thus, the coupling system will measure the activities of both glutamine synthetase and adenylate kinase. It will then be useful to run a blank without ammonium (or conversely glutamate). However, the effect of ammonium (or glutamate) on the activity of adenylate kinase has to be checked.

Therefore, it is sometimes useful to add specific inhibitors to block interfering enzymes. P^1,P^5-di (adenosine-5')-pentaphosphate is a strong inhibitor of adenylate kinase [29] and can be included into the assay mixture for the determination of glutamine synthetase. A blank without one of the substrates is however still necessary, as the inhibition of the interfering activity may be incomplete.

Another way to diminish interferences from the extracts is to dilute them until interferences become negligible. As mentioned above, purification steps including desalting of extracts are time consuming and may provoke losses of activities. Nevertheless, the fact that many enzymes become unstable when they are diluted [42] has to be taken into account. Dilution experiments can be performed in order to determine the optimal dilution of the extract in the assay. Interestingly, various enzymes from leaves of various species could be measured at strong dilutions without losses in activity ([16] and unpublished results). When stopped assays are used, linearity with time should always be checked, as well as the recovery of the product of the enzyme under investigation. This is best achieved by spiking the extraction buffer with various amounts of the product, below and above the range expected to be produced by the enzyme. Alternatively, extracts under study can be mixed with an extract with a known activity.

4.3.2 Logistics

Building a microplate-based platform enabling the determination of dozens of enzyme activities in parallel requires several points to be taken into account.

4.3.2.1 Type of Assay

Stopped assays are usually preferred to continuous assays, as they provide several major advantages. First, they offer more flexibility because the determination of the products of the enzymatic reactions under study can be performed separately, while the continuous assays require as many temperature-controlled photometers as enzymes being measured. Incubators, including automated hotels, where microplates can be incubated at predetermined temperatures for set periods of time before reactions are stopped, are indeed less expensive than photometers and can easily be included as part of an automated pipetting station. Secondly, stopped assays can provide a much higher sensitivity, when products of enzymes are being measured with kinetic or fluorimetric methods. This allows routine determination of enzymes with low activities from raw extracts, without the need for sophisticated time-consuming purification and concentration procedures. Thirdly, the first step of stopped assays can be performed with low volumes (e.g., 20 µl in 96-well microplates), while for optical reasons, continuous assays require a minimal volume (e.g., 100 µl in 96-well flat bottom microplates). This can lead to a substantial lowering of the costs, assuming the reagents used for the determination of the products are less expensive than those used in the first step. The major disadvantage of stopped assays is that they require more pipetting steps, which implies that they are more time consuming and more prone to error. Time and error can however be considerably reduced when using electronic multichannel pipettes or liquid handling robots. The use of continuous assays will be usually restricted to enzymes with high activities and those requiring inexpensive reagents, like triose phosphate isomerase, malate dehydrogenases, or phosphoglucomutase.

4.3.2.2 Reagents

As by definition, each enzyme activity requires unique conditions to be determined; a multiparallel platform will require a large variety of reagents, which implies well-organized logistics. Typically, microplates have to be prepared in advance, so that enzyme reactions can be started right after extracts have been prepared. However, assay mixes are generally stable for only a few hours, so that they have to be prepared right before starting extractions. It is very useful to build a "bank" of reagents that can be organized as ready-to-use kits. Whenever possible, stock solutions should be prepared in advance and stored at adequate temperatures (e.g., $-80°C$ when containing enzymes and/or coenzymes). Pipetting schemes used for the preparation of assay mixes should be kept as simple as possible in order to decrease the time needed, for example, by adjusting concentrations of reagents in such a way that only a few, easy to manage, pipetting volumes will be used.

Another important issue is that more and more reagents are no longer commercially available, probably due to the fact that many enzyme-based analytical procedures used to determine metabolites have been replaced by mass spectrometry-based methods. For example, yeast glycerokinase, used as a coupling enzyme in a range of assays measuring ATP- or UTP-producing enzymes and exploiting the glycerol-3-phosphate cycling system [16, 18], cannot be replaced by its homologues from bacteria, as these have a much weaker affinity for ATP and do not react with UTP. Heterologous expression systems can be used to produce such enzymes, but imply extra costs and can be time consuming. Substrates like xylulose-5-phosphate or sedoheptulose-1,7-bisphosphate, necessary for the determination of important enzymes from the Calvin cycle or the oxidative pentose phosphate pathway, also became unavailable recently. In these cases, skills in organic chemistry will be welcome. Alternatively, private or public laboratories can produce such compounds on demand, as relevant protocols are often available, but this is usually very expensive.

4.3.2.3 Sample Handling

Samples are prepared by quenching tissues into liquid nitrogen immediately after harvesting. If labile posttranslational modifications are studied, sampling should happen within seconds. Furthermore, several enzymes that are regulated via light-dependent redox mechanism, like fructose-1,6-bisphosphase (EC 3.1.3.11;[9]), may activate/deactivate very quickly. It is then crucial to plunge the tissues into liquid nitrogen in the light. Samples should always be stored at $-80°C$ and processed at very low temperatures (grinding and weighing of aliquots for analysis) until extraction.

4.3.2.4 Preparation of Extracts

The optimal dilution varies from enzyme to enzyme, in large part because enzymes from various pathways cover 4–5 orders of magnitude in terms of activity. Thus, depending on the enzymes being measured, it will be necessary to achieve several dilutions of the extracts. This is best achieved when extracts are prepared in 96-well format.

4.3.2.5 Stability of Enzymes

Many enzymes are not stable once extracted and do not resist a freezing/thawing cycle, even in the presence of glycerol. The assay must therefore be performed as quickly as possible once the extracts have been prepared. This implies that a compromise has to be found between number of extractions and number of enzymes to be measured.

4.3.2.6 Temperature

Kinetic properties of enzymes vary with temperature, it is thus important to keep it constant, by using incubators and/or temperature controlled photometers.

4.3.2.7 Timing

Time management is essential for conducting stopped assays. When several enzymes are assayed in parallel in many extracts, manual timing becomes very difficult. This can be simplified by using the automated timing available in standard programs driving liquid handling robots.

4.3.2.8 Automation

The need to process more samples faster is a continuing trend in academic and industrial research. The wide adoption of microplates in laboratory routines has significantly influenced the development of a huge diversity of labware and automation solutions. Almost everything dealing with enzyme activities can now be processed with the help of robots in this format, from preparation of samples to detection. Depending on needs and means, the best balance between man and machine has nevertheless to be found in the jungle of laboratory robotics.

Based on the desired throughput, and assuming the labor of 2–4 people, we can roughly estimate the needs:

- Low throughput, below 500 activity determinations a week, robotics is not an absolute requirement. Standard microplate equipment including multichannel pipettes and a spectrophotometer might be adequate.
- Mid throughput, between 500 and 50,000, at least one liquid handling robot is needed, ideally a 96-channel robot equipped with a gripper to transport microplates, a shaker, temperature control, and several microplate readers. Pipetting robots working in the range of 0.5–50 µl are the most adequate; in addition to the throughput, they usually provide a very good accuracy at low volumes. A cryogenic grinding/weighing robot (Labman, Stokesley, UK) may also be very useful to process samples prior to extraction, as these steps are highly time consuming. A laboratory information management system may also be implemented in order to decrease time and error in calculations.
- High throughput, above 50,000, requires fully automated solutions. A high degree of sophistication will be required to include steps such as centrifugation, adhering, or removing adhesive lids, integration of microplate readers, and so on, implying an exponential increase of the costs. A further consequence is a strong decrease in flexibility.

4.3.3 Determination of Enzyme Properties in Raw Extracts

The purification of enzymes from living organisms, even partial, is a time-consuming process eventually leading to losses in activity and/or alterations in the actual catalytic properties. In consequence, its use is generally restricted to detailed biochemical studies. Highly purified enzymes are needed to determine important kinetic parameters like K_m and k_{cat}, to search for inhibitors or activators, or to obtain crystals.

If k_{cat} is known, it is then theoretically possible to evaluate enzyme concentrations in complex extracts. This is however biased by possible changes due to isoform composition or to posttranslational regulation events. This is also not a priority, as advances in proteomics are likely to become more adequate for such purpose and able to deal with a much larger number of analytes in parallel [8]. Apparent activities of enzymes should therefore be considered as integrating various levels of regulation, each of them being potentially subject to environmental or genetic inputs.

We believe that the collection of large sets of activity data obtained from various genotypes, organs, or tissues and from various growth conditions could be useful to modeling scientists. It would therefore be advantageous to determine the activities in standardized conditions, like temperature, pH (depending on the subcellular compartmentation and assuming that most enzymes from a compartment will have similar pH optima), or buffers. Metadata consisting in precise documentations of the assay condition for each enzyme should also be documented.

4.3.3.1 Measurement of Total Activity

When assay conditions are optimized in such a way that a given activity from a raw extract is maximized, we will consider that v_{total} is being measured. Under conditions at which enzymes are by far more diluted than their substrates, most of them obey the law of Michaelis–Menten. As a consequence and assuming that assay conditions, including concentrations of substrates, are kept nearly constant, rates of reactions will be dependent solely on the enzyme concentration, due to the establishment of pseudo-zero-order reactions (see Section 4.2.1). Thus, measurement of v_{total} is an estimate of the amount of enzyme present.

4.3.3.2 Measurement of Apparent Kinetic Constants

Various linearization methods have been established to determine such constants (see Section 4.2.5 and Fig. 4.5), but as previously stated [12], computer-based methods should be preferred, assuming some understanding of the underlying calculations. In particular, the structure of the experimental error may drive the choice of the method being used, as each method handles the error in a different way (see Section 4.2.5). It is possible to determine kinetic constants in raw extracts that are close to kinetic data obtained with purified enzymes and that can be found in literature or in databases [47, 48]. As shown in Fig. 4.10, the affinity of rubisco for ribulose-1,5-bisphosphate and its total activity were determined by fitting the Michaelis–Menten

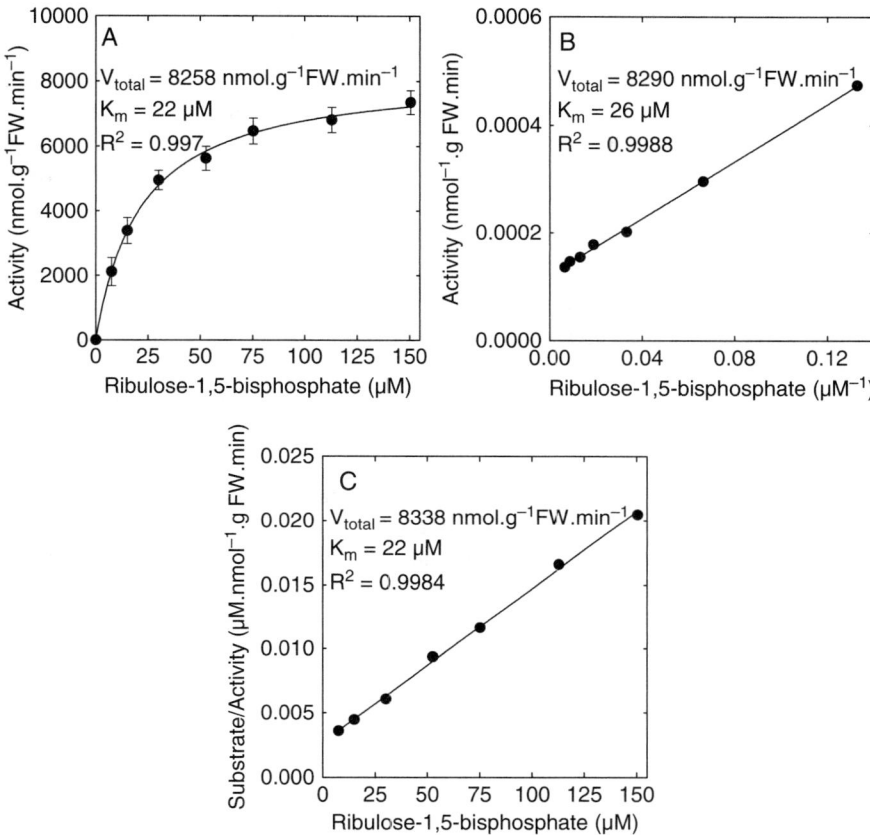

Fig. 4.10 Change in velocity with the concentration of ribulose-1,5-bisphosphate for the reaction catalyzed by ribulose-1,5-carboxylase/oxygenase (EC 4.1.1.39) and determination of K_m and V_{total} with hyperbola fitting (**A**), Lineweaver and Burk (**B**), and Hanes (**C**) methods. Data are expressed as means ± SD ($n = 6$). The fitting of the hyperbola was achieved using the Sigma Plot software [48]

equation (Eq. 4.26) and by using the methods of Lineweaver and Burk, and Hanes. Values obtained with the three methods were very similar, with the exception of the apparent K_m for ribulose-1,5-bisphosphate which was found to be higher when using the method of Lineweaver and Burk, probably due the overweighting of the 2 points obtained with the two lowest substrate concentration (see Section 4.2.5). The apparent K_m or $K_{0.5}$ for ribulose-1,5-bisphosphate was found to have a value of about 20 µM, which is close to values obtained with the purified enzyme from various species of higher plants (http://www.brenda.uni-koeln.de/).

Kinetic properties obtained with raw extracts should always be considered with caution as various sources of error are possible. Effectors present in the extracts may inhibit or activate the enzyme under study and thus lead to erroneous results. Artifacts may also result from the destabilizing effect of the dilution of the substrate, especially when the enzyme is already highly diluted [42]. Such an effect can lead to an erroneous interpretation, as the dose–response curve may have a sigmoid shape and thus evoke cooperativity (see Section 4.2.7). It may therefore be useful to repeat measurements with different concentrations of both extract and substrate. Furthermore, the use of K_m might be misleading and should be replaced with apparent K_m or more generally $K_{0.5}$, i.e., the condition at which the enzyme reaches 50% of its total velocity in the extract.

Inhibition types and constants may also be determined using raw extracts. Therefore, a large number of determinations have to be performed, i.e., various substrate concentrations at various inhibitor concentrations. Such determinations are probably prone to error, due to the complexity of raw extracts. If alterations in inhibition constants are to be searched across a large range of genotypes and/or growth conditions, it seems more adequate to determine the $K_{0.5}$ corresponding to the inhibitors or activators under study first, by using the same approach described above.

It is important to note that both low and high concentrations of substrate should be used anyway. For example, competitive inhibitors would not exert any visible effect at high substrate concentrations. Once $K_{0.5}$ has been determined in conditions that are satisfactory in terms of accuracy and reproducibility, a high throughput screen can be designed in "saturating" and "half saturating" conditions. Any shift in the ratio between $v_{0.5}$ and v_{total} would indicate a possible variation in the properties of the enzyme under study.

4.4 Conclusion

Enzyme activity integrates information from several levels of biological organization and in that respect the information is perhaps more valuable than relying on assumptions made from, for example gene transcript abundance alone, and unlike metabolite or gene transcript data, enzyme activity also provides information about flux, which is key to understanding metabolic networks. However, parallel determination of, for example, gene transcript abundance, metabolite and protein levels in conjunction with enzyme activity will provide rich data sets where integration of information is likely to be of greater value than the sum of the parts.

Traditional methods for analyzing enzyme activity are laborious and not comparable to the high throughput "omics" approaches currently being used to investigate the levels of gene transcripts, proteins, and metabolites. The enzyme analysis platform described here is a step toward the type of high throughput tool that we have become familiar with in the "omics" arena. However, because the biochemistry of enzyme activity analysis is considerably more complex than other high throughput technologies, true high throughput profiling on the scale of genomics is unlikely. Nevertheless, high throughput enzyme activity analysis is now a reality.

Acknowledgments A.R. acknowledges support from the U.S. Department of Energy (DOE), Office of Science, Biological and Environmental Research (BER) program as part of its Program for Ecosystem Research (PER) and contract No. DE-AC02-98CH10886 to Brookhaven National Laboratory. Y.G. acknowledges Mark Stitt, Melanie Höhne, Jan Hannemann, John Lunn, Hendrik Tschoep, Ronan Sulpice, Marie-Caroline Steinhauser, and support from the Max Planck Society and the German Ministry for Research and Technology (GABI 0313110).

References

1. Ballicora MA, Frueauf JB, Fu YB, Schurmann P, Preiss J (2000) Activation of the potato tuber ADP-glucose pyrophosphorylase by thioredoxin. J Biol Chem 275:1315–1320.
2. Bergmeyer H (1987) Methods of Enzymatic Analysis. VCH Weinheim, Germany.
3. Bieniawska Z, Barratt DHP, Garlick AP, Thole V, Kruger NJ, Martin C, Zrenner R, Smith AM (2007) Analysis of the sucrose synthase gene family in *Arabidopsis*. Plant J 49:810–828.
4. Bisswanger H (2002) Enzyme Kinetics, Principals and Methods. Wiley-VCH Weinheim, Germany.
5. Blasing OE, Ernst K, Streubel M, Westhoff P, Svensson P (2002) The non-photosynthetic phosphoenolpyruvate carboxylases of the C4 dicot *Flaveria trinervia* – implications for the evolution of C4 photosynthesis. Planta 215:448–456.
6. Briggs G, Haldane J (1925) A note on the kinetics of enzyme action. Biochem J 19:338–339.
7. Burrin D (1993) Spectroscopic techniques. In: Wilson K, Goulding K (eds). A Biologist's Guide to Principals and Techniques of Practical Biochemistry. University Press, Cambridge, UK.
8. Chen SX, Harmon AC (2006) Advances in plant proteomics. Proteomics 6:5504–5516.
9. Chiadmi M, Navaza A, Miginiac-Maslow M, Jacquot JP, Cherfils J (1999) Redox signalling in the chloroplast: structure of oxidized pea fructose-1,6-bisphosphate phosphatase. EMBO J 18:6809–6815.
10. Ching TM (1982) A sensitive and simple assay of starch synthase activity with pyruvate-kinase and luciferase. Anal Biochem 122:139–143.
11. Cole HA, Wimpenny JW, Hughes DE (1967) ATP pool in *Escherichia coli* .I. Measurement of pool using a modified luciferase assay. Biochim Biophys Acta 143:445–453.
12. Cornish-Bowden A (2004) Fundamentals of Enzyme Kinetics. Portland Press, London, UK.
13. Estevez M, Skarda J, Spencer J, Banaszak L, Weaver TM (2002) X-ray crystallographic and kinetic correlation of a clinically observed human fumarase mutation. Prot Sci 11: 1552–1557.
14. Fan F, Wood KV (2007) Bioluminescent assays for high-throughput screening. Assay Drug Dev Technol 5:127–136.
15. Fridman E, Carrari F, Liu YS, Fernie AR, Zamir D (2004) Zooming in on a quantitative trait for tomato yield using interspecific introgressions. Science 305:1786–1789.
16. Gibon Y, Blaesing OE, Hannemann J, Carillo P, Hohne M, Hendriks JHM, Palacios N, Cross J, Selbig J, Stitt M (2004) A robot-based platform to measure multiple enzyme activities in *Arabidopsis* using a set of cycling assays: Comparison of changes of enzyme activities and transcript levels during diurnal cycles and in prolonged darkness. Plant Cell 16: 3304–3325.
17. Gibon Y, Usadel B, Blaesing OE, Kamlage B, Hoehne M, Trethewey R, Stitt M (2006) Integration of metabolite with transcript and enzyme activity profiling during diurnal cycles in *Arabidopsis rosettes*. Genome Biology 7:R76.
18. Gibon Y, Vigeolas H, Tiessen A, Geigenberger P, Stitt M (2002) Sensitive and high throughput metabolite assays for inorganic pyrophosphate, ADPGlc, nucleotide phosphates, and glycolytic intermediates based on a novel enzymic cycling system. Plant J 30:221–235.
19. Gomes A, Fernandes E, Lima J (2006) Use of fluorescence probes for detection of reactive nitrogen species: A review. J Fluorescence 16:119–139.

20. Greenberg LJ (1962) Fluorometric measurement of alkaline phosphatase and aminopeptidase activities in order of 10–14 mole. Biochem Biophys Res Comm 9:430–435.
21. Greis KD (2007) Mass spectrometry for enzyme assays and inhibitor screening: an emerging application in pharmaceutical research. Mass Spec Rev 26:324–339.
22. Gronwald JW, Plaisance KL (1998) Isolation and characterization of glutathione S-transferase isozymes from sorghum. Plant Physiol 117:877–892.
23. Gutteridge A, Thornton JM (2005) Understanding nature's catalytic toolkit. Trends Biochem Sci 30:622–629.
24. Hausler RE, Fischer KL, Flugge UI (2000) Determination of low-abundant metabolites in plant extracts by NAD(P)H fluorescence with a microtiter plate reader. Anal Biochem 281: 1–8.
25. Hendriks JHM, Kolbe A, Gibon Y, Stitt M, Geigenberger P (2003) ADP-glucose pyrophosphorylase is activated by posttranslational redox-modification in response to light and to sugars in leaves of *Arabidopsis* and other plant species. Plant Physiol 133:838–849.
26. Kaiser WM, Huber SC (2001) Post-translational regulation of nitrate reductase: mechanism, physiological relevance and environmental triggers. J Exp Bot 52:1981–1989.
27. Kappes T, Hauser PC (2000) Recent developments in electrochemical detection methods for capillary electrophoresis. Electroanalysis 12:165–170.
28. Koshland DE, Nemethy G, Filmer D (1966) Comparison of experimental binding data and theoretical models in proteins containing subunits. Biochemistry 5:365–385.
29. Lienhard GE, Secemski, II (1973) P1,P5-Di(Adenosine-5′)Pentaphosphate, a Potent Multisubstrate Inhibitor of Adenylate Kinase. J Biol Chem 248:1121–1123.
30. Lorimer GH, Badger MR, Andrews TJ (1977) D-Ribulose-1,5-Bisphosphate Carboxylase-Oxygenase – Improved Methods for Activation and Assay of Catalytic Activities. Anal Biochem 78:66–75.
31. Lowry CV, Kimmey JS, Felder S, Chi MMY, Kaiser KK, Passonneau PN, Kirk KA, Lowry OH (1978) Enzyme patterns in single human muscle-fibers. J Biol Chem 253:8269–8277.
32. Lowry OH, Rock MK, Schulz DW, Passonneau JV (1961) Measurement of pyridine nucleotides by enzymatic cycling. J Biol Chem 236: 2746–2755.
33. McIntosh CA, Oliver DJ (1992) NAD+-linked isocitrate dehydrogenase – isolation, purification, and characterization of the protein from pea mitochondria. Plant Physiol 100:69–75.
34. Michaelis L, Menton M (1913) Die kinetik der invertinwirkung. Biochemische Zeitschrift 49:333–369.
35. Monod J, Wyman J, Changeux JP (1965) On nature of allosteric transitions – a plausible model. J Mol Biol 12:88–118.
36. Morcuende R, Bari R, Gibon Y, Zheng WM, Pant BD, Blasing O, Usadel B, Czechowski T, Udvardi MK, Stitt M, Scheible WR (2007) Genome-wide reprogramming of metabolism and regulatory networks of *Arabidopsis* in response to phosphorus. Plant Cell Environ 30:85–112.
37. Outlaw WH, Manchester J (1980) Conceptual error in determination of NAD+-malic enzyme in extracts containing NAD+-malic dehydrogenase. Plant Physiol 65:1136–1138.
38. Schaffer AA, Levin I, Oguz I, Petreikov M, Cincarevsky F, Yeselson Y, Shen S, Gilboa N, Bar M (2000) ADPglucose pyrophosphorylase activity and starch accumulation in immature tomato fruit: the effect of a *Lycopersicon hirsutum*-derived introgression encoding for the large subunit. Plant Sci 152:135–144.
39. Scheible WR, GonzalezFontes A, Laurer M, MullerRober B, Caboche M, Stitt M (1997) Nitrate acts as a signal to induce organic acid metabolism and repress starch metabolism in tobacco. Plant Cell 9:783–798.
40. Schmidt O, Bassler M, Kiesel P, Knollenberg C, Johnson N (2007) Fluorescence spectrometer-on-a-fluidic-chip. Lab on a Chip 7:626–629.
41. Schwarz MA, Hauser PC (2001) Recent developments in detection methods for microfabricated analytical devices. Lab on a Chip 1: 1–6.
42. Selwyn MJ (1965) A simple test for inactivation of an enzyme during assay. Biochim Biophys Acta 105:193–195.

43. Shen JB, Ogren WL (1992) Alteration of Spinach Ribulose-1,5-Bisphosphate Carboxylase Oxygenase Activase Activities by Site-Directed Mutagenesis. Plant Physiol 99:1201–1207.
44. Slade AJ, Fuerstenberg SI, Loeffler D, Steine MN, Facciotti D (2005) A reverse genetic, nontransgenic approach to wheat crop improvement by TILLING. Nature Biotechnol 23: 75–81.
45. Stitt M, Gibon Y, Lunn JE, Piques M (2007) Multilevel genomics analysis of carbon signalling during low carbon availability: coordinating the supply and utilisation of carbon in a fluctuating environment. Funct Plant Biol 34:526–549.
46. Storer AC, Cornishb.A (1974) Kinetics of coupled enzyme reactions – applications to assay of glucokinase, with glucose-6-phosphate dehydrogenase as coupling enzyme. Biochem J 141:205–209.
47. Studart-Guimaraes C, Gibon Y, Frankel N, Wood CC, Zanor MI, Fernie AR, Carrari F (2005) Identification and characterisation of the alpha and beta subunits of succinyl CoA ligase of tomato. Plant Mol Biol 59:781–791.
48. Sulpice R, Tschoep H, Von Korff M, Bussis D, Usadel B, Hohne M, Witucka-Wall H, Altmann T, Stitt M, Gibon Y (2007) Description and applications of a rapid and sensitive non-radioactive microplate-based assay for maximum and initial activity of D-ribulose-1,5-bisphosphate carboxylase/oxygenase. Plant Cell Environ 30:1163–1175.
49. Tcherkez GGB, Farquhar GD, Andrews TJ (2006) Despite slow catalysis and confused substrate specificity, all ribulose bisphosphate carboxylases may be nearly perfectly optimized. Proc Nat Acad Sci U S A 103:7246–7251.
50. Teipel JW, Hill RL (1971) Subunit interactions of fumarase. J Biol Chem 246: 4859–4865.
51. Tiessen A, Hendriks JHM, Stitt M, Branscheid A, Gibon Y, Farre EM, Geigenberger P (2002) Starch synthesis in potato tubers is regulated by post-translational redox modification of ADP-glucose pyrophosphorylase: a novel regulatory mechanism linking starch synthesis to the sucrose supply. Plant Cell 14:2191–2213.
52. Usadel B, Blasing OE, Gibon Y, Poree F, Hohne M, Gunter M, Trethewey R, Kamlage B, Poorter H, Stitt M (2008) Multilevel genomic analysis of the response of transcripts, enzyme activities and metabolites in *Arabidopsis rosettes* to a progressive decrease of temperature in the non-freezing range. Plant Cell Environ 31:518–547.
53. Warburg O, Christian W, Griese A (1935) Wasserstoffubertragendes co-ferment seine zusammensetzung und wirkungsweise. Biochemische Zeitschrift 282:157–165.
54. Zhang Q, Xu JJ, Chen HY (2006) Glucose microfluidic biosensors based on immobilizing glucose oxidase in poly(dimethylsiloxane) electrophoretic microchips. J Chromatography A 1135:122–126.

Chapter 5
Quantification of Isotope Label

D.K. Allen and R.G. Ratcliffe

5.1 Introduction

The use of stable and radioactive isotopes of low natural abundance has a long history in plant research. At a metabolic level, isotope labeling leads to the discovery of new pathways (e.g., [48, 70]) and to detailed descriptions of the fluxes that underpin the metabolic phenotype [93, 108]. In particular, and as described in detail elsewhere in this book, ^{13}C-labeling experiments provide the inputs for generating the large-scale flux maps that emerge from network flux analysis. The availability of robust, accurate methods for the analysis of the redistribution of label is central to the success of this approach, and so this chapter focuses on the nuclear magnetic resonance (NMR) and mass spectrometry (MS) techniques that make this possible.

Typically, a ^{13}C-labeling experiment will lead to the production of a range of isotopomers for any given metabolite. Isotopomers differ in the number and position of the incorporated isotopes, and the resulting labeling pattern will reflect the pathways that led to the incorporation of the label into the metabolite. It turns out that the quantification of these isotopomers is crucial for the analysis of typical metabolic networks. While net intracellular fluxes can be deduced from measurements of metabolic inputs and outputs using flux-balancing techniques, this method requires additional assumptions [12], and a cellular objective such as optimal cell growth, to predict the flux patterns (see Chapter 8). In contrast, labeling experiments [97] permit the calculation of net and exchange flux values for reversible reactions [146], the analysis of parallel fluxes in subcellular compartments [44], and the measurement of cyclic fluxes for nonlinear pathways like the tricarboxylic acid cycle [109]. The success of the labeling approach arises because variations in fluxes at branch points can lead to differences in isotopomer distribution [103], reflecting the carbon rearrangements of the different pathways. As a result, investigations of label enrichment are invaluable for network-based studies of metabolism [60, 138].

R.G. Ratcliffe (✉)
Department of Plant Sciences, University of Oxford, South Parks Road, Oxford OX1 3RB, UK
e-mail: george.ratcliffe@plants.ox.ac.uk

Both gas chromatography/mass spectrometry (GCMS) and ^{13}C NMR can provide large amounts of labeling information, and their use is universal in network flux analysis. As GCMS and ^{13}C NMR evaluate chemical compounds differently, the information content of the spectra is complementary and sometimes redundant. Figure 5.1 compares the spectra and labeling information obtained from MS and ^{13}C NMR. In principle, both techniques are capable of completely resolving all of the isotopomers of a given compound, though for larger compounds (>4 carbons) determination of even the majority of the isotopomers is time consuming, laborious [10], and rarely attempted. However, partial identification of isotopomer distributions is often sufficient for flux estimation [27].

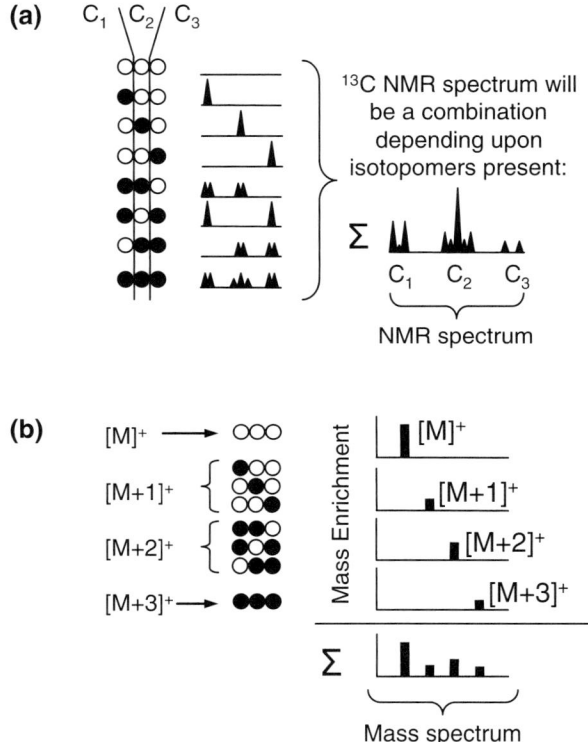

Fig. 5.1 Spectroscopic signatures for a chemical structure containing three carbon atoms that can each be labeled (^{13}C, •) or unlabeled (^{12}C, ○): (**a**) NMR; (**b**) MS. All $2^n = 2^3 = 8$ positional isotopomers (n = total number of atoms that can be either unlabeled or labeled) and their spectroscopic signatures are shown for both instruments. The NMR spectrum shows three complex signals, corresponding to the three carbon atoms, and the relative intensities of the signals contain information on the relative abundance of seven of the eight positional isotopomers. The MS spectrum shows four signals, corresponding to the four mass isotopomers, and their relative intensity defines the mass enrichment for each mass isotopomer

MS characterizes the labeling of a molecular fragment in terms of molecular mass, and fragments that only differ in the number of labeled atoms are referred to as mass isotopomers. The *mass enrichment* of a mass isotopomer is defined as the abundance of the isotopomer (A_i) as a fraction of the total abundance of the fragment (i.e., $A_i/\sum A$, $A_{i+1}/\sum A$, etc.). Note that it is also common to represent the abundance A_i for a particular mass fraction (e.g., M^+, $M+1^+$, etc.) as the bracketed form ($[M]^+$, $[M+1]^+$, etc.). The brackets signify the relative abundance or concentration of the mass isotopomer. NMR techniques characterize the labeling of intact molecules in terms of the abundance of *positional isotopomers*, defined as molecules with the same chemical structure that differ in isotopic (^{13}C) composition. NMR can also detect the presence or absence of adjacent ^{13}C atoms, and this is referred to as *bond connectivity* information. These terms are discussed in more detail elsewhere [93].

A number of factors, including time, accessibility to instruments, sample number, sample size, instrument sensitivity, label(s) selection, and label costs will affect the choice of analysis by MS or NMR. Furthermore, GCMS requires derivatization of metabolites to convert them into volatile forms that can pass through the GC column (see also Chapter 3). Accordingly, if both methods are used, it may be advantageous to perform the non-destructive NMR measurements first, and then to recover the sample for subsequent derivatization and MS processing.

5.2 Sample Preparation

Sample preparation is an essential first step in analyzing the redistribution of label by either MS or NMR, and indeed it has been suggested that advances in sample preparation, including derivatization techniques for particular compounds, will result in more significant progress than the use of increasingly sophisticated and expensive instrumentation [54]. Moreover, the complexity of compartmented plant metabolism increases the need for high-quality information on the redistribution of the label, and comprehensive flux maps can only be obtained from large datasets. Fortunately, careful fractionation before NMR analysis or derivatization for GCMS can increase both the range and the number of metabolites available for analysis. Some of the extraction procedures that are commonly used in flux analysis are summarized in the following paragraphs.

Lipid extraction is often the first step in sample processing because the oil fraction can complicate subsequent processing steps if it is present in significant quantities. Extraction with hexane:isopropanol or chloroform:methanol mixtures can be followed by acid-catalyzed esterification with boron trifluoride or methanolic:HCl [13], base-catalyzed hydrolysis with sodium or potassium hydroxide or methoxide [39], or butyl amide formation [1, 62]. If the oil fraction is the only component of interest, then transmethylation can be carried out in situ to aid recovery, but the tissue is then unavailable for further analysis. More general instructions on lipid handling and analysis are given in [18].

Low molecular weight polar components can be extracted with 80% ethanol, 70% methanol, or perchloric acid in water. The sugars, free amino acids, and organic acids can be further fractionated with cation and anion exchange columns (e.g., [64, 151]).

Proteins are extracted using buffered aqueous solutions that also contain surfactants and denaturants, precipitated with trichloroacetic acid (TCA), and hydrolyzed with 6 M HCl at elevated temperatures (above 100°C) for up to 24 h to liberate the amino acids [106]. The hydrolysis process is detrimental to several amino acids. The basic amino acids glutamine and asparagine are largely converted to glutamate and aspartate, and tryptophan and cysteine are destroyed [42]. Also the side chain of arginine is labile and may only be recognized in GCMS as a smaller product [57, 85]. After liberation, the amino acids can be processed through a cation exchange column and dried for subsequent derivatization (see Section 5.4.3).

Following protein extraction, starch can be solubilized by heating with an alkaline solution, and after adjusting the pH to 4.5–5 using acetic acid, the soluble starch can be degraded with amyloglucosidase/amylase [45]. Glucose can be converted to the alditol acetate derivative for label quantitation with or without prior reduction [11, 88].

Finally, the recalcitrant material that remains is likely to be composed in large part of cell wall, including pectins, hemicellulose, and cellulose. By treating the cell wall with 2 M trifluoroacetic acid for several hours at 120°C [11], the pectin and hemicellulose and to a far lesser extent the cellulose are broken down to their monomeric forms. Protein glycans segregated with the protein extraction can be handled in a similar fashion [1]. These monosaccharides can also be derivatized with or without prior reduction. Alternatively, during protein hydrolysis, glycans are converted to levulinic acid and hydroxyacetone which can then be analyzed directly [117].

5.3 Nuclear Magnetic Resonance (NMR)

NMR is a property of isotopes with non-zero nuclear magnetic moments. It provides a mechanism for detecting such isotopes in a wide range of circumstances, and it gives rise to a method, NMR spectroscopy, that is arguably one of the most versatile analytical techniques available for the detection and identification of chemical compounds. The general principles of NMR spectroscopy are fully described in many textbooks (e.g., [43, 51, 69]). Moreover, given that the versatility of NMR spectroscopy as an analytical technique stems in large part from the plethora of experiments that exist for manipulating the nuclear spin system prior to detection, guides to the implementation of comprehensive selections of the many different NMR experiments are also available (e.g., [7]).

The specific application of NMR spectroscopy relevant here is the analysis of the redistribution of label that occurs when an isotopically labeled substrate is introduced into the plant metabolic network. NMR is well suited to this task, since it is readily applied to the analysis of plants and plant metabolites [92, 94], and it is also well established that NMR is a powerful method for analyzing the complex mixtures of metabolites that are routinely obtained in tissue extracts [36, 63]. Accordingly after a brief description of some general considerations, this section will focus on the NMR analysis of fractional enrichments and isotopomeric composition.

5.3.1 General Considerations

The NMR-detectable isotopes that are most commonly used in isotopic labeling experiments are ^{2}H, ^{13}C, and ^{15}N. These isotopes have low natural abundance (0.015, 1.1, and 0.37%, respectively), and supplying a plant tissue with a suitably labeled substrate will generally lead to the selective enhancement of the NMR signals from a range of metabolites. The redistribution of the label reflects the metabolic activity of the tissue, and thus provides information on the pathways that are present and the fluxes they support. However, while all three isotopes are used extensively in pathway delineation, network flux analysis is largely based on measuring the redistribution of ^{13}C [93] (see Chapter 9). It follows that ^{1}H and ^{13}C NMR methods are the most important for flux analysis, although notable exceptions can be found, for example, a ^{15}N NMR analysis of the time course of [^{15}N]ammonium assimilation in *Corynebacterium glutamicum* [124].

Typically, flux analysis requires an analysis of the labeling of a mixture of metabolites in one or more fractions obtained by tissue extraction (Section 5.2). The sensitivity of the chosen NMR experiment and the extent to which the NMR signals overlap are key factors in defining the scope of the analysis. Adequate sensitivity can be achieved by extracting sufficient labeled material – typically a few grams fresh weight – and in general, the NMR analysis will always need a larger quantity of labeled material than the corresponding MS analysis. In compensation, NMR can usually provide a wealth of information on the positional isotopomers within the mixture, thus ensuring that NMR is used extensively in flux analysis [121].

The other important consideration is the extent to which the signals overlap in the NMR spectrum. ^{13}C NMR signals are dispersed over an intrinsically large spectral window ($\sim$200 ppm), but the signals from the isotopomers that contribute to the labeling of a specific carbon atom will invariably overlap. Similarly in the much more crowded ^{1}H NMR spectrum, it may be difficult to detect the informative ^{13}C satellites because of overlap with other signals. Fortunately, the limitations that sometimes arise in one-dimensional (1D) NMR analysis (Section 5.3.2) can often be solved by switching to a two-dimensional (2D) NMR experiment (Section 5.3.3), and in general, the success of an NMR flux analysis is not limited by the availability of sufficient data on the redistribution of the label.

Several other considerations and experimental options have a bearing on the overlap problem. First, the NMR signals from low-molecular weight compounds are usually narrower than those from macromolecules. It follows that it is more informative to analyze hydrolysates of macromolecules, for example starch [28] and protein

[120], than to attempt to record spectra from the macromolecules themselves. This generalization holds true even though several amino acids (Asn, Cys, Gln, and Trp) are lost during acid hydrolysis of protein. Secondly, the resolution of some NMR signals is adversely affected by the presence of paramagnetic ions in aqueous tissue extracts, and this problem can be avoided by the addition of small amounts of a chelating agent such as EDTA. Thirdly, the extent to which a tissue extract is fractionated before analysis is entirely at the discretion of the investigator. Since absolute pool sizes are not required for steady-state analysis, the inevitable losses that will occur in a multi-step procedure are of no consequence, provided there is sufficient label for analysis in the final fraction. Moreover for time-course analysis, the same consideration holds true if NMR is only used to measure the isotopomeric composition of the pool, with some other technique being used to measure pool size at an earlier stage in the fractionation procedure. In practice, many NMR investigations are based on rather crude fractionation of the tissue extract, relying on the peak separation attributes of 2D NMR to reduce spectral overlap to manageable levels. Finally, some readily detected metabolites exist in solution as a mixture of stereoisomers, for example, the α- and β-anomers of glucose. This increases the number of peaks in the NMR spectrum without increasing the information content. Depending on the investigation, this may turn out to be only a minor complication that can be minimized spectroscopically; but on other occasions, it may be worthwhile finding a chemical route to eliminate the problem. For example, glucose can be converted to monoacetone glucose, a procedure that halves the number of glucose signals because the derivative only exists as a single stereoisomer [1, 56]. Note that the redundancy arising from the detection of the α- and β-anomers of glucose can be useful because a comparison between the two sets of signals provides a check on the analysis [32]. However, this advantage is probably outweighed by the reduced overlap and improved sensitivity achievable, when the signals from the two anomers are combined through the formation of a suitable derivative.

5.3.2 One-Dimensional NMR Methods

5.3.2.1 Fractional Enrichments

One way of characterizing the isotopic composition of a metabolite pool is to measure the fractional enrichment of specific carbon atoms. A metabolite with n carbon atoms has n fractional enrichments – also known as positional enrichments – and there are several ways of extracting this information from 1D NMR spectra [121]. Note that every method depends on comparing signal intensities, and it is therefore important to ensure that the signals are indeed comparable. Thus, if the signals are not fully relaxed, then the degree of saturation must be identical or a suitable correction factor must be determined. Similarly if ^{1}H-decoupled ^{13}C NMR signals have different nuclear Overhauser enhancement (NOE) effects [51], then this must be allowed for in the calculation.

5 Quantification of Isotope Label

All carbon-bonded hydrogen atoms give ^{1}H NMR signals with ^{13}C-satellites. Only 1.1% of the total signal intensity will be in the satellites at natural abundance, but this fraction increases if the carbon atom is selectively labeled. Accordingly, measuring the relative intensities of the satellite and central resonances provides a direct route to the fractional enrichment of the specific carbon atom that is bonded to the observed hydrogen atom. The fractional enrichments of the other carbon atoms in the metabolite can then be obtained by comparing the intensities of the signals in the ^{13}C NMR spectrum (Fig. 5.2). A useful refinement to the basic method is to compare ^{1}H NMR spectra recorded with or without ^{13}C-decoupling [113]. The difference between these two spectra allows accurate quantification of satellites that would otherwise be obscured by signals from hydrogen atoms bonded to unlabeled carbon moieties. Even with this refinement, the method is largely restricted to metabolites that give rise to readily resolvable signals in the less crowded regions of the ^{1}H NMR spectra of tissue extracts, such as glucose, the glucosyl moiety of sucrose, and alanine (e.g., [28]). However, the scope of the method increases considerably if specific metabolites are separated from the tissue extract by chromatography before analysis [76].

A ^{1}H NMR method has also been developed for measuring the fractional enrichment of non-protonated carbon atoms [137]. The method exploits the weaker coupling that occurs between the carbon of interest and the proton(s) attached to the

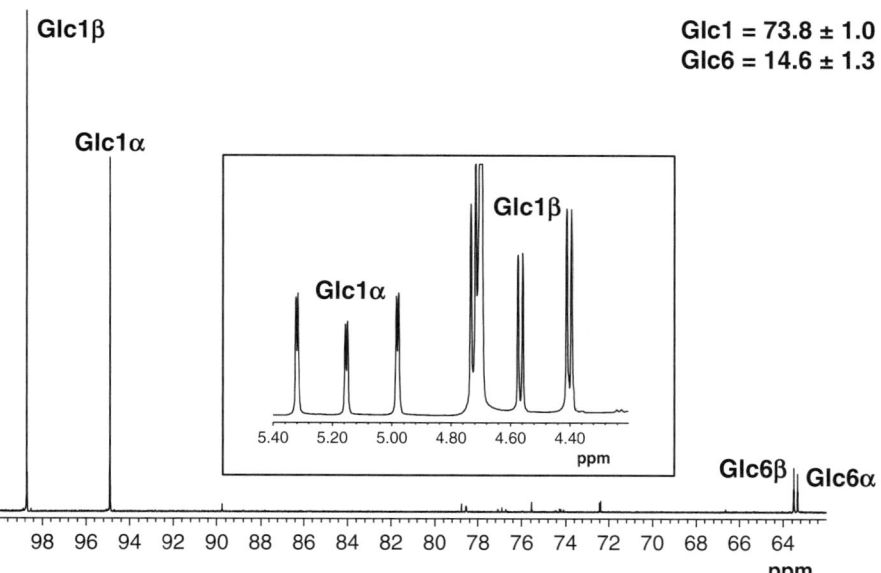

Fig. 5.2 ^{1}H (inset) and ^{13}C NMR spectra of glucose extracted from maize root tips after labeling with [1-^{13}C]glucose. Each carbohydrate ^{1}H signal consists of a central resonance flanked by two satellite peaks. The former derives from H bonded to ^{12}C and the latter to H bonded to ^{13}C. The relative intensities can be used to obtain the fractional enrichments listed. The figure is adapted from Alonso et al. [2] with permission from the American Society of Plant Biologists

neighboring carbon atom(s). ^{1}H spin-echo spectra are recorded with and without selective ^{13}C inversion of the neighboring carbon atom(s) and the difference between these spectra reveals the extent to which the neighboring non-protonated carbon atom is labeled. The method requires careful calibration, but it works well and it was shown to give more reliable estimates of the fractional enrichment of the non-protonated carbon atoms in glutamate than a method based on quantifying the ^{13}C NMR spectrum [137].

Fractional enrichments can also be deduced directly from ^{13}C NMR spectra in particular circumstances. For example, if there are grounds for considering that one or more carbon atoms in a metabolite will not be enriched in the labeling experiment, then the intensity of the corresponding signals represents natural abundance, allowing the fractional enrichment of the labeled carbon atoms in the metabolite to be determined by direct comparison of their intensities. For example, in an analysis of a [1-^{13}C]glucose labeling experiment on transgenic tobacco (*Nicotiana tabacum*) cells, it was assumed that C3 of fructose would be negligibly enriched, allowing the redistribution of label between C1 and C6 to be quantified directly [38]. Similarly if a labeled metabolite is derivatized with an unlabeled reagent, then comparison of the natural abundance signals from the unenriched part of the derivative with the signals derived from the metabolite leads directly to the fractional enrichments for the labeled metabolite [5, 56].

Although measurements of fractional enrichments provide only a limited characterization of the isotopic composition of a labeled metabolite pool, such measurements can provide sufficient constraints to generate large-scale flux maps of central metabolism. Examples include an early demonstration of the approach in *C. glutamicum* [76] and a series of papers on heterotrophic plant tissues [28, 29, 96]. NMR-determined fractional enrichments have also been used in studies that focus on specific features of central metabolism in heterotrophic plant cells, including carbon entry into the tricarboxylic acid cycle [30], recycling between triose phosphates and hexose phosphates [38] glucose resynthesis [2], and cold-induced sweetening of potato tubers [75].

5.3.2.2 Isotopomer Analysis

The most informative way of characterizing the isotopic composition of a metabolite pool is to analyze the entire isotopomer content. A metabolite with n carbon atoms has 2^n positional isotopomers, each of which can be specified using a notation in which 0 and 1 represent ^{12}C and ^{13}C, respectively. So for a four-carbon compound, the fractional enrichment of carbon 2 will depend on the relative abundance of the isotopomers labeled in carbon 2 (0100, 1100, 0110, 1110, 0101, 1101, 0111, 1111) and those that are unlabeled (0000, 1000, 0010, 1010, 0001, 1001, 0011, 1011). Many of these isotopomers give a recognizable spectroscopic signature in the ^{13}C NMR spectrum as a result of the fine structure that arises when two chemically inequivalent ^{13}C atoms are bonded to each other. This fine structure arises from the ^{13}C–^{13}C scalar coupling interaction, and it is characterized by a coupling constant J. For directly bonded carbon atoms, the one-bond coupling constant ($^1J_{CC}$) tends to

5 Quantification of Isotope Label

be in the range 30–60 Hz giving rise to a readily detectable splitting of the signals in 1D ^{13}C NMR spectra. Unfortunately, the interactions between carbon atoms linked by two or more bonds are much weaker, and rarely detectable, with the result that a complete isotopomer analysis is unlikely to be an option for molecules with more than three carbon atoms. However, even the incomplete analysis that is possible for larger molecules is very informative, and NMR is frequently the method of choice for steady-state flux analysis [121].

The fine structure caused by ^{13}C–^{13}C interactions is a conspicuous feature in 1D ^{13}C NMR spectra of multiply labeled compounds (Fig. 5.3), and there is a long history of using this information for metabolic analysis [71, 102]. The spectra are invariably recorded with ^{1}H-decoupling, to remove the additional fine structure that would otherwise arise from the scalar interactions between ^{1}H and ^{13}C atoms, and the contribution of specific isotopomers, and groups of isotopomers, to the intensity of the signal for a particular carbon atom is determined by integrating the peaks within the fine structure.

This procedure is generally robust, but as discussed elsewhere [71, 121], it is necessary to be aware of the several factors that can complicate the analysis. First, it is essential to avoid acquisition conditions that might distort the relative intensities through relaxation effects. Long relaxation delays are necessary if the analysis is to include carboxyl and carbonyl group carbon atoms with long relaxation times; and small variations in NOE, particularly for carbon atoms with long relaxation times, may make it preferable to record spectra without the NOE, despite the penalty in sensitivity. Secondly, the appearance of the expected multiplets becomes increasingly distorted as the chemical shift separation decreases between the signals of the scalar coupled atoms. This problem is less severe at higher magnetic fields, but if necessary, multiplets can be analyzed by simulating the pattern using measured

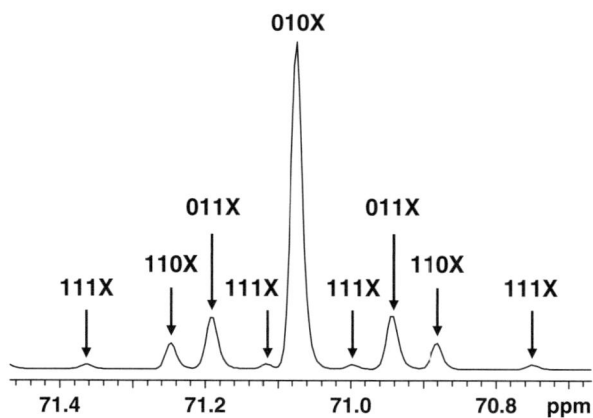

Fig. 5.3 ^{1}H-decoupled ^{13}C NMR spectrum of malate extracted from an *Arabidopsis thaliana* cell culture after labeling with [1-^{13}C]glucose. The spectrum shows the C2 signal of malate, with the fine structure arising from multiply labeled isotopomers. The signals are assigned according to the following scheme: 0, ^{12}C; 1, ^{13}C; and X, ^{12}C or ^{13}C

chemical shift and coupling constant data. Finally, ^{13}C isotope effects lead to small shifts in the signals of carbon atoms bonded to one or more ^{13}C atoms, and this leads to a slight asymmetry in the separation of the signals of the expected multiplets.

A good example of the scope of isotopomer analysis by 1D ^{13}C NMR can be found in a comprehensive analysis of the isotopomers of ^{13}C-labeled glucose, extracted from the leaves of tobacco plants grown on a medium supplemented with [U-^{13}C$_6$]glucose (Fig. 5.4; [32]). Glucose was isolated chromatographically from a tissue extract, and the complex multiplets in the resolution-enhanced 1D ^{13}C NMR spectrum were analyzed by numerical deconvolution using a genetic algorithm [31]. In this procedure, the aim is to identify the contributions of different positional isotopomers to the observed multiplets, which in turn requires an accurate knowledge of the observable ^{13}C–^{13}C coupling constants and the ^{13}C isotope shifts for each carbon atom in both anomers. Inevitably, it is not possible to observe a unique spectroscopic signature for every isotopomer because scalar coupling is only detectable between carbon atoms separated by at most two, or in some cases three, bonds. The labeling of carbon atoms that do not contribute to the multiplet structure is

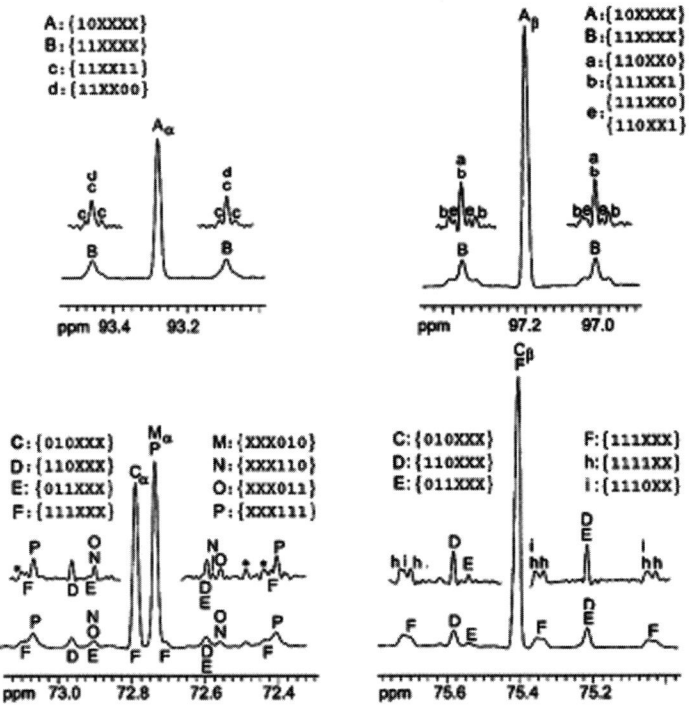

Fig. 5.4 ^{1}H-decoupled ^{13}C NMR signals of glucose extracted from tobacco leaves after labeling with [U-^{13}C$_6$]glucose. Detailed analysis of the fine structure allows intensity measurements to be made for many different isotopomers. The spectra have been resolution enhanced with a Gaussian function, and asterisks (∗) indicate contaminants. Adapted from Ettenhuber et al. [32] with permission from Elsevier

unknown: it is either 0 or 1 in the nomenclature used above, and this uncertainty can be represented by the symbol X.

To take a specific example, analysis of the C1 signal of the α-anomer of glucose allowed the detection of four sets of isotopomers ("X groups"): 10XXXX; 11XXXX; 11XX11; 11XX00. In the tobacco leaf analysis, it was possible to obtain intensity measurements for 29 of these X groups, and numerical deconvolution of these measurements yielded the relative abundances of the 21 positional isotopomers that could have been expected to contribute to the NMR signals. Note that by supplementing the growth medium with only a small amount of [U-^{13}C$_6$]glucose, it was possible to argue that the 42 isotopomers with non-contiguous labeling would have made a negligible contribution to the observed X groups, thus converting an under-determined problem (29 measurements, 63 variables) into an over-determined one (29 measurements, 21 variables). This procedure has been successfully applied to both *Drosophila melanogaster* [31] and tobacco leaves [32], and the relative abundance of the observed isotopomers can be used to model the relative contribution of the fluxes that are responsible for the recycling of glucose.

The extent to which 1D ^{13}C NMR can provide a complete analysis of the isotopomer composition of a ^{13}C-labeled metabolite is limited by two factors. First, as mentioned above, the strength of the interaction that causes the all-important splitting of the signals falls off very rapidly as the number of bonds separating the coupled carbon atoms increases. The coupling constant for directly bonded carbon atoms is typically 30–60 Hz leading to well-resolved splitting; whereas the coupling constants for carbon atoms linked by two or more bonds ($^2J_{CC}$, $^3J_{CC}$, etc.) are invariably less than 5 Hz with the result that two and three bond coupling is only resolved in favorable cases. Secondly, it is not uncommon for different pairs of carbon atoms to have identical coupling constants leading to overlapping signals. For example, in glutamate, $^1J_{C2C3}$ and $^1J_{C3C4}$ have the same value (35 Hz) making it impossible to distinguish the contributions of [2,3-^{13}C$_2$]glutamate and [3,4-^{13}C$_2$]glutamate to the C3 signal [65]. It follows from these inherent limitations that it is not usually possible to determine the complete isotopomer composition of a metabolite with more than three carbon atoms by NMR. In this respect, the determination of 21 of the 63 independent isotopomers of glucose in the studies described above is a notable achievement. However, it should also be emphasized that a complete analysis of the isotopomer composition is not usually necessary – measurements of isotopomer subsets (unresolved X groups) still provide constraints on the redistribution of the label – and the flux mapping problem is in any case usually over-determined, with many more isotopomer measurements than free fluxes.

A ^{1}H NMR method that allows a more complete analysis of isotopomeric composition has been described [24]. The method is based on heteronuclear spin-echo difference spectroscopy, in which ^{1}H spin-echo spectra are recorded with and without selective ^{13}C inversion of carbon atoms $^2J_{CH}$ or $^3J_{CH}$ coupled to the observed proton. This ^{13}C editing of the ^{1}H NMR spectrum reveals weak interactions that would usually give only poorly resolved fine structure in the ^{1}H spectrum, allowing the quantitative determination of the corresponding isotopomers. The method was demonstrated on an aspartate sample that was shown to contain only unlabeled,

[1-^{13}C]-, [4-^{13}C]- and [1,4-^{13}C$_2$]-labeled aspartate. The method is sensitive and generates isotopomer data that would otherwise be inaccessible, and it can provide key information in specific instances [24]. However, it has not been used extensively, most probably because it is quite intricate to implement and because flux analysis is not usually limited by a shortage of isotopomer measurements.

In general, 1D NMR is a convenient approach for generating isotopomer data, and such measurements are frequently used for flux analysis in plant and microbial systems. For example, in a notable study, 58 isotopomer measurements, almost all determined by 1D NMR, and 15 fractional enrichment measurements were used as constraints in an analysis of anaplerosis in *C. glutamicum* [87].

5.3.3 Two-Dimensional NMR Methods

5.3.3.1 Fractional Enrichments

Many of the ^{13}C satellite signals that can provide information on the specific enrichment of particular carbon atoms are masked by overlapping signals in the 1D ^{1}H NMR spectrum of a typical tissue extract, and while it is possible to circumvent this problem by purifying individual metabolites, the procedure is time consuming. The alternative is to use 2D ^{1}H,^{1}H correlation experiments to reduce overlap by distributing the signals along two ^{1}H frequency axes. In principle, two commonly used 2D NMR experiments, COSY and TOCSY (Fig. 5.5), could be useful, and a detailed comparison of the implementation and reliability of these experiments for measuring fractional enrichments is available [78]. This comparison showed that the 2D double quantum filtered COSY experiment was incapable of producing accurate measurements of positional enrichment and that the TOCSY experiment was much better, provided zero quantum filters were applied during the mixing period.

The unreliability of the COSY experiment stemmed from the distortion of the lineshapes in the magnitude corrected spectra and the effects of long-range heteronuclear couplings ($^2J_{CH}$, etc.). Neither of these problems occurs in the TOCSY experiment, and this experiment is capable of determining fractional enrichments with good precision [66, 77]. It appears that the best approach is to implement a version of the TOCSY experiment with zero quantum filters since this avoids signal distortions that would otherwise arise from zero quantum coherences. These unwanted coherences lead to dispersive components in the signals, reducing the accuracy of peak integration and compromising measurements of fractional enrichment [77]. The efficiency of this approach for quantifying fractional enrichments in complex mixtures has been demonstrated for a protein hydrolysate, where it was possible to obtain reliable measurements for 35 protonated carbon atoms simultaneously [78].

5.3.3.2 Isotopomer Analysis

2D NMR spectroscopy can also be used for isotopomer analysis. Here, the objective is to perform a ^{1}H,^{13}C correlation experiment that produces a spectrum with the ^{1}H

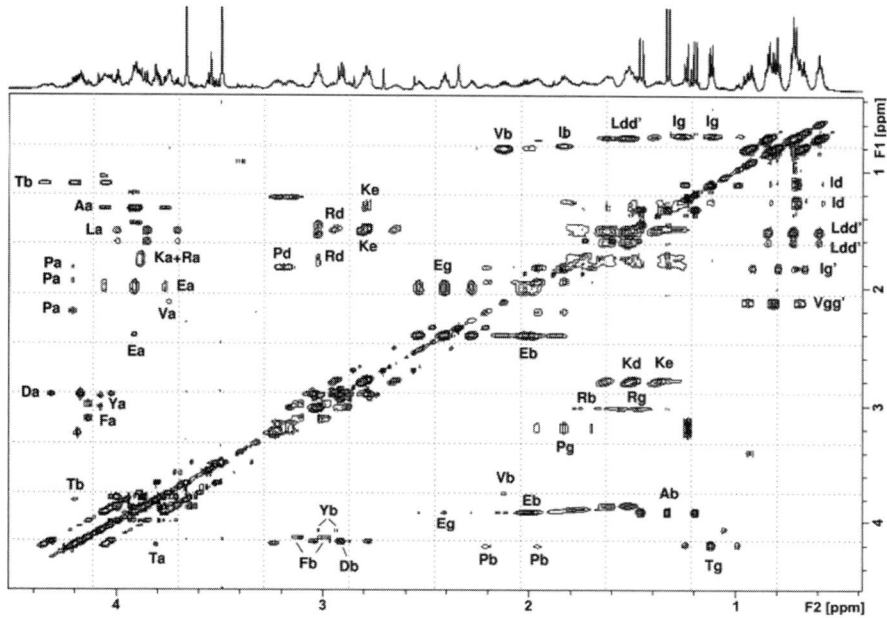

Fig. 5.5 2D zero quantum filtered TOCSY spectrum of a biomass hydrolysate from *E. coli* cells grown on 20% [U-^{13}C$_6$]glucose +80% [1-^{13}C]glucose. The annotated cross-peaks derive from the labeled amino acids, and the greatly improved resolution in the 2D spectrum allows direct measurement of the fractional enrichment of numerous carbon atoms. In contrast, the severe overlap in the 1D spectrum makes the analysis of the ^{13}C fine structure impracticable in most cases. Reprinted from Massou et al. [78] with permission from Elsevier

and ^{13}C chemical shifts along the two axes. Several experiments are available, and while there are occasional applications of HMQC in flux analysis (e.g., [149, 150]), the most commonly implemented method is HSQC (often referred to as 2D-[^{13}C,^{1}H] COSY). The increased spectral dispersion in the 2D spectrum has the immediate advantage of reducing spectral overlap in the ^{13}C dimension; there is also a sensitivity advantage because the ^{13}C signals are detected via the more sensitive ^{1}H nucleus. HSQC is particularly useful for experiments involving uniformly labeled substrates, since these lead to NMR signals with a rich multiplet structure, allowing the relative abundance of the corresponding isotopomers to be deduced from the relative intensities of the peaks within a particular signal and avoiding the difficult task of comparing HSQC intensities from different carbon atoms. Note that the method reports only on the carbon atoms directly bonded to hydrogen in most instances and so it provides no information on the labeling of the carboxyl groups in amino acids and organic acids.

The power of HSQC for isotopomer analysis was demonstrated in a comprehensive analysis of the ^{13}C-labeling of the amino acids in a protein hydrolysate, obtained from *E. coli* [120]. This study showed that 48 of the 50 aliphatic ^{13}C resonances expected for the 16 amino acids present in the hydrolysate (recall that Asn,

Cys, Gln, and Trp are lost during hydrolysis) were well resolved in the 2D spectrum, and 42 signals were useful for the subsequent metabolic analysis. Note that it was not possible to encompass the entire ^{13}C chemical shift range in a single spectrum, and it was necessary to run a second HSQC spectrum with the ^{13}C carrier frequency centered on the aromatic region to capture further data for Tyr and His.

Subsequently, HSQC was adopted as a routine tool for analyzing fluxes through the pathways of central metabolism in microbes [101], and the usefulness of the approach was extended by the development of parameter fitting methods for analyzing the data [104, 139]. Several refinements of the basic HSQC experiment have been proposed on the basis of a detailed analysis of the way the data are used for network flux analysis [130]. First, a line-fitting tool was developed for the accurate determination of peak areas, together with a method for estimating the errors in the relative intensities. Secondly, it was also shown that improved spectral resolution could be obtained for protein hydrolysates by derivatizing the amino acids and recording spectra in a non-polar solvent. This derivatization procedure is not often implemented since it adds a further step to the analysis, and the benefit is usually modest. Thirdly, the paper describes a method for correcting HSQC data for incomplete isotopic equilibration in a continuous flow culture, thus facilitating more cost-effective experiments in which the labeled substrate is deliberately supplied for periods shorter than those that would approximate to an isotopic steady state. Finally, it is shown that additional, non-redundant labeling information, arising from long-range coupling, can be extracted with the line-fitting tool from some of the tyrosine and histidine signals [130].

HSQC has been used to analyze protein hydrolysates from plant cell extracts (Fig. 5.6; [115, 116]). Non-overlapping multiplets were quantified using NMRView (http://onemoonscientific.com/nmrview), and overlapping multiplets were analyzed using software based on the peak-fitting model proposed by van Winden et al. [130]. Sriram et al. [115] investigated the accuracy of the HSQC experiment by recording spectra from mixtures containing known quantities of natural abundance, [2,3-$^{13}C_2$]- and [U-$^{13}C_3$] alanine, and obtained good agreement between the measured and expected relative abundances of the multiplet signals. NMR signal intensities can be easily manipulated by the spectroscopist, and verification of the validity of the measured intensities is a sensible precaution prior to investing time and effort in analyzing the data. Note that the isotopomer abundances derived from these spectra can either be used directly [115] or be reduced to a smaller number of bondomers [114, 131] before starting the parameter fitting [116]. Bondomers are defined according to whether the carbon–carbon bonds in a molecule are derived intact from a uniformly labeled substrate or whether they have been formed ab initio through biosynthesis. Qualitatively, this focus on connectivity provides a convenient method for interpreting labeling experiments, while quantitatively it leads to computationally efficient modeling.

Overlap in the indirect (^{13}C) dimension of an HSQC spectrum can be further reduced by using J-scaling [140]. This technique increases the apparent coupling constants between ^{13}C atoms, improving the separation of signals in multiplets by a defined factor. This method has been used to good effect in the analysis of soybean embryo labeling experiments [117].

5 Quantification of Isotope Label

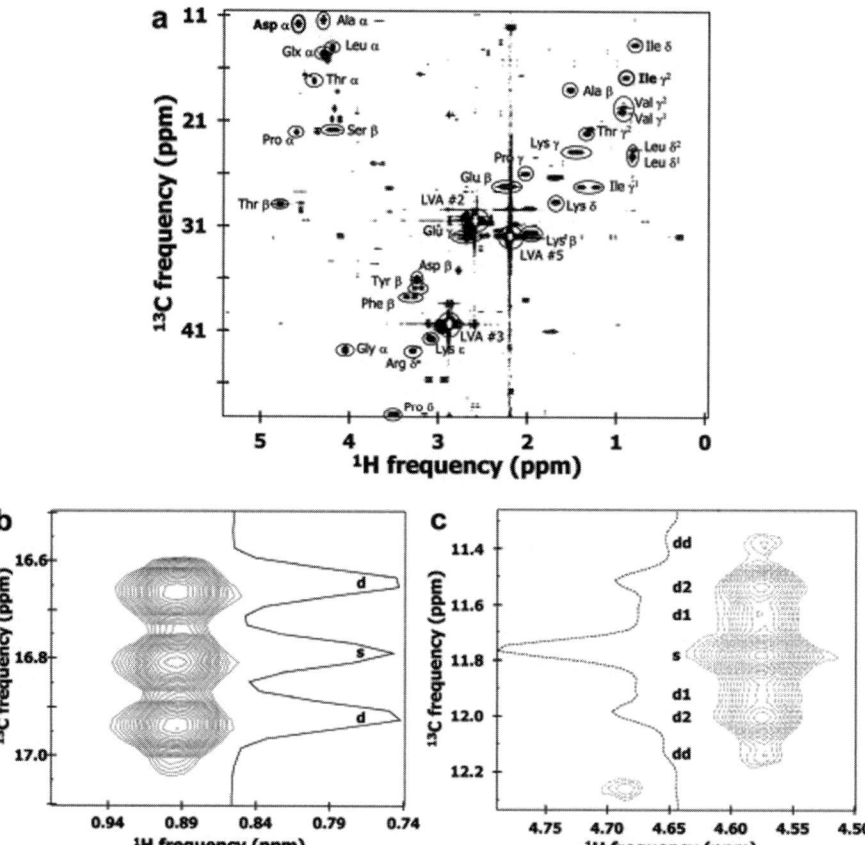

Fig. 5.6 (a) HSQC spectrum of hydrolysate from *Catharanthus roseus* hairy roots grown on 5% [U-$^{13}C_6$]glucose. Expansion of the signals for (**b**) the Ile γ methyl group and (**c**) the Asp α carbon atom, in each case showing a 1D slice alongside the 2D multiplet. The relative intensities of the multiplet signals provide information on the relative abundance of the contributing isotopomers. Reprinted from Sriram et al. [116] with permission from Elsevier

While HSQC is the main 2D NMR method used in isotopomer analysis, other methods may be used to decipher complex coupling patterns. For example, the 2D INADEQUATE pulse sequence was used to identify multiply labeled glucose isotopomers derived from starch in ^{13}C-labeling experiments on maize (*Zea mays*) kernel cultures [45]. This double quantum coherence experiment was particularly useful in analyzing the results of [U-$^{13}C_6$]glucose labeling, and, for example, could be used to show the presence of [1,2-$^{13}C_2$]glucose, [1,2,3-$^{13}C_3$]glucose, but not [2,3-$^{13}C_2$]glucose isotopomers. Similarly, it provided a clear demonstration that the contiguous labeling of C3 and C4 in the original [U-$^{13}C_6$]glucose source did not survive metabolic rearrangement to any significant extent prior to incorporation of glucose into starch. Ultimately, the value of this experiment lies in the way in which it correlates pairs of contiguous carbon atoms, thus removing the uncertainties in the interpretation of the 1D spectrum arising from the similarity of most of the J_{CC} values.

5.3.4 In Vivo NMR

In contrast to MS, NMR signals can be detected directly from living tissues, and this in vivo NMR approach has found many applications in the analysis of plants [91, 94]. In vivo NMR is particularly useful when it can supply information that would be lost on tissue extraction; for example, in some circumstances, in vivo NMR can provide information on the subcellular distribution of ions and metabolites. In vivo NMR can also provide a convenient and statistically robust way of monitoring changes in a tissue over a time course without the need for serial extraction. This latter approach is potentially useful in labeling studies since it allows in vivo NMR to be used to monitor the kinetics of label redistribution and to obtain direct evidence for the establishment of the isotopic and metabolic steady states that are required for steady-state network flux analysis. However, it should be noted that the resolution of in vivo spectra is usually less good than in the corresponding extracts, and in vivo NMR requires careful attention to the maintenance of the tissue under relevant physiological conditions during data acquisition [94]. These two considerations probably explain why the in vivo approach has not been commonly used in flux studies on plant and microbial tissues, where it is relatively straightforward to analyze multiple extracts from replicate experiments. In contrast, in vivo NMR is used extensively in flux analysis in animals systems, including the human brain [50].

The potential of in vivo NMR in this area for plants can be judged from a recent study on excised linseed (*Linum usitatissimum*) embryos [126]. The large pools of sucrose and lipids were readily detected in vivo, and time-course data showed that the intermediates of the central carbon metabolism reached isotopic equilibrium over a time scale of 3 h, that the sucrose pool required 6 h, and that it took 18 h to reach a complete isotopic and metabolic steady state. It was also possible to extract the rates of lipid and sucrose synthesis from the time course, and a detailed analysis led to the expected conclusion that sucrose synthesis was largely the result of sucrose phosphate synthase activity. Overall in vivo NMR analysis is not a substitute for the detailed NMR and MS analysis of tissue extracts, but it does provide an efficient and direct assessment of the time scales necessary to achieve isotopic and metabolic steady state, thus facilitating the design of steady-state experiments.

5.4 Mass Spectrometry

From the time of its experimental conception by the "father of mass spectrometry" J. J. Thomson nearly a century ago, the mass spectrometer has become widely recognized as an instrument of choice for identification of chemical structures. A mass spectrometer measures abundances for given mass-to-charge ratios (*m/z*) of gas phase ions. The charge (z) is almost always equal to one so that the measurement of abundance will be on an atomic mass unit (amu) basis. Each compound is fragmented during MS and the abundances of ions within each fragment are monitored. The comparison of ion abundances within a fragment provides a relative set of mass enrichments. Together, all mass isotopomers of a fragment represent the complete isotopomer pool for that combination of atoms. Therefore, mass enrichments are reported on a fractional basis with their sum totaling one.

The MS is usually linked in series behind a chromatograph that separates compounds as a function of their chemical and physical properties. Given the analytical capabilities of these two techniques to resolve complex mixtures and precisely identify compounds at very sensitive levels, it is not surprising that the two are frequently used in tandem with biological samples that are diverse in composition and present at low concentrations.

MS has recently received much attention as a tool in flux analysis and network studies for several reasons. First, it has been shown that the mass isotopomer distributions given by MS measurements reflect the fluxes through metabolic pathways and are therefore diagnostic of cellular behavior [23, 103, 143]. Secondly, mass isotope ratios can be measured accurately and precisely, to within 0.4 mol% for many amino acid fragments [3], and the precision of these measurements is matched by a sensitivity that greatly exceeds the sensitivity of NMR.

The number of isotopomer measurements can be maximized by analyzing different carbon fragments (Figs 5.7 and 5.8). A compound of n carbons contains 2^n isotopomers and can have up to 2^n-1 different fragments. In practice, only a subset of the possible fragments is observable at sufficiently high intensities for quantitative analysis. For example, the three-atom molecule in Fig. 5.7 has seven possible fragmentation products, but molecular rearrangement to obtain the two-atom fragment containing atoms one and three would require both bond-breaking and bond-forming steps, whereas the other two-atom products only require bond breaking. Although some molecular rearrangements are characteristic of the ion collision processes (e.g., TBDMS-derived $[M-85]^+$), multi-step rearrangements are generally limited by energetics and proximity of other reactive groups. Furthermore, the label information for one-atom fragments is likely to be obscured by the contaminating ions that are higher in number at lower molecular weights.

The ionization of a molecule results in fragments that contain carbon atoms from different positions in the molecule (Fig. 5.8). It is essential to establish the origin of the carbon atoms in each fragment, and this is usually done by using labeled standards. Note that fragment identification can be complicated when multiple breakdown products are formed with different carbon compositions but the same masses [3]. The measurement of the mass enrichments can also be confused by the presence of contaminating peaks with the same mass. This may lead to inconsistencies with the expected relationship between the labeling of the different fragments (Fig. 5.8), and if this is observed, then further investigation of the fragmentation pattern would certainly be necessary before proceeding with the quantitative analysis.

Multiple fragment measurements can also provide positional labeling information. Thus as well as direct measurement of the completely unlabeled (000) and

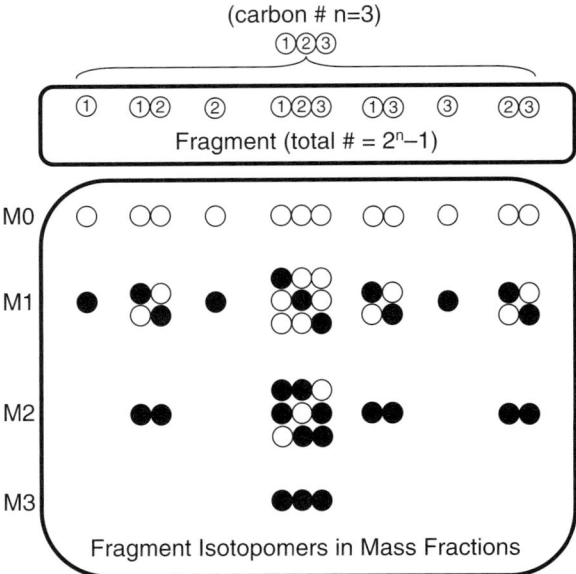

Fig. 5.7 Theoretical fragmentation and mass distribution of isotopomers for a three-carbon compound. A statistical analysis shows that there are 2^n-1 possible fragments, each of which can contain both ^{12}C and ^{13}C atoms

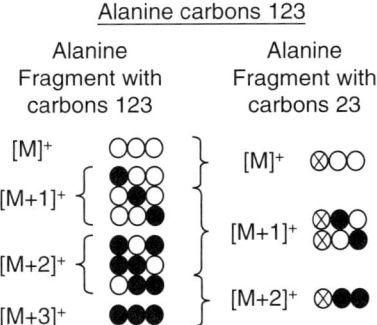

Fig. 5.8 Mass isotopomers for two fragments from alanine. The labeling of the 23 fragment will necessarily reflect the labeling of the 123 fragment, as indicated by the brackets, and in the absence of any underlying contaminant signals, $[M]^+_{123} \leq [M]^+_{23} \leq ([M]^+_{123} + [M+1]^+_{123})$ and $[M+3]^+_{123} \leq [M+2]^+_{23} \leq ([M+2]^+_{123} + [M+3]^+_{123})$

completely labeled (111) isotopomers from the Ala123 fragment, the abundance of the 100 and 011 isotopomers can be deduced by comparing the Ala123 and Ala23 measurements (Fig. 5.8). In favorable instances, it is possible to deduce the abundance of all the positional isotopomers for molecules such as serine or glycine, depending on the number of measurements made [23].

Overall, accurate quantification of label by MS relies on (i) adequate separation of metabolites through sample preparation and chromatography; (ii) the detection of multiple fragmentation patterns that are readily identifiable and well separated from other fragments; and (iii) proper corrections for naturally abundant isotopes. Each of these topics is considered in the following sections.

5.4.1 Chromatographic Parameters

The use of chromatography to separate molecules dates back to the separation of plant pigments by Tswett in the early 1900s [46]. Today, the separation of the compounds is efficiently achieved using either high performance liquid chromatography (HPLC) or capillary GC, the methodological descendants of the earlier work. Much work has been done on establishing the theory associated with chromatographic separations, and this is discussed elsewhere (e.g., [46]).

Several considerations influence the choice between liquid and gas chromatography. LC analyses do not require volatilization, since they exploit properties such as ion exchange, size exclusion, and hydrophobicity, and so may not require derivatization. However, liquids are more viscous and therefore require equipment (e.g., seals, pumps, valves, lines, injectors) with higher pressure tolerances. Moreover finding the right column for LC can be an expensive and time-consuming process, and it may require substantial method development. It is also usually necessary to use tandem MS/MS to generate daughter ions because the initial ionization of the compounds coming off the LC column does not lead to extensive fragmentation.

Recent applications of LCMS in flux analysis have focused on the analysis of intracellular intermediates [58, 61, 73, 128, 133]. The lack of a derivatization step is an advantage for unstable intermediates, but there are some limitations in the extent to which the positional isotopomers of the amino acids can be analyzed [90, 100], and the use of LCMS analysis requires higher amounts of labeled material than GCMS [21].

Capillary electrophoresis (CE) provides another option for compound separation, and this has recently been explored as an alternative to GC or LC for isotopomer analysis [125] This study reported improvements in sensitivity and mass resolution when CE was coupled to time-of-flight MS, suggesting that this approach may be an emerging technology for flux analysis. However, CE and LC are currently used less frequently for isotopomer analysis than GC, and it is GCMS that is emphasized here.

Accurate isotopomer analysis by GCMS starts with a good separation of metabolites in the labeled mixture. GC parameters such as column length, the choice of column packing material for the stationary phase, the time and temperature profile for the column (i.e., ramp profile), sample injection volume, sample concentration, and purity of the sample can all have dramatic consequences on the separation process and need to be chosen with care [14]. Some of the instrumental components and parameters are well established. For example, helium gas, because of its inert, non-reactive, non-toxic, non-flammable, and affordable nature, is commonplace.

Splitless injection of 0.1–2 μl avoids compound partitioning from differences in volatility and is therefore frequently used. Finally, oven temperature profiles are set to vary over the range 5–400°C for optimized separation of compounds as well as sample throughput.

The power of GC with careful attention to these parameters is evident in the clean separation of many peaks. Free amino acids or those resulting from protein hydrolysis are handled with ease (Fig. 5.9), and this is routinely exploited in biological studies of metabolic networks (e.g., [16, 41, 60, 61, 147]). However, in the chromatography field, far more elaborate separations have been achieved. For example, the modified GC column construction by Berger [8] resulted in separation of 970 components from standard gasoline by serially connecting nine 50-m columns.

In fact, the superior separation qualities of GC can also lead to errors in quantitative analysis. GC resolution is sufficient in some cases to partially separate isotopes by weight. Isotope fractionation has been the study of many investigations (reviews: [40, 129]) and was apparently first reported with deuterium labeling studies by Wilzbach and Riesz [141]. Generally, ions elute in a weight-decreasing fashion, with the highest weight ions eluting first (Fig. 5.10), reflecting reduced molecular interactions with the stationary phase for the larger atoms. The degree of separation is not enough to cause peak splitting for most deuterated amino acids [135], though it can be more pronounced with deuterated (e.g., [86]), tritiated (e.g., [110]), or ^{13}C-labeled (e.g., [81]) fatty acid methyl esters and sugars (e.g., [6, 119]). To avoid bias from isotopic discrimination, the entire peak must be integrated, sacrificing some of the sensitivity at each measurement.

5.4.1.1 Column Stationary Phase

Adequate separation of most compounds is provided by a relatively small subset of the available column stationary phases [142]. The 100% dimethyl polysiloxane

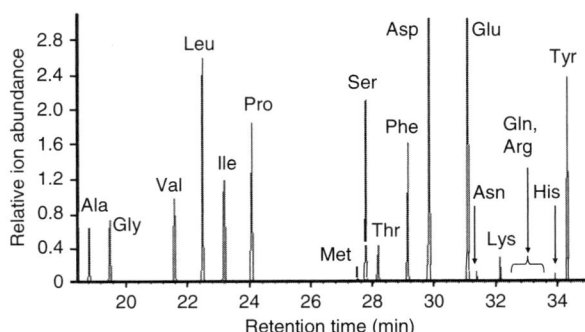

Fig. 5.9 Gas chromatogram of TBDMS-derivatized amino acids from a protein hydrolysate monitored by selected ion mode (SIM) MS. Cysteine and tryptophan are absent because they are oxidatively destroyed; glutamine and asparagine are barely detectable because of deamidation. Arginine is often not monitored because the detectable ions for SIM result from more complex fragmentation patterns (Section 5.4.3)

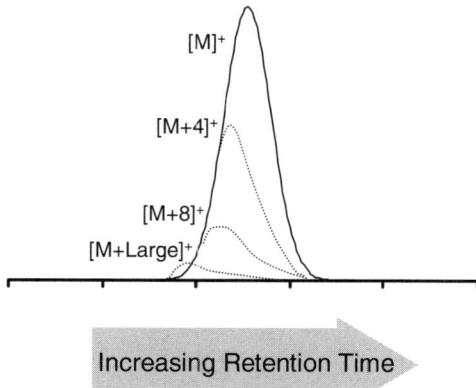

Fig. 5.10 Isotope fractionation during gas chromatography. The dispersion in the peak depends on the nature of the compound and its size, so the relative peak positions in the figure are only indicative of a general trend

and 5% methyl-replaced phenyl groups known commercially as DB-1®, DB-5®, HP-1®, HP-5® (Agilent J&W), or SPB-1®, SPB-5® (Supelco) are the most nonpolar and very resilient to repeated use. Changing the stationary phase will change the binding interactions between the column and the mobile phase and affect separation. In some instances, failure to adequately resolve the metabolites of interest necessitates a longer column or a different column choice. The phenomenon is exemplified by fatty acid methyl esters or butyl amides, where the separation of 18-carbon compounds with different degrees of unsaturation cannot be resolved by a DB-1®, and instead a DB-23® (Agilent J&W) is used (Fig. 5.11). See Niessen [83] for a more thorough description of column packing compounds.

5.4.1.2 Derivative Groups

Low volatility, thermal instability, and complications arising from covalent and hydrogen bond formation can all interfere with the separation of compounds containing carboxyl, mercaptan, hydroxyl, amino, imino, phosphate, sulfoxide, and carbonyl groups by GC. Fortunately, proton substitution with less polar acyl, isopropyl, silyl, or other functional groups is routine. The choice of derivative is important because derivatives are not universally reactive, and the added group can enhance sensitivity, selectivity, resolution, improve peak symmetry, and when coupled to the MS can result in distinct carbon fragments that provide unique isotopomer information. A suitable derivative should bestow improved volatility and thermal stability, and result in high yields, with few if any side reactions, while being cost effective and simple to use. Table 5.1 lists the most commonly used derivatization methods for biomolecules and the functional groups they block. TBDMS is especially useful because it reacts with most functional groups and because the resulting derivatives are stable, have high molecular weight, and have several common breakdown fragments that lose only derivative components, making them ideal for quantifying isotopomers. See Section 5.4.3 for further information.

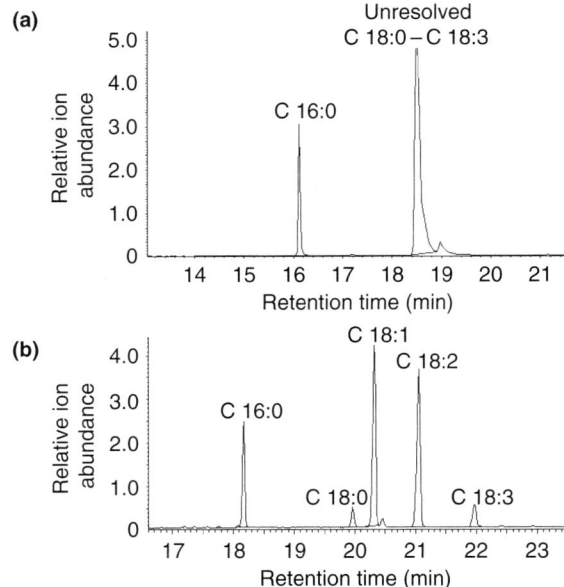

Fig. 5.11 Separation of derivatives of 16- and 18-carbon fatty acids on different columns: (**a**) A DB-1® column fails to separate the 18-carbon fatty acids with 18:0, 18:1, and 18:2, completely unresolved at a retention time of 18.5 min; (**b**) A more polar DB-23 column (50% cyanopropyl) results in complete separation of the same compounds

5.4.2 Spectrometric Detection

5.4.2.1 Types of Mass Spectrometer

Mass spectrometers come in different forms, including ion traps, time of flight, Fourier transform, and magnetic-sector and quadrupole instruments. All instruments rely on the same basic concept: interaction of charged particles with electrical or magnetic fields, but vary in dynamic range, sensitivity, resolution, and throughput. Ion traps and quadrupoles offer a lot of the same benefits: high sensitivity, ease of use, and low cost, but with a primary disadvantage of limited mass range. However, ion traps generally have a smaller linear dynamic range than quadrupoles. Time-of-flight and Fourier transform instruments can measure a large range of masses, and along with magnetic sector analyzers offer the highest resolution (up to 1 part in 10,000) but require significant expertise and can be quite costly. Moreover, magnetic sensor instruments have relatively poor sensitivity. The use of the quadrupole remains the scientific and industrial workhorse in network studies (e.g., [16, 41]). Overviews describing the physical construction and principles of each instrument are given elsewhere (e.g., [112]). Aside from quadrupoles, several notable network studies have used ion traps [60], MALDI-TOF [144, 145], or LCMS-TOF [148].

Table 5.1 Most common derivatization methods for GCMS

Method	Common Reagent	Before	After	Gain in MW	Application	Common Losses	References
Methylation	Methanolic HCl BF_3, NaOH Diazomethane	-COOH	-COO-CH_3	14	Lipids Organic acids	$[M-31]^+$ $[M-59]^+$ $[M-73]^+$ $[M-59]^+$	[33, 106]
Acetylation	Acetic anhydride[1]	-OH -SH -NH_2	-O-$COCH_3$ -S-$COCH_3$ -NH-$COCH_3$	42	Sugars	$[M-73]^+$ $[M-101]^+$	[1, 88, 107][2]
TMS	MSA^3 BSA MSTFA BSTFA	-OH =NH -SH -COOH -NH_2 -OH =NH	-O-Si$(CH_3)_3$ =N-Si$(CH_3)_3$ -S-Si$(CH_3)_3$ -COO-Si$(CH_3)_3$ -NH-Si$(CH_3)_3$ -O-Si$(CH_3)_2$-C$(CH_3)_3$ =N-Si$(CH_3)_2$-C$(CH_3)_3$	72	Sugars Amino acids Organic acids Intermediates	$[M-15]^+$ $[M-43]^+$ $[M-117]^+$	[52, 74[4], 95, 111, 119, 134]
TBDMS	MTBSTFA	-COOH -NH_2 -SH	-COO-Si$(CH_3)_2$-C$(CH_3)_3$ -NH-Si$(CH_3)_2$-C$(CH_3)_3$ -S-Si$(CH_3)_2$-C$(CH_3)_3$	114	Amino acids Organic acids	$[M-15]^+$ $[M-57]^+$ $[M-85]^+$ $[M-159]^+$	[23, 25, 85, 147]

Abbreviations: MSA, N-methyl1-N-TMS-acetamide; BSA, N,O-bis-TMs-acetamide; MSTFA, N-methyl1-N-TMS-trifluoroacetamide; BSTFA, N,O-bis-TMS-trifluoroacetamide; MIBSTFA, N-(tert-butyldimethylsilyl)-N-methyltrifluoroacetamide.

[1] Other less frequently used derivatives that are well suited to sugars are reviewed in Price [8], and MacLeod et al. [74].
[2] Ethylchloroformate has also been used to derivatize amino acids, bearing some similarity to acetylation see Yang et al. [149] and Christensen and Nielsen [17].
[3] A more comprehensive list of reagents is given in Wittmann [142].
[4] Describes methoxime-TMS derivatization that can be used to gain additional labeling information.

5.4.2.2 Quadrupole Design

Quadrupole mass filters are designed with four rods running parallel to the stream of ions. The rods impose different radiofrequency (rf) and direct current (dc) voltages to separate the ions as they travel toward the detector. A small m/z window of ions reaches the detector (an electron multiplier), where they strike metal surfaces, emitting electrons. The electron signal is amplified through more collisions with metal surfaces, until sufficiently amplified to be accurately recorded digitally. By alternating radiofrequency voltages, each m/z is monitored in succession, generating a spectrum of intensities of different ions (y-axis) for the range of m/z values monitored (x-axis) by the spectrometer. Mechanically adjusting the ion beam and the source and detector slits through which the beam passes alters the resolution and sensitivity in an inverse fashion. Masses within the instrument's range are distinguishable to levels of 0.1 amu [14]. However since there is a trade-off between resolution and sensitivity, detection sensitivity can be maximized by limiting the quantitative resolution to 1 Da, since this is sufficient to separate the mass isotopomers that arise in labeling experiments.

Mass spectrometers can also be operated in tandem, an approach referred to as MS/MS. For triple quadrupole instruments, this is done by separating two linked MS by a third set of quadrupole rods that operate on a radiofrequency only. By adjusting the voltages on the first quadrupole, particular masses are transmitted to the second rf-only quadrupole. In the second quadrupole, they collide with argon atoms that are neutral leading to further ion fragmentation, producing daughter ions that are measured by the third quadrupole. MS/MS monitoring of the parent and daughter spectra provides further information on structure and labeling [55, 58, 128], and it is particularly important for LC-based MS because the usual ionization method, electrospray, imparts less energy to the molecular ion reducing the tendency to fragment.

5.4.2.3 Ionization Methods

There are three main approaches to ionization: desorption, evaporative, and gas-phase ionization. Desorption methods include fast atom bombardment (FAB), plasma desorption, matrix-assisted laser desorption (MALDI), and field desorption. These methods are used with compounds that are poorly volatilized or of high molecular weight (e.g., proteins). Evaporative ionization methods include electrospray and thermospray. In both approaches, liquid particles from an LC column pass through a capillary tube and are nebulized into aerosols as they enter the ion source. These methods are frequently used for peptides, proteins, and non-volatiles but have found fewer applications in analyzing metabolite labeling. This may change with the increasing availability of LC-MS/MS spectrometers. Finally gas-phase ionization methods, particularly electron impact (EI), and to a lesser extent, chemical ionization (CI), are both well suited for volatile, nonionic compounds of molecular weight less than 1,000 Da.

EI is the most common ionization technique for GCMS. Briefly, electrons released from a hot filament are accelerated at 70 V through a small opening toward the effluent from the GC column. As the electrons bombard the sample, they impart an excess of energy to strip off the outer shell electrons, generating positive ions. The remaining energy results in covalent bond cleavage to produce characteristic fragmentation patterns composed of cations and neutral radicals. The fragments contain different subsets of the original molecule and therefore provide unique labeling information. The molecular ion is often present at low abundance, and it may be difficult to measure accurately. However, fragments that result from small losses of part of the derivative group are usually abundant, diagnostic for the compound under study, and provide the same mass isotopomer measurements as the molecular ion. The electron voltage can be reduced if greater abundance of molecular ions is desired, but this will come at the expense of the fragmentation, as well as the overall sensitivity of the instrument.

CI is a "softer" method that results in less fragmentation and would appear to be particularly well suited for label quantification in the molecular ion but has received little attention. In this approach, the volatile sample is subjected to an ionized gas (usually methane that has itself been ionized by electron bombardment). The collisions result in ionized sample molecules through proton transfer events to generate $[M+H]^+$ ions. Labeled sugars have been analyzed in this way [26, 27, 49]. Negative ion chemical ionization is also possible, but it has been used infrequently and is not discussed here. For further details on ionization techniques and the voltage set points that are adjusted to control ionization for optimal sensitivity and resolution, see [4].

5.4.2.4 Ion Monitoring

An important distinction in using MS for label quantification versus structural identification is that the compound(s) of interest are known. This allows the instrument to be set up for selective ion monitoring (SIM) rather than total ion monitoring (TIM). Derivatized standards, both labeled and unlabeled, provide a quick assessment of the fragmentation patterns and masses of interest, and the conclusions can be confirmed by calculation of expected losses and by comparison with existing MS libraries (e.g., National Institute of Standards and Technology (NIST) Mass Spectral Libraries, or Palisade Complete MS Library 600 K^{TM}). For example, glucose standards labeled at different carbon atoms provide direct evidence for the fragmentation of the original derivative (Fig. 5.12). Note that while it is standard practice to use a single calibrant, and single fragments, the resulting abundances include not only the species ion enrichment but also the effects of mass discrimination, contaminants, background noise, and electrometer offset errors [105]. It follows that it is advisable to run multiple standards.

SIM allows the entire instrumental monitoring time to be focused on a selected number of diagnostically important ions, reducing the workload of the spectrometer, and improving sensitivity. The use of SIM was pioneered by Sweeley et al. [119] and provides optimal signal and minimal interference [89, 142]. In a quadrupole system, the ions are selected by altering the rod voltage ("ion beam switching"), and accurate monitoring/quantifying of the chosen ions is limited by the user-defined dwell times [79]. The dwell time is the amount of time in each pass that is spent monitoring a particular ion. It is limited by the GC peak width that defines the total time that

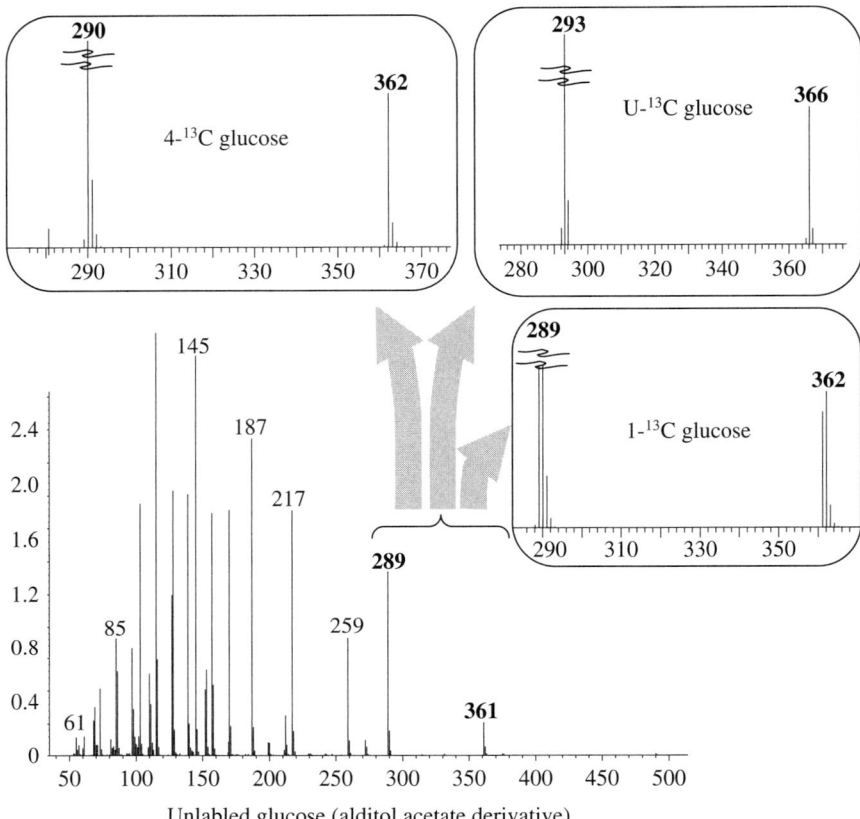

Fig. 5.12 The spectrum of the glucitol hexa-acetate derivative of glucose shows peaks at m/z 289 and 361 that are well removed from other peaks. The 361-amu base peak was hypothesized to represent the loss of CH$_2$OCOCH$_3$ (M-73) that would occur by cleavage between either carbon 1 and 2 or carbon 5 and 6 of the symmetric derivative. This was verified by running spectra of labeled glucose derivatives. The boxes show the expected shifts of the 361 base peak to higher molecular weights, with a splitting of the peak when the carbon is labeled at C1 or C6 because of the ambiguity in the cleavage of the symmetric derivative

ion information will be collected for a particular compound. The dwell time should be sufficiently small that all of the monitored ions within the compound are counted at least 15–20 times [79, 118]. Obviously the more peaks that are monitored the less time that can be devoted to each ion, and decreasing the amount of a compound shortens the monitoring time for each ion. By measuring each ion abundance multiple times, isotope discrimination events from the GC separation process are allowed for through averaging [23].

Even so, electrometer offsets, poor ion monitoring, and nonlinear gain characteristics or recording can still result in significant systematic errors [79, 105] and may require further study. Given a peak retention time window, the selection of the ions to be monitored is driven by the nature of the investigation. Higher molecular weight fragments are the most diagnostic because there are fewer possible breakdown products with high molecular weights. The smaller number of fragments reduces the opportunity for overlaps in the spectrum, although usually these high molecular weight fragments are less abundant and more difficult to measure accurately. In terms of sensitivity, SIM-GCMS allows analysis of nanograms to picograms of material even when chromatography does not resolve compounds completely. For isotope measurements, the resolution of the peaks for adjacent ions serves to define measurement precision. More general considerations on SIM-GCMS can be found in the literature (e.g., [72]).

Accurate analysis also depends on operating within the linear range of abundance measurement for the spectrometer [9, 34]. At low abundance levels, the background chemical interference becomes significant, while at very high levels some of the most abundant fragments may saturate the detector and result in an inaccurate ceiling value. This problem is particularly important when quantifying isotopomers because of large disparities between the most abundant and least abundant ions. Acceptable conditions are established by running multiple concentrations (Fig. 5.13).

The ratio of the mass isotopomers within a compound should be constant over the linear concentration range [9], and Dauner and Sauer [23] noted no impact on the ratios of isotopomers for amino acids over a concentration range of three orders of magnitude. Evidence for saturation can usually be found on modern instruments by extracting individual ion chromatograms, and if sample dilution is not possible then saturation can also be avoided by reducing the electron multiplier voltage [142]. A nonlinear response can also arise for other reasons [33, 85, 127], including sample volume, repeller voltage, or changes in chemical compound size/length [34,127]. Differences in gas phase chemistry [35], and operating conditions such as sample

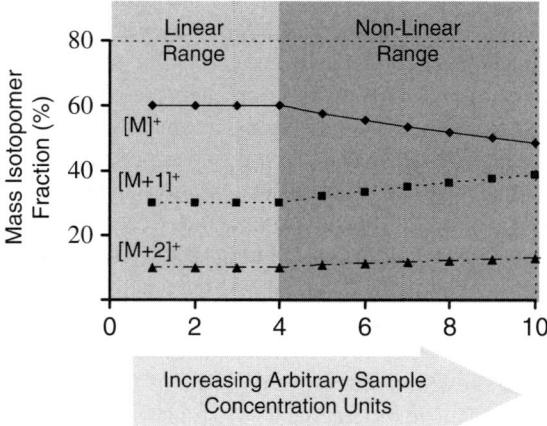

Fig. 5.13 Effect of sample concentration on relative mass isotopomer abundance. The relative abundances are usually constant over several orders of magnitude, but at high concentrations the detector becomes saturated for the mass present in greatest abundance, leading to an underestimate of the saturating fraction

pressure [47], can also affect the response of the instrument. Some of these issues have been addressed elsewhere [3, 9] and running sets of standards can help to identify whether these concerns are justified. In all cases, background noise should be minimized by subtracting a baseline signal from integrated ion chromatograms during the processing of data [3].

5.4.3 Analysis of Biomolecules

All the major classes of biomolecule can be analyzed by GCMS, and methods for storage lipids, amino acids, organic acids, sugars, and storage carbohydrates are described in this section. In general, quantitative analysis of the label distribution in these compounds requires (i) clear chromatographic separation of the metabolites within a class; (ii) well-documented fragmentation patterns; and (iii) detection of abundant ions representing multiple fragments of the carbon skeleton. While numerous protocols are described in the literature, it is essential to verify the reproducibility of the chosen method using appropriate standards before implementing a new method.

5.4.3.1 Storage Lipids

Biosynthesis of lipids can be regarded as a polymeric addition of repeating acetyl-CoA building blocks. From a modeling perspective, there is a direct relation between the labeling of the acetyl-CoA pool and the labeling of the entire fatty acyl chain, if the acetyl CoA pool is in an isotopic steady state. Fatty acid methyl ester (FAME) derivatives are easily quantified by flame ionization detection (GC-FID), making them a frequent choice. FAME produces two-carbon McLafferty fragments [80, 122] as base peaks that represent the labeling of carbons 1 and 2 of the fatty acid and presumably describe carbons within one acetate group (Fig. 5.12).

However, the GCMS evaluation of McLafferty product labeling is problematic for several reasons. First, while the abundance and carbon composition of the McLafferty fragments make them ideal for analysis, their molecular weight of 74 lies in a region of the spectrum that has overlapping fragments (Fig. 5.14). For example, the C_5H_9 alkane radical (m/z 69) also represents a significant (up to $\sim$20% of base peak) breakdown product of fatty acids. In a highly labeled fatty acid, this alkyl fragment may contaminate the McLafferty region. Secondly, FAME labeling can be accurately quantified only for saturated fatty acids, so plant oils that are usually high in unsaturated oil content must first be hydrogenated [106] involving extra processing steps. Thirdly, well-documented proton transfer events [35] and the bias of operational parameters [33, 85, 127] also make interpretation of FAME spectra more challenging. An alternative that circumvents some of these problems is to use the butyl amide approach to move the fragments to a higher weight (m/z 115) [1, 62].

5.4.3.2 Amino and Organic Acids

Numerous options are available for the derivatization of amino and organic acids, and silyl derivatives are particularly useful (Table 5.1). The choice of derivative

5 Quantification of Isotope Label

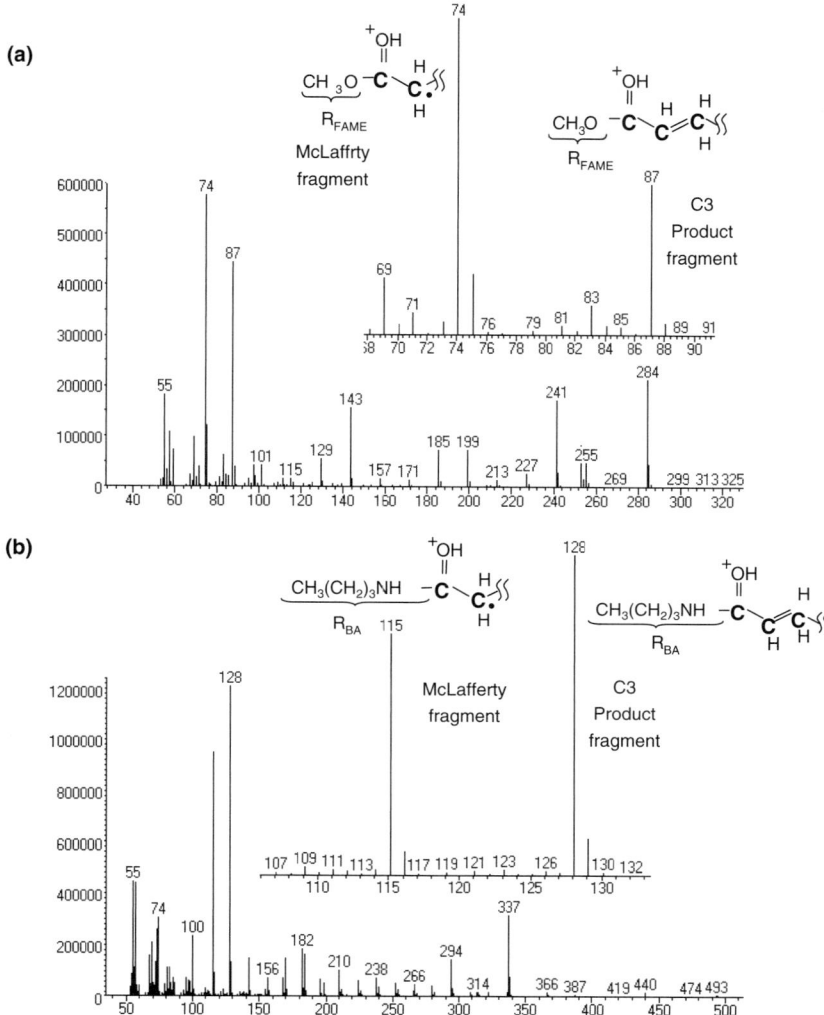

Fig. 5.14 Comparison of FAME and butyl amide McLafferty fragments. (**a**) FAME shows overlapping/nearby peaks in the McLafferty region for the derivatized heptadecanoate. (**b**) Butyl amide fragments are shifted to higher molecular weights away from other fragment products, and, unlike FAME, can produce accurately quantified label measurements for both unsaturated (octadecenoate shown) and saturated fatty acids, and therefore do not require hydrogenation. Reprinted from Allen et al. [1] with permission from Elsevier

has implications for derivative stability, molecular weight, volatility, and ion fragmentation and therefore needs to be considered carefully. The derivatization of amino acids is particularly challenging because of the presence of carbamide, imino, hydroxyl, disulfidic, and extra amino and carboxyl groups. The derivatization and GC analysis of amino acids has received much attention with a review published over 30 years ago citing 415 publications that describe approximately 100 different approaches to chemical derivatization [53]. Today, the literature frequently reports the use of TBDMS derivatives because of the good reactivity with multiple functional groups, good volatility, and good peak resolution.

TBDMS derivatives are synthesized by combining equal volumes of N-(tert-butyldimethylsilyl)-*N*-methyltrifluoroacetamide (MTBSTFA) and dimethylformamide (DMF) to the dried sample and heating to a temperature above 60°C for 60 min [59, 82]. Longer times or increased temperatures are used to prevent incomplete derivatization. The use of MTBSTFA is advantageous because of limited byproduct generation and interference. The quantitative isotopomer analysis of amino acids in protein hydrolysates has been the subject of many recent investigations. Common fragments for TBDMS-amino acids are the $[M-15]^+$ and $[M-57]^+$ signals that represent the entire carbon backbone and the $[M-85]^+$ and $[M-159]^+$ signals that reflect carbon cleavage between the first and second carbons. Using this approach, Dauner and Sauer [23] identified 125 independent mass isotopomer measurements that were deemed acceptable. Acceptance was based on the correction of an unlabeled BSA standard. Signals for $[M+1]^+$ and $[M+2]^+$ that combined to more than 7% of the $[M]^+$ peak were discarded. No fragments were accepted for histidine or threonine due to their low abundance, and arginine was rejected because of difficulties defining the fragmentation patterns. However, arginine has been analyzed by others [57, 85, 142] (and references within), and Fig. 5.15 shows one of the several TBDMS-substituted arginine products with $m/z = 499$, following the loss of NH_3 from the guanidino group. It has also been suggested [85] that the product formed after the NH_3 loss could be a side-chain nitrile in which two derivatizing groups are attached to the amino group at carbon 2 of the amino acid, although this would be sterically less favored.

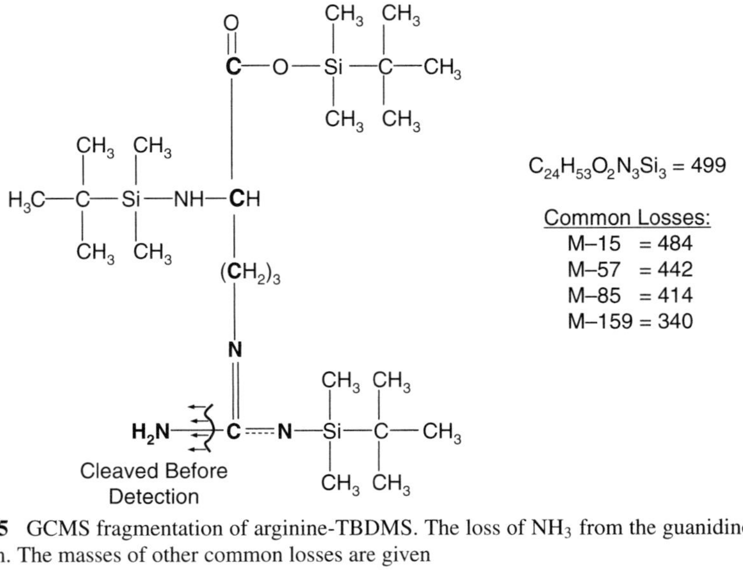

Fig. 5.15 GCMS fragmentation of arginine-TBDMS. The loss of NH_3 from the guanidino group is shown. The masses of other common losses are given

Other derivatizing agents, such as DMFDMA and DMFDBA, have been used but do not lead to particular differences in measurable amino acids, fragment numbers, or overall information [16]. Husek and Simek [54] have summarized recent work in this area.

TBDMS, TMS, and methyl ester formation have also been used for Krebs cycle intermediates (e.g., [20, 26, 27]). Due to the symmetry of some organic acid metabolites, the total number of isotopomers and thus label descriptions are reduced [143].

5.4.3.3 Sugars and Storage Carbohydrates

Plant tissue may contain considerable starch and cell wall, along with simple sugars such as glucose and sucrose. For MS, glycans can be broken down to provide information on the labeling of the cytosolic and plastidic pools of hexose phosphates. Cell wall components report on the labeling of cytosolic UDP-glucose [19], and a comparison of this pool with the labeling of the plastidic glucose units in starch [22] allows an assessment of the degree of equilibration of the upper glycolytic pathways between the plastid and cytosol. This analysis establishes whether there is sufficient information to identify fluxes for parallel pathways in multiple compartments and thus to allow compartmental modeling. Comparison of the monosaccharide units from starch, cell wall, and protein glycans using GCMS has been considered in the context of embryo labeling studies of soybean [1]. Additionally, starch labeling has been considered in flux models of *Brassica* [107, 109], and soybean [115].

Sugars offer a high ratio of reactive groups to the total number of atoms, and in the open-chain form, all six carbons of a hexose molecule can be derivatized. While most derivatives are suitable for carbohydrate label analysis (Table 5.1), often the metabolic interest lies in establishing the extent of bond breaking and reforming between carbons 1 and 2 since this reflects the relative contribution of glycolysis and OPPP to carbohydrate oxidation. In this regard, it is useful that sugars with a reducing carbon, for example glucose, can be derivatized uniquely because of the aldehyde group at carbon 1, allowing this atom to be distinguished from the other backbone carbons (see [74, 88] for more details). The aldehyde group can also be converted directly to an alcohol, and for small sample sizes this has the benefit of grouping α and β anomers that would otherwise elute from the GC differently. The ready interconversion of the anomers means that analyzing both sets of peaks would only yield redundant information. Prior reduction to sugar alcohols therefore increases the sensitivity of analysis by combining two peaks into one, leading to higher signal-to-noise ratios. Unfortunately, reduction can also introduce symmetry, for example in the case of glucose, and this restricts the analysis of the mass isotopomers.

Reduction prior to acetylation is performed by first suspending carbohydrates in 2 M NH_4OH and then reducing through the addition of $NaBH_4$ (e.g., [1, 11]). Peracetylation follows after water removal by resuspension in 1-methyl-imidazole or pyridine and subsequent addition of acetic anhydride [1, 88]. The reaction is neutralized with water and extracted multiple times with methylene chloride. The soluble fraction can be directly loaded into the GCMS and analyzed as described elsewhere [1, 10, 88].

Additional isotopomer information can be abstracted from formation of deuterioalditol acetates, aldonitrile acetates, or dialkyldithioacetal acetate derivatives (see [88] for review). The analysis of most of these derivatives is relatively straightforward, with [M-59]$^+$, arising from the loss of acetate, being one of the most informative fragments. For the peracetate derivative, other prominent losses include cleavage of the C1–C2 or C2–C3 bonds generating m/z 317 and 242 products respectively. Deuterioalditol acetates give the symmetrical cleavage of C2–C3 or C4–C5 bonds resulting in a peak at m/z 289, while the aldononitrile acetates result in significant cleavage between C5–C6.

5.4.4 Corrections for Natural Abundance

MS is unable to discriminate between ^{13}C atoms derived from a labeled precursor and the naturally abundant ^{13}C found at low levels in all carbon compounds. Failure to account for the natural abundance of ^{13}C can lead to significant errors in flux analysis [143] and so it is necessary to apply corrections for natural abundance to MS data.

Ultimately, the correction has to take account of every atom in the detected fragment of the derivatized metabolite, and in making the correction it is important to make a distinction between the atoms that could have been labeled in the experiment, i.e., the carbon atoms of the detected metabolite for a ^{13}C-labeling experiment, and the atoms that can only be labeled by virtue of the natural isotopic abundance. In a commonly used approach, the experimental measurements are first corrected for the atoms in the latter category only, and then the labeling of the backbone is simulated using the correct isotopomer abundances for the input substrate during the modeling of the fluxes that redistribute the label [67, 68, 136, 152].

The correction procedure can be conveniently implemented using a matrix approach. The first step is to define a mass isotopomer distribution vector (MDV) for the detected fragment:

$$\mathrm{MDV_{obs}} = \begin{bmatrix} [M]^+ \\ [M+1]^+ \\ \vdots \\ [M+n]^+ \end{bmatrix} \quad \text{where}: \sum_{i=0}^{n}[M+i]^+ = 1 \quad (5.1)$$

Here $[M]^+$, $[M+1]^+$, etc. represent the abundances of the signals observed for the fragment. The corrected MDV ($\mathrm{MDV_{cor}}$) is calculated by multiplying $\mathrm{MDV_{obs}}$ by an inverted correction matrix (CM):

$$\mathrm{MDV_{cor}} = \mathrm{CM_{CHNOSiS}^{-1}} \cdot \mathrm{MDV_{obs}} \quad (5.2)$$

where $\mathrm{CM_{CHNOSiS}}$ is the product of individual correction matrices for each element [132]:

5 Quantification of Isotope Label

$$CM_{CHNOSiS} = CM_C \cdot CM_H \cdot CM_N \cdot CM_O \cdot CM_{Si} \cdot CM_S \qquad (5.3)$$

Note that the dimensions of the MDV vectors in Eq 5.2 establish the dimensions of the overall correction matrix. The row dimension in MDV_{obs} represents the measured mass isotopomers, and the row dimension in MDV_{cor} contains all the mass isotopomers that arise from the labeling experiment. As the number of atoms within a molecule increases so too does the number of possible masses that can be measured, reflecting the contribution of the naturally abundant isotopes. However, the probability of obtaining higher masses due to the presence of multiple heavy isotopes at natural abundance becomes very small, and the number of rows in MDV_{obs} is determined by the detection sensitivity of the instrument.

Consider a compound containing a single atom of each of the three elements (X, Y, Z), each of which has a heavy isotope (^1X, ^1Y, ^1Z) with a mass 1 unit higher than the naturally more abundant isotope (^0X, ^0Y, ^0Z). If ^1X is supplied during a labeling experiment, then the observed mass isotopomer abundances for XYZ must be corrected for the natural abundance of the Y and Z isotopes. The observed mass isotopomer abundances have contributions from the following isotopomers:

$$[M]^+ \qquad ^0X\,^0Y\,^0Z$$

$$[M+1]^+ \qquad ^0X^1Y^0Z + {}^0X^0Y^1Z + {}^1X^0Y^0Z$$

$$[M+2]^+ \qquad ^0X^1Y^1Z + {}^1X^1Y^0Z + {}^1X^0Y^1Z$$

$$[M+3]^+ \qquad ^1X^1Y^1Z$$

The relative contributions of these isotopomers to the observed abundances depend on the natural abundances of the Y and Z isotopes — $p(^0X)$, etc. — and the relative abundance of ^0X and ^1X in the labeled molecule. The latter corresponds to the values of $[M]^+$ and $[M+1]^+$ after correction for the natural abundance of Y and Z. This leads to a matrix relationship:

$$\begin{bmatrix} [M]^+ \\ [M+1]^+ \\ [M+2]^+ \\ [M+3]^+ \end{bmatrix}_{obs} = \begin{pmatrix} p(^0Y)p(^0Z) & 0 \\ p(^1Y)p(^0Z)+p(^0Y)p(^1Z) & p(^0Y)p(^0Z) \\ p(^1Y)p(^1Z) & p(^1Y)p(^0Z)+p(^0Y)p(^1Z) \\ 0 & p(^1Y)p(^1Z) \end{pmatrix} \begin{bmatrix} [M]^+ \\ [M+1]^+ \end{bmatrix}_{cor}$$

(5.4)

This equation can be rearranged as shown in Eq. (5.2), allowing MDV_{cor} to be calculated from MDV_{obs} and the generalized inverse of the correction matrix.

Note that the size of the matrix is determined by the number of detectable mass isotopomers and the number of X atoms that can be labeled [23, 136]. It is advantageous to make use of all reliable measurements, even if they provide redundant information, but making extra measurements for a given compound may limit the number of different fragments that can be measured during the dwell time – most current mass spectrometers allow for 30 ions to be monitored and the time associated with each ion will be reduced as more are monitored. So if the $[M+2]^+$ and $[M+3]^+$ mass isotopomers are undetectable, Eq. (5.4) would reduce to

$$\begin{bmatrix} [M]^+ \\ [M+1]^+ \end{bmatrix}_{obs} = \begin{pmatrix} p(^0Y)p(^0Z) & 0 \\ p(^1Y)p(^0Z) + p(^0Y)p(^1Z) & p(^0Y)p(^0Z) \end{pmatrix} \begin{bmatrix} [M]^+ \\ [M+1]^+ \end{bmatrix}_{cor} \quad (5.5)$$

This can be re-written as

$$\begin{bmatrix} [M]^+ \\ [M+1]^+ \end{bmatrix}_{obs} = \begin{pmatrix} p(^0Y) & 0 \\ p(^1Y) & p(^0Y) \end{pmatrix} \begin{pmatrix} p(^0Z) & 0 \\ p(^1Z) & p(^0Z) \end{pmatrix} \begin{bmatrix} [M]^+ \\ [M+1]^+ \end{bmatrix}_{cor} \quad (5.6)$$

emphasizing that the correction matrix can be broken down into elemental matrices, C_Y and C_Z, as indicated in Eq. (5.3).

The elemental correction matrices in Eq. (5.6) apply to a fragment with only a single atom of each element, Y and Z. If there is more than one atom of each element, then the fractional abundance or probabilities of the naturally abundant isotopomers that contribute to the observed mass isotopomers can be calculated from the combinatorial probability equation [23, 132]:

$$\text{Fractional abundance} = N! \cdot \prod_{i=1}^{n} \left(\frac{p(I_i)^{f(I_i)}}{f(I_i)!} \right) \quad (5.7)$$

Here, n is the number of naturally occurring isotopes, $I_1,\ldots,I_n$ of the element under consideration, $p(I_i)$ is the natural abundance of the isotope I_i (Table 5.2), $f(I_i)$ is the number of atoms of I_i in the fragment that is being analyzed, and N is the total number of atoms of the element in the fragment.

The correction for natural abundance can be illustrated with a specific example. Assume that the observed ion abundances for the M-57 fragment of TBDMS-serine (Fig. 5.16) are $[M]^+ = 225,000$; $[M+1]^+ = 100,000$; $[M+2]^+ = 53,000$; $[M+3]^+ = 90,000$.

Table 5.2 Natural abundances for some commonly encountered elements [99]

Element	$[M+0]^+$	$[M+1]^+$	$[M+2]^+$	$[M+4]^+$
Carbon	0.9893	0.0107		
Hydrogen	0.999885	0.000115		
Nitrogen	0.99632	0.00368		
Oxygen	0.99757	0.00038	0.00205	
Silicon	0.922297	0.046832	0.03087	
Sulfur	0.9493	0.0076	0.0429	0.0002

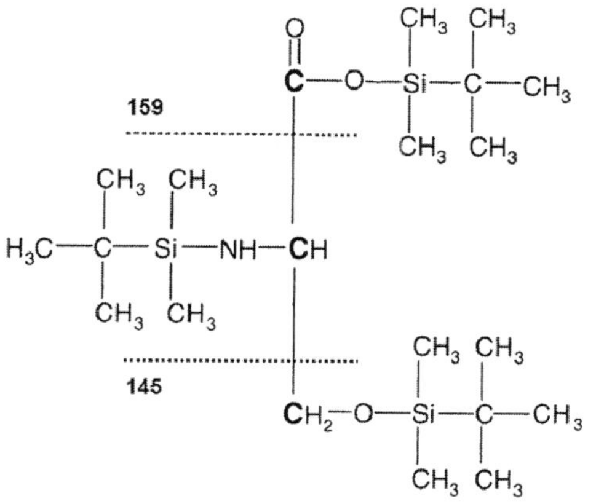

5 Quantification of Isotope Label

Fig. 5.16 The chemical structure of serine-TBDMS showing fragmentation of the amino acid backbone. Fragments M-15 and M-57 correspond to losses of methyl and t-butyl groups respectively; while fragments M-145 and M-159 correspond to the loss of carbon 3 and carbon 1 respectively of the serine backbone. Here the loss of the amino acid side chain corresponds to a loss of 145

Structure annotations: $C_{21}H_{49}O_3N_1Si_3 = 447.3$

Common Losses:
- M−15 = 432.3
- M−57 = 390.2
- M−85 = 362.2
- M−side chain = 302.2
- M−159 = 288.2

These values define an MDV_{obs} in which the relative abundances sum to 1:

$$MDV_{obs} = \begin{bmatrix} 0.4808 \\ 0.2137 \\ 0.1132 \\ 0.1923 \end{bmatrix} \tag{5.8}$$

The M-57 fragment contains 14 carbon atoms from the derivatizing agent and since the observed fragment ions extend to $[M+3]^+$, it is necessary to construct a carbon matrix that allows for contributions to the ion abundances from up to three ^{13}C atoms:

$$CM_C = \begin{pmatrix} ^{12}C_{14} & 0 & 0 & 0 \\ ^{12}C_{13}{}^{13}C_1 & ^{12}C_{14} & 0 & 0 \\ ^{12}C_{12}{}^{13}C_2 & ^{12}C_{13}{}^{13}C_1 & ^{12}C_{14} & 0 \\ ^{12}C_{11}{}^{13}C_3 & ^{12}C_{12}{}^{13}C_2 & ^{12}C_{13}{}^{13}C_1 & ^{12}C_{14} \end{pmatrix} \tag{5.9}$$

The probabilities of these combinations can be calculated from (5.7) using the fractional abundances of ^{12}C and ^{13}C given in Table 5.2:

$$CM_C = \begin{pmatrix} 0.8602 & 0 & 0 & 0 \\ 0.1302 & 0.8602 & 0 & 0 \\ 0.0092 & 0.1302 & 0.8602 & 0 \\ 0.0004 & 0.0092 & 0.1302 & 0.8602 \end{pmatrix} \tag{5.10}$$

Correction matrices can be generated for each of the other elements in the fragment in a similar fashion. The overall correction matrix is the product of each elemental matrix and this turns out to be

$$\mathrm{CM_{CHNOSi}} = \begin{pmatrix} 0.6644 & 0 & 0 & 0 \\ 0.2081 & 0.6644 & 0 & 0 \\ 0.1003 & 0.2081 & 0.6644 & 0 \\ 0.0213 & 0.1003 & 0.2081 & 0.6644 \end{pmatrix} \quad (5.11)$$

Finally, the $\mathrm{MDV_{cor}}$ is calculated from the inverse of the correction matrix and $\mathrm{MDV_{obs}}$ using (5.2):

$$\mathrm{MDV_{cor}} = \begin{bmatrix} 0.6626 \\ 0.0870 \\ 0.0288 \\ 0.2216 \end{bmatrix} \quad (5.12)$$

Note that the sum of the values in $\mathrm{MDV_{cor}}$ has been rescaled to 1, to allow for round-off errors caused by truncating the natural abundance values in the calculation of the probabilities for the correction matrix. The more precise the mathematical accounting and the estimate of natural abundance (see [99]), the less is the need for scaling. However, scaling is always recommended, particularly if the data are to be used for flux modeling.

The assumptions behind this correction strategy are described by Lee et al. [67]. In particular, it is assumed that (i) the relative abundance of any isotopomer is equivalent to the probability of finding that isotopomer in the entire population of possible isotopomers and (ii) the occurrence of a particular isotope at a particular position in a molecule is independent of the presence/absence of all other isotopes elsewhere in the molecule. These assumptions allow the distribution of label to be treated as a multinomial probability problem (5.7).

Corrections for natural abundance are routinely and systematically implemented through the use of matrix algebra developed in the literature [23, 37, 84, 132, 143]. The correction for all isotopes of a particular element should be performed in a single step [132], and the corrections can be implemented through the use of a standard spreadsheet or mathematical software packages available upon request [37, 123, 136, 152].

Generally, natural abundance correction is applied only to the atoms that cannot be labeled during the experiment. The natural abundance associated with other atoms, for example the backbone carbon atoms in a ^{13}C-labeling experiment, is built into the model that describes the metabolic redistribution of the label. However, several authors [37, 98] have developed approaches to correct for natural abundance in the atoms that are labeled in an experiment, as well as those that cannot be labeled. This strategy entails the use of extra labeled standards that can be costly, and it is usually unnecessary in flux analysis when the fluxes have to be simulated anyway.

After correction for natural abundance, it may also be necessary to adjust for the contribution of unlabeled original biomass to the MDV [82, 130, 150]. Suppose that the analytical sample used to analyze the labeling of serine above was contaminated with 5% of the original biomass. The MDV ($\mathrm{MDV_{biomass}}$) for this original biomass can be calculated using (5.7):

5 Quantification of Isotope Label

$$\text{MDV}_{\text{biomass}} = \begin{bmatrix} {}^{12}C_3 \\ {}^{12}C_2{}^{13}C \\ {}^{12}C{}^{13}C_2 \\ {}^{13}C_3 \end{bmatrix} \cdot 5\% = \begin{bmatrix} 0.9682 \\ 0.0314 \\ 0.0003 \\ 0.0000 \end{bmatrix} \cdot 0.05 = \begin{bmatrix} 0.0484 \\ 0.0016 \\ 0.0000 \\ 0.0000 \end{bmatrix} \quad (5.13)$$

$\text{MDV}_{\text{biomass}}$ can then be subtracted from MDV_{cor}, to give, after scaling, the MDV that is corrected for both natural abundance and original biomass ($\text{MDV}_{\text{final}}$):

$$\text{MDV}_{\text{final}} = \begin{bmatrix} 0.6626 \\ 0.0870 \\ 0.0288 \\ 0.2216 \end{bmatrix} - \begin{bmatrix} 0.0484 \\ 0.0016 \\ 0.0000 \\ 0.0000 \end{bmatrix} = \begin{bmatrix} 0.6142 \\ 0.0854 \\ 0.0288 \\ 0.2216 \end{bmatrix} \Rightarrow \begin{bmatrix} 0.6465 \\ 0.0899 \\ 0.0303 \\ 0.2333 \end{bmatrix} \quad (5.14)$$

Note that the biomass correction can be applied before the correction for natural abundance if preferred, but it is then necessary to include atoms other than carbon in $\text{MDV}_{\text{biomass}}$.

Finally, particular care is required if the mass ion is affected by the loss of an ion, for example the loss of a proton leading to the detection of $[M-1]^+$. Correction for natural abundance in this situation is discussed elsewhere [60, 72]. Sometimes it is better to avoid using the mass isotopomer data from such ions [23], athough the loss of a proton is characteristic of a number of compounds (e.g., sugars) that contain very useful label information.

5.5 Concluding Remarks

Although many steady-state flux analyses are based on quantifying the redistribution of stable isotopes using either NMR or MS, it should be emphasized that the two approaches provide complementary information. In fact, it has been repeatedly demonstrated that combining NMR and MS measurements leads to more reliable flux analysis [15, 16, 23, 60, 61, 149]. In particular, a combined approach invariably provides a more complete picture of the isotopomer distribution at steady state and thus helps to improve the definition of the flux map. Confirmation of this can be found in a recent flux analysis of *Saccharomyces cerevisiae* using GCMS, LCMS, and NMR data on the redistribution of ^{13}C-label [61]. This study found that three methods varied significantly in the extent to which they could define the flux distribution around specific metabolic nodes, with each of the methods being the best approach in particular instances.

While there are clear advantages in combining NMR and MS for flux analysis, it is likely that personal preferences, access to equipment, and the availability of expertise will continue to hinder the adoption of such an analytical strategy. This is not necessarily a disadvantage, since either technique is fully capable of providing useful information on its own and a sensitivity analysis will demonstrate the quality of the deduced flux map. NMR will continue to have the advantages of minimal sample handling and direct insight into labeling pathways via the readily obtainable information on positional labeling; while MS methods will continue to have the advantage of sensitivity and rapid data collection. Sensitivity will be a key issue as the experimental focus switches from steady-state analysis to dynamic analysis of time courses, and for this purpose MS will be the method of choice in most instances.

References

1. Allen DK, Shachar-Hill Y, Ohlrogge JB (2007) Compartment-specific labeling information in ^{13}C metabolic flux analysis of plants. Phytochemistry 68:2197–2210.
2. Alonso AP, Vigeolas H, Raymond P, Rolin D, Dieuaide-Noubhani M (2005) A new substrate cycle in plants. Evidence for a high glucose-phosphate-to-glucose turnover from in vivo steady-state and pulse-labeling experiments with [^{13}C]glucose and [^{14}C]glucose. Plant Physiol 138:2220–2232.
3. Antoniewicz M, Kelleher JK, Stephanopoulos G (2007) Accurate assessment of amino acid mass isotopomer distributions for metabolic flux analysis. Anal Chem 79:7554–7559.
4. Ashcroft AE (1997) Ionization Methods in Organic Mass Spectrometry. Royal Society of Chemistry, Cambridge UK.
5. Bacher A, Le Van Q, Keller PJ, Floss HG (1983) Biosynthesis of riboflavin. Incorporation of ^{13}C-labeled precursors into the xylene ring. J Biol Chem 258: 13431–13437.
6. Bentley R, Saha NC, Sweeley CC (1965) Separation of protium and deuterium forms of carbohydrates by gas chromatography. Anal Chem 37:1118–1122.
7. Berger S, Braun S (2004) 200 and More NMR Experiments. Wiley-VCH Verlag, Weinheim, Germany.
8. Berger TA (1996) Separation of a gasoline on an open tubular column with 1.3 million effective plates. Chromatographia 42:63–71.
9. Bergner EA, Lee WNP (1995) Testing gas chromatographic/mass spectrometric systems for linearity of response. J Mass Spectrom 30:778–780.
10. Beylot M, David F, Brunengraber H (1993) Determination of the ^{13}C-labeling pattern of glutamate by gas chromatography-mass spectrometry. Anal Biochem 212:532–536.
11. Blakeney AB, Harris PJ, Henry RJ, Stone BA (1983) A simple and rapid preparation of alditol acetates for monsaccharide analysis. Carbohydrate Res 113:291–299.
12. Bonarius HPJ, Schmid G, Tramper J (1997) Flux analysis of underdetermined metabolic networks: the quest for the missing constraints. Trends Biotechnol 15:308–314.
13. Browse J, McCourt PJ, Somerville CR (1986) Fatty-acid composition of leaf lipids determined after combined digestion and fatty-acid methyl-ester formation from fresh tissue. Anal Biochem 152:141–145.
14. Brunengraber H, Kelleher JK, Des Rosiers C (1997) Applications of mass isotopomer analysis to nutrition research. Annu Rev Nutr 17:559–596.
15. Chatham JC, Bouchard B, Des Rosiers C (2003) A comparison between NMR and GCMS ^{13}C-isotopomer analysis in cardiac metabolism. Mol Cell Biochem 249:105–112.
16. Christensen B, Nielsen J (1999) Isotopomer analysis using GC-MS. Met Eng 1: 282–290.
17. Christensen B, Nielsen J (2000) Metabolic network analysis of *Penicillium chrysogenum* using ^{13}C-labeled glucose. Biotechnol Bioeng 68:652–659.
18. Christie WW (2003) Lipid Analysis. Oily Press, Bridgwater.
19. Coates SW, Gurney Jr T, Sommers LW, Yeh M, Hirschberg CB (1980) Subcellular localization of sugar nucleotide synthetases. J Biol Chem 255:9225–9229.
20. Comte B, Vincent G, Bouchard B, Jette M, Cordeau S, Des Rosiers C (1997) A ^{13}C mass isotopomer study of anaplerotic pyruvate carboxylation in perfused rat hearts. J Biol Chem 272:26125–26131.

21. Costenoble R, Muller D, Barl T, van Gulik WM, van Winden WA, Reuss M, Heijnen JJ (2007) ^{13}C-Labeled metabolic flux analysis of a fed-batch culture of elutriated *Saccharomyces cerevisiae* FEMS Yeast Res 7:511–526.
22. da Silva PMFR, Eastmond PJ, Hill LM, Smith AM, Rawsthorne S (1997) Starch metabolism in developing embryos of oilseed rape. Planta 203:480–487.
23. Dauner M, Sauer U (2000) GC-MS analysis of amino acids rapidly provides rich information for isotopomer balancing. Biotechnol Prog 16:642–649.
24. de Graaf AA, Mahle M, Möllney M, Wiechert W, Stahmann P, Sahm H (2000) Determination of full ^{13}C isotopomer distributions for metabolic flux analysis using heteronuclear spin echo difference spectroscopy. J Biotechnol 77:20–35.
25. Des Rosiers C, Montgomery JA, Descrochers S, Garneau M, David F, Mamer OA, Brunegraber H (1988) Interference of 3-hydroxyisobutyrate with measurements of ketone body concentration and isotopic enrichment by gas chromatography-mass spectrometry. Anal Biochem 173:96–105.
26. Des Rosiers C, Di Donato L, Comte B, Laplante A, Marcoux C, David F, Fernandez CA, Brunengraber H (1995) Isotopomer analysis of citric acid cycle and gluconeogenesis in rat liver. Reversibility of isocitrate dehydrogenase and involvement of ATP-citrate lyase in gluconeogenesis. J Biol Chem 270:10027–10036.
27. Di Donato L, Des Rosiers C, Montgomery JA, David F, Garneau M, Brunengraber H (1993) Rates of gluconeogenesis and citric acid cycle in perfused livers, assessed from the mass spectrometric assay of the ^{13}C labeling pattern of glutamate. J Biol Chem 268:4170–4180.
28. Dieuaide-Noubhani M, Raffard G, Canioni P, Pradet A, Raymond P (1995) Quantification of compartmented metabolic fluxes in maize root tips using isotope distribution from ^{13}C- or ^{14}C-labeled glucose. J Biol Chem 270: 13147–13159.
29. Dieuaide-Noubhani M, Canioni P, Raymond P (1997) Sugar-starvation-induced changes of carbon metabolism in excised maize root tips. Plant Physiol 115:1505–1513.
30. Edwards S, Nguyen BT, Do B, Roberts JKM (1998) Contribution of malic enzyme, pyruvate kinase, phospho*enol*pyruvate carboxylase, and the Krebs cycle to respiration and biosynthesis and to intracellular pH regulation during hypoxia in maize root tips observed by nuclear magnetic resonance and gas chromatography-mass spectrometry. Plant Physiol 116: 1073–1081.
31. Eisenreich W, Ettenhuber C, Laupitz R, Theus C, Bacher A (2004) Isotopolog perturbation techniques for metabolic networks: metabolic recycling of nutritional glucose in *Drosophila melanogaster*. Proc Natl Acad Sci USA 101:6764–6769.
32. Ettenhuber C, Radykewicz T, Kofer W, Koop HU, Bacher A, Eisenreich W (2005) Metabolic flux analysis in complex isotopolog space. Recycling of glucose in tobacco plants. Phytochemistry 66:323–335.
33. Fagerquist CK, Schwarz JM (1998) Gas-phase acid-base chemistry and its effects on mass isotopomer abundance measurements of biomolecular ions. J Mass Spectrom 33:144–153.
34. Fagerquist CK, Neese RA, Hellerstein MK (1999) Molecular ion fragmentation and its effects on mass isotopomer abundances of fatty acid methyl esters ionized by electron impact. J Am Soc Mass Spectrom 10:430–439.
35. Fagerquist CK, Hellerstein MK, Faubert D, Bertrand MJ (2001) Elimination of the concentration dependence in mass isotopomer abundance mass spectrometry of methyl palmitate using metastable atom bombardment. J Am Soc Mass Spectrom 12:754–761.
36. Fan TWM (1996) Metabolite profiling by one- and two-dimensional NMR analysis of complex mixtures. Prog Nucl Magn Reson Spectrosc 28:161–219.
37. Fernandez CA, Des Rosiers C, Previs SF, David F, Brunengraber H (1996) Correction of 13C mass isotopomer distributions for natural stable isotope abundance J Mass Spectrom 31:255–262.
38. Fernie AR, Roscher A, Ratcliffe RG, Kruger NJ (2001) Fructose 2,6-bisphosphate activates pyrophosphate: fructose-6-phosphate 1-phospho-transferase and increases triose phosphate to hexose phosphate cycling in heterotrophic cells. Planta 212:250–263.

39. Feuge RO, Gros AT (1949) Modification of vegetable oils. 7. Alkali catalyzed interesterification of peanut oil with ethanol. J Amer Oil Chem Soc 3:97–102.
40. Filer CN (1999) Isotopic fractionation of organic compounds in chromatography. J Labeled Cpd Radiopharm 42:169–197.
41. Fischer E, Sauer U (2003) Metabolic flux profiling of *Escherichia coli* mutants in central carbon metabolism using GC-MS. Eur J Biochem 270:880–891.
42. Fountoulakis M, Lahm HW (1998) Hydrolysis and amino acid composition analysis of proteins. J Chromatog A 826:109–134.
43. Freeman R (2003) Magnetic Resonance in Chemistry and Medicine. Oxford University Press, Oxford.
44. Frick O, Wittmann C (2005) Characterization of the metabolic shift between oxidative and fermentative growth in *Saccharomyces cerevisiae* by comparative ^{13}C flux analysis. Microb Cell Fact 4:Art. 30.
45. Glawischnig E, Girl A, Tomas A, Bacher A, Eisenreich W (2002) Starch biosynthesis and intermediary metabolism in maize kernels. Quantitative analysis of metabolite flux by nuclear magnetic resonance. Plant Physiol 130:1717–1727.
46. Grob RL (1985) Modern Practice of Gas Chromatography. John Wiley & Sons, New York, Chichester, Brisbane, Toronto, Singapore.
47. Harrison AG, Cotter RJ (1990) Methods of ionization. Meth Enzymol 193:3–37.
48. Hatch MD, Slack CR (1966) Photosynthesis by sugar cane leaves – a new carboxylation reaction and pathway of sugar formation. Biochem J. 101:103–111.
49. Hellerstein MK, Neese RA, Linfoot P, Christiansen M, Turner S, Letscher A (1997) Hepatic gluconeogenic fluxes and glycogen turnover during fasting in humans. A stable isotope study. J Clin Invest 100:1305–1319.
50. Henry PG, Adriany G, Deelchand D, Gruetter R, Marjanska M, Öz G, Seaquist ER, Shestov A, Uğurbil K (2006) In vivo ^{13}C NMR spectroscopy and metabolic modelling in the brain: a practical perspective. Magn Reson Imaging 24:527–539.
51. Hore PJ (1995) Nuclear Magnetic Resonance. Oxford University Press, Oxford.
52. Huege J, Sulpice R, Gibon Y, Lisec J, Koehl K, Kopka J (2007) GC-EI-TOF-MS analysis of in vivo carbon-partitioning into soluble metabolite pools of higher plants by monitoring isotope dilution after ^{13}CO$_2$ labeling. Phytochemistry 68:2258–2272.
53. Husek P, Macek K (1975) Gas-chromatography of amino acids. J Chromatog 113:139–230.
54. Husek P, Simek P (2001) Advances in amino acid analysis. LC GC North America 19: 986–999.
55. Jeffrey FMH, Roach JS, Storey CJ, Sherry AD, Malloy CR (2002) ^{13}C Isotopomer analysis of glutamate by tandem mass spectrometry. Anal Biochem 300:192–205.
56. Jin ES, Jones JG, Merritt M, Burgess SC, Malloy CR, Sherry AD (2004) Glucose production, gluconeogenesis, and hepatic tricarboxylic acid cycle fluxes measured by nuclear magnetic resonance analysis of a single glucose derivative. Anal Biochem 327: 149–155.
57. Jin H, Pfeffer PE, Douds DD, Piotrowski E, Lammers PJ, Shachar-Hill Y (2005) The uptake, metabolism, transport and transfer of nitrogen in an arbuscular mycorrhizal symbiosis. New Phytol 168:687–696.
58. Kiefer P, Nicolas C, Letisse F, Portais JC (2007) Determination of carbon labeling distribution of intracellular metabolites from single fragment ions by ion chromatography tandem mass spectrometry. Anal Biochem 360:182–188.
59. Kitson FG, Larsen BS, McEwen CN (1996) Gas Chromatography and Mass Spectrometry. Academic Press, San Diego.
60. Klapa MI, Aon JC, Stephanopoulos G (2003) Systematic quantification of complex metabolic flux networks using stable isotopes and mass spectrometry. Eur J Biochem 270:3525–3542.
61. Kleijn RJ, Geertman JMA, Nfor BK, Ras C, Schipper D, Pronk JT, Heijnen JJ, van Maris AJA, van Winden WA (2007) Metabolic flux analysis of a glycerol-overproducing

Saccharomyces cerevisiae strain based on GC-MS, LC-MS and NMR-derived ^{13}C-labelling data. FEMS Yeast Res 7:216–231.
62. Kopka J, Ohlrogge JB, Jaworski JG (1995) Analysis of in vivo levels of acyl-thioesters with gas chromatography/mass spectrometry of the butyl amide derivative. Anal Biochem 224:51–60.
63. Krishnan P, Kruger NJ, Ratcliffe RG (2005) Metabolite fingerprinting and profiling in plants using NMR. J Exp Bot 56:255–265.
64. Kruger NJ, Huddleston JE, Le Lay P, Brown ND, Ratcliffe RG (2007) Network flux analysis: Impact of ^{13}C-substrates on metabolism in *Arabidopsis thaliana* cell suspension cultures. Phytochemistry 68:2176–2188.
65. Künnecke B, Cerdan S, Seelig J (1993) Cerebral metabolism of [1,2-^{13}C$_2$]glucose and [U-^{13}C$_4$]3-hydroxybutyrate in rat brain as detected by ^{13}C NMR spectroscopy. NMR Biomed 6:264–277.
66. Lane AN, Fan TWM (2007) Quantification and identification of isotopomer distributions of metabolites in crude cell extracts using ^{1}H TOCSY. Metabolomics 3:79–86.
67. Lee WNP, Byerley LO, Berger EA (1991) Mass isotopomer analysis: theoretical and practical considerations. Biol Mass Spectrom 20:451–458.
68. Lee WNP, Bergner EA, Guo ZK (1992) Mass isotopomer pattern and precursor product relationship. Biol Mass Spectrom 21:114–122.
69. Levitt MH (2002) Spin Dynamics. Basics of Nuclear Magnetic Resonance. Wiley-VCH Verlag, Weinheim.
70. Lichtenthaler HK, Schwender J, Disch A, Rohmer M (1997) Biosynthesis of isoprenoids in higher plant chloroplasts proceeds via a mevalonate independent pathway. FEBS Lett 400:271–274.
71. London RE (1988) ^{13}C labelling in studies of metabolic regulation. Prog Nucl Magn Reson Spectrosc 20:337–383.
72. Low IA, Liu RH, Barker SA, Fish F, Settine RL, Piotrowski EG, Damert WC, Liu JY (1985) Selected ion monitoring mass spectrometry: parameters affecting quantitative determination. Biomed Mass Spectrom 12:633–637.
73. Luo B, Groenke K, Takors R, Wandrey C, Oldiges M (2007) Simultaneous determination of multiple intracellular metabolites in glycolysis, pentose phosphate pathway and tricarboxylic acid cycle by liquid chromatography-mass spectrometry. J Chromatog A 1147:153–164.
74. MacLeod JK, Flanigan IL, Williams JF, Collins JG (2001) Mass spectrometric studies of the path of carbon in photosynthesis: positional isotopic analysis of ^{13}C-labelled C$_4$ to C$_7$ sugar phosphates. J Mass Spectrom 36:500–508.
75. Malone JG, Mittova V, Ratcliffe RG, Kruger NJ (2006) The response of carbohydrate metabolism in potato tubers to low temperature. Plant Cell Physiol 47:1309–1322.
76. Marx A, de Graaf AA, Wiechert W, Eggeling L, Sahm H (1996) Determination of the fluxes in the central metabolism of *Corynebacterium glutamicum* by nuclear magnetic resonance spectroscopy combined with metabolite balancing. Biotechnol Bioeng 49:111–129.
77. Massou S, Nicolas C, Letisse F, Portais J-C (2007) Application of 2D-TOCSY NMR to the measurement of specific ^{13}C-enrichments in complex mixtures of ^{13}C-labeled metabolites. Metab Eng 9:252–257.
78. Massou S, Nicolas C, Letisse F, Portais J-C (2007) NMR-based fluxomics: quantitative 2D NMR methods for isotopomer analysis. Phytochemistry 68:2330–2340.
79. Matthews DE, Hayes JM (1976) Systematic errors in gas chromatography-mass spectrometry isotope ratio measurements. Anal Chem 48:1375–1382.
80. McLafferty FW (1959) Mass spectrometric analysis-molecular rearrangements. Anal Chem 31:82–87.
81. Meier-Augenstein W, Watt PW, Langhans CD (1996) Influence of gas chromatographic parameters on measurement of ^{13}C/^{12}C isotope ratios by gas-liquid chromatography-combustion isotope ratio mass spectrometry. I. J Chromatog 752:233–241.

82. Nanchen A, Fuhrer T, Sauer U (2006) Determination of metabolic flux ratios from ^{13}C-experiments and gas chromatography-mass spectrometry data: protocol and principles. In: Weckwerth W (ed.) Methods in Molecular Biology, Vol. 358, Metabolomics: Methods and Protocols. Humana Presss, Totowa, NJ, pp. 177–198.
83. Niessen WMA (2001) Current Practice of Gas Chromatography-Mass Mpectrometry. Marcel Dekker, New York Basel.
84. Park SM, Sinskey AJ, Stephanopoulos G (1997) Metabolic and physiological studies of *Corynebacterium glutamicum* mutants. Biotechnol Bioeng 55:864–879.
85. Patterson BW, Carraro F, Wolfe RR (1993) Measurement of ^{15}N enrichment in multiple amino acids and urea in a single analysis by gas chromatography/mass spectrometry. Biol Mass Spectrom 22:518–523.
86. Pawlosky RJ, Sprecher HW, Salem N (1992) High-sensitivity negative-ion GC-MS method for detection of desaturated and chain-elongated products of deuterated linoleic and linolenic acids. J Lipid Res 33:1711–1717.
87. Petersen S, de Graaf AA, Eggeling L, Möllney M, Wiechert W, Sahm H (2000) In vivo quantification of parallel and bidirectional fluxes in the anaplerosis of *Corynebacterium glutamicum*. J Biol Chem 275:35932–35941.
88. Price NPJ (2004) Acyclic sugar derivatives for GC/MS analysis of ^{13}C-enrichment during carbohydrate metabolism. Anal Chem 76:6566–6574.
89. Prinsen E, Van Dongen, W, Esmans EL, Van Onckelen HA (1997) HPLC linked electrospray tandem mass spectrometry: a rapid and reliable method to analyzed indole-3-acetic acid metabolism in bacteria. J Mass Spectrom 32:12–22.
90. Rantanen A, Rousu J, Kokkonen JT, Tarkiainen V, Ketola RA (2002) Computing positional isotopomer distributions from tandem mass spectrometric data. Met Eng 4:285–294.
91. Ratcliffe RG (1994) In vivo NMR studies of higher plants and algae. Adv Bot Res 20: 43–123.
92. Ratcliffe RG, Shachar-Hill Y (2001) Probing plant metabolism with NMR. Annu Rev Plant Physiol Plant Mol Biol 52:499–526.
93. Ratcliffe RG, Shachar-Hill Y (2006) Measuring multiple fluxes through plant metabolic networks. The Plant J 45:490–511.
94. Ratcliffe RG, Roscher A, Shachar-Hill Y (2001) Plant NMR spectroscopy. Prog Nucl Magn Reson Spectrosc 39:267–300.
95. Roessner U, Wagner C, Kopka J, Trethewey RN, Willmitzer L (2000) Simultaneous analysis of metabolites in potato tuber by gas chromatography-mass spectrometry. Plant J 23: 131–142.
96. Rontein D, Dieuaide-Noubhani M, Dufourc EJ, Raymond P, Rolin D (2002) The metabolic architecture of plant cells. Stability of central metabolism and flexibility of anabolic pathways during the growth cycle of tomato cells. J Biol Chem 277: 43948–43960.
97. Roscher A, Kruger NJ, Ratcliffe RG (2000) Strategies for metabolic flux analysis in plants using isotope labelling. J Biotechnol 77:81–102.
98. Rosenblatt J, Chinkes D, Wolfe M, Wolfe RR (1992) Stable isotope tracer analysis by GC-MS, including quantification of isotopomer effects. Am J Physiol 263:E584–596.
99. Rosman KJR, Taylor PDP (1998) Isotopic composition of the elements. Pure Appl Chem 70:217–235.
100. Rousu J, Rantanen A, Ketola RA, Kokkonen JT (2005) Isotopomer distribution computation from tandem mass spectrometric data with overlapping fragment spectra. Spectroscopy 19:53–67.
101. Sauer U, Hatzimanikatis V, Bailey JE, Hochuli M, Szyperski T, Wüthrich K (1997) Metabolic fluxes in riboflavin-producing *Bacillus subtilis*. Nature Biotechnol 15: 448–452.
102. Schaefer J, Stejskal EO, Beard CF (1975) Carbon-13 nuclear magnetic resonance analysis of metabolism in soybeans labelled by $^{13}CO_2$. Plant Physiol 55:1048–1053.

103. Schmidt K, Carlsen M, Nielsen J, Villadsen J (1997) Modeling isotopomer distributions in biochemical networks using isotopomer mapping matrices. Biotechnol Bioeng 55: 831–840.
104. Schmidt K, Nielsen J, Villadsen J (1999) Quantitative analysis of metabolic fluxes in *Escherichia coli*, using two-dimensional NMR spectroscopy and complete isotopomer models. J Biotechnol 71:175–190.
105. Schoeller DA (1980) Model for determining the influence of instrumental variations on the long-term precision of isotope dilution analyses. Biomed Mass Spectrom 7:457–463.
106. Schwender J, Ohlrogge JB (2002) Probing in vivo metabolism by stable isotope labeling of storage lipids and proteins in developing *Brassica napus* embryos. Plant Physiol 130: 347–361.
107. Schwender J, Ohlrogge JB, Shachar-Hill Y (2003) A flux model of glycolysis and the oxidative pentose phosphate pathway in developing *Brassica napus* embryos. J Biol Chem 278:29442–29453.
108. Schwender J, Ohlrogge J, Shachar-Hill Y (2004) Understanding flux in plant metabolic networks. Curr Opin Plant Biol 7:309–317.
109. Schwender J, Shachar-Hill Y, Ohlrogge JB (2006) Mitochondrial metabolism in developing embryos of Brassica napus. J Biol Chem 281:34040–34047.
110. Sgoutas DS (1966) Isotope fractionation of methyl esters by gas-liquid chromatography. Nature 211:296–297.
111. Shastri AA, Morgan JA (2007) A transient isotopic labeling methodology for ^{13}C metabolic flux analysis of photoautotropic microorganisms. Phytochemistry 68:2302–2312.
112. Silverstein RM, Webster FX, Kiemle D (2005) Spectrometric identification of organic compounds. Academic Press, New York.
113. Sonntag K, Eggeling L, de Graaf AA, Sahm H (1993) Flux partitioning in the split pathway of lysine synthesis in *Corynebacterium glutamicum*. Quantification by ^{13}C- and ^{1}H-NMR spectroscopy. Eur J Biochem 213:1325–1331.
114. Sriram G, Shanks JV (2004) Improvements in metabolic flux analysis using carbon bond labelling experiments: bondomer balancing and Boolean function mapping. Metab Eng 6:116–132.
115. Sriram G, Fulton DB, Iyer VV, Peterson JM, Zhou R, Westgate ME, Spalding MH, Shanks JV (2004) Quantification of compartmented metabolic fluxes in developing soybean embryos by employing biosynthetically directed fractional ^{13}C labelling, two dimensional [^{13}C, ^{1}H] nuclear magnetic resonance, and comprehensive isotopomer balancing. Plant Physiol 136:3043–3057 (Erratum: Plant Physiol 142:1771).
116. Sriram G, Fulton DB, Shanks JV (2007) Flux quantification in central carbon metabolism of *Catharanthus roseus* hairy roots by ^{13}C labelling and comprehensive bondomer balancing. Phytochemistry 68:2243–2257.
117. Sriram G, Iyer VV, Fulton DB, Shanks JV (2007) Identification of hexose hydrolysis products in metabolic flux analytes: A case study of levulinic acid in plant protein hydrolysate. Metab Eng 9:442–451.
118. Stellaard F, Paumgartner G (1985) Measurement of isotope ratios in organic compounds at picomole quantities by capillary gas chromatography/quadrupole electron impact mass spectrometry. Biomed Mass Spectrom 12:560–564.
119. Sweeley CC, Elliott WH, Fries I, Ryhage R (1966) Mass spectrometric determination of unresolved components in gas chromatographic effluents. Anal Chem 38:1549–1553.
120. Szyperski T (1995) Biosynthetically directed fractional ^{13}C-labeling of proteinogenic amino acids. An efficient analytical tool to investigate intermediary metabolism. Eur J Biochem 232:433–448.
121. Szyperski T (1998) ^{13}C-NMR, MS and metabolic flux balancing in biotechnology research. Quart Rev Biophys 31:41–106.
122. Takayama M (1995) Metastable McLafferty rearrangement reaction in the electron impact ionization of stearic acid methyl ester. Int J Mass Spectrom Ion Proc 144:199–204.

123. Talwar P, Wittmann C, Lengauer T, Heinzle E (2003) Software tool for automated processing of ^{13}C labeling data from mass spectrometric spectra. Biotechniques 35:1214–1215.
124. Tesch M, de Graaf AA, Sahm H (1999) In vivo fluxes in the ammonium-assimilatory pathways in *Corynebacterium glutamicum* studied by ^{15}N nuclear magnetic resonance. Appl Environ Microbiol 65:1099–1109.
125. Toya Y, Ishii N, Hiasawa T, Naba M, Hirai K, Sugawara K, Igarashi S, Shimizu K, Tomita M, Soga T (2007) Direct measurement of isotopomer of intracellular metabolites using capillary electrophoresis time-of-flight mass spectrometry for efficient metabolic flux analysis. J Chromatog A 1159:134–141.
126. Troufflard S, Roscher A, Thomasset B, Barbotin JN, Rawsthorne S, Portais JC (2007) In vivo ^{13}C NMR determines metabolic fluxes and steady state in linseed embryos. Phytochemistry 68:2341–2350.
127. Tulloch AP, Hogge LR, (1985) Investigation of the formation of MH$^+$ and other ions in the mass spectrum of methyl decanoate using specifically deuterated decanoates. Chem Phys Lipids 37:271–281.
128. van Dam JC, Eman MR, Frank J, Lange HC, van Dedem GWK, Heijnen JJ (2002) Analysis of glycolytic intermediates in *Saccharomyces cerevisiae* using anion exchange chromatography and electrospray ionization with tandem mass spectrometric detection. Anal Chim Acta 460:209–218.
129. Van Hook WA (1969) Isotope separation by gas chromatography. Adv Chem Ser 89:99–118.
130. van Winden W, Schipper D, Verheijen P, Heijnen J (2001) Innovations in generation and analysis of 2D [^{13}C, ^{1}H] COSY NMR spectra for metabolic flux analysis purposes. Metab Eng 3:322–343.
131. van Winden W, Heijnen JJ, Verheijen PJT (2002) Cumulative bondomers: A new concept in flux analysis from 2D [^{13}C, ^{1}H] COSY NMR data. Biotechnol Bioeng 80:731–745.
132. van Winden WA, Wittmann C, Heinzle E, Heijnen JJ (2002) Correcting mass isotopomer distributions for naturally occurring isotopes. Biotechnol Bioeng 80:477–479.
133. van Winden WA, van Dam, JC, Ras C, Kleijn RJ, Vinke JL, van Gulik WM, Heijnen JJ (2005) Metabolic-flux analysis of *Saccharomyces cerevisiae* CEN.PK113-7D based on mass isotopomer measurements of ^{13}C-labeled primary metabolites. FEMS Yeast Res 5:559–568.
134. Vandenheuvel WJA, Cohen JS (1970) Gas-liquid chromatography-mass spectrometry of C-13 enriched amino acids as trimethylsilyl derivatives. Biochim Biophys Acta 208:251–259.
135. Vandenheuvel WJA, Smith JL, Putter I, Cohen JS (1970) Gas-liquid chromatography and mass spectrometry of deuterium-containing amino acids as their trimethylsilyl derivatives. J Chromatog 50:405–412.
136. Wahl SA, Dauner M, Wiechert W (2004) New tools for mass isotopomer data evaluation in 13C flux analysis: mass isotope correction, data consistency checking and precursor relationships. Biotechnol Bioeng 259–268.
137. Wendisch VF, de Graaf AA, Sahm H (1997) Accurate determination of ^{13}C enrichments in nonprotonated carbon atoms of isotopically enriched amino acids by ^{1}H nuclear magnetic resonance. Anal Biochem 245:196–202.
138. Wiechert W, Möllney M, Isermann N, Wurzel W, de Graaf AA (1999) Bidirectional reaction steps in metabolic networks: III. Explicit solution and analysis of isotopomer labeling systems. Biotechnol Bioeng 66:69–85.
139. Wiechert W, Möllney M,, Petersen S, de Graaf AA (2001) A universal framework for ^{13}C metabolic flux analysis. Metab Eng 3:265–283.
140. Willker W, Flögel U, Leibfritz D (1997) Ultra-high-resolved HSQC spectra of multiple ^{13}C-labeled biofluids. J Magn Reson 125:216–219.
141. Wilzbach KE, Riesz P (1957) Isotope effects in gas-liquid chromatography. Science 126:748–749.
142. Wittmann C (2007) Fluxome analysis using GC-MS. Microbiol Cell Fact 6:Art. 6.
143. Wittmann C, Heinzle E (1999) Mass spectrometry for metabolic flux analysis. Biotechnol Bioeng 62:739–750.

144. Wittmann C, Heinzle E (2001) Application of MALDI-TOF MS to lysine-producing *Corynebacterium glutamicum*. Eur J Biochem 268:2441–2455.
145. Wittmann C, Heinzle E (2001) MALDI-TOF MS for quantification of substrates and products in cultivations of *Corynebacterium glutamicum*. Biotechnol Bioeng 72:642–647.
146. Wittmann C, Heinzle E (2001) Modeling and experimental design for metabolic flux analysis of lysine-producing Corynebacteria by mass spectrometry. Met Eng 3:173–191.
147. Wittmann C, Hans M, Heinzle E (2002) In vivo analysis of intracellular amino acid labelings by GC/MS. Anal Biochem 307:379–382.
148. Wu ZP (2004) Determination of phenylalanine isotope ratio enrichment by liquid chromatography/time-of-flight mass spectrometry. Eur J Mass Spectrom 10:619–623.
149. Yang C, Hua Q, Shimizu K (2002) Quantitative analysis of intracellular metabolic fluxes using GC-MS and two dimensional NMR spectroscopy. J Biosci Bioeng 93:78–87.
150. Yang C, Hua Q, Shimizu K (2002) Metabolic flux analysis in *Synechocystis* using isotope dilution from ^{13}C-labeled glucose. Metab Eng 4:202–216.
151. Yazdi-Samadi B, Rinne RW, Seif RD, (1977) Components of developing soybean seeds: oil, protein, sugars, starch, organic acids, and amino acids. Agron J 69:481–486.
152. Zamboni N, Fischer E, Sauer U (2005) Fiat Flux: a software for metabolic flux analysis from ^{13}C-glucose experiments. BMC Bioinformatics 6:209.

Chapter 6
Data Integration

Aaron Fait and Alisdair R. Fernie

6.1 Introduction

In the last decade, an unprecedented amount of post-genomic experimental information has become available. Datasets originating from transcriptomic analysis, metabolite profiling, and proteomics can be produced faster, with ever increasing accuracy and decreasing cost. However, putting the pieces together is not trivial. Our understanding of cellular phenomena based on "omics" data depends on – and is limited by – our capability to implement appropriate analysis tools able to integrate the different "omics" approaches [75, 78]. Bringing together such disparate datasets presents a considerable challenge [76]. Such analysis is time consuming and prone to both error and speculation. Consequently, there is a substantial need to consider both the methods currently being used and the statistical principles involved in the analysis of post-genomic experimental data.

Metabolic profiling has become a commonly used tool to characterize a plant under different environmental, developmental, or genetic conditions. The metabolome, the complete set of small-molecule metabolites in a system of interest,[1] is a dynamic entity. Its quantitative composition results from the combined actions of gene regulation, gene expression, and enzyme activity. It thus can be seen as a manifestation of the biological information flow. Several studies have shown that small changes in transcript and enzyme levels have significant effect on the concentration of the metabolic intermediates but only limited effects on the metabolic fluxes [39]. Thus, the metabolome more sensitively responds to genetic and environmental changes than fluxes. Furthermore, a substantial part of the regulation of the metabolic steady state is being processed at the posttran-

---

[1]Common-used instruments for metabolomics can identify only a fraction of the whole metabolic variability within a cell at once. This is mainly due to the extreme diversity in the chemical nature and abundance exposed by biological systems. Specifically in the plant kingdom, roughly hundred thousands of metabolites, primary and secondary, are estimated to occur.

A. Fait (✉)
Ben-Gurion University of the Negev, Jacob Blaustein Insts. for Desert Research, French Associates Institute for Agriculture & Biotechnology of Drylands, Midreshet Ben-Gurion, 84990, Israel
e-mail: fait@bgu.ac.il

scriptional and posttranslational levels. As a result, the regulation of the metabolic network cannot be predicted solely from the genomic sequence or transcriptome analysis [21, 44, 83]. For instance, when the tricarboxylic acid (TCA) cycle in tomato plants is impaired by reducing the activity of one of its enzymes, succinyl coA ligase, 2-oxoglutarate is diverted to a metabolic bypass known as the γ-aminobutyrate (GABA) shunt [74]. Increased GABA shunt activity is eventually measured. However, when a broad gene expression analysis of the transgenic lines was performed, the authors measured only a minor increase in transcripts of the GABA shunt key enzyme glutamate decarboxylase (GAD). On the other hand, high-throughput analysis of primary metabolism revealed a major accumulation of its product, GABA. Enzymatic essays and stable isotope-based feeding experiments confirmed the increase in the activity of GAD and glutamate dehydrogenase (GDH) and in the enrichment of Glu-derived ^{13}C-GABA [74]. These results corroborate the metabolite profile results and reflect the posttranslational nature of the regulation of GABA biosynthesis [5].

Nowadays, the accuracy of analysis and metabolic coverage of the metabolome is constantly increasing, and a systems approach to metabolism is becoming realistic [7, 17, 60, 86 see also Chapter 3]. Parallel high-throughput analysis of transcripts, metabolites, and proteins from one sample has become routine [4, 51]. Data integration across these different data types can reveal system-wide regulatory principles. This has necessitated the development of new algorithms to analyze complex data matrices as well as databases, annotation tools, and pathway mappings, in order to facilitate maximal knowledge acquisition [55]. In the following sections, we will discuss methodologies and possible pitfalls of complex data management, including data standardization, the problems of outliers and missing data, and non-linear features of metabolism. We will additionally highlight clustering tools, co-response analysis as well as network and pathway reconstruction.

6.2 Overcoming the Noise

6.2.1 Data Normalization

Non-biological variability in experimental data can be reduced by data normalization. For example, running samples at different days on an analytical instrument will contribute to technical variability, which may, e.g., be caused by changes in detector sensitivity of the instrument. To reduce bias of the data analysis by this kind of technical variability, randomization of the order in which the samples are measured as well as inclusion of reference samples is important. During subsequent data elaboration, normalization of the elements building the data matrix aims at filtering out the non-biological contribution to the variance of the dataset. For a variable x with n elements, *median normalization* can be achieved by dividing each element x_i ($i = 1 \ldots n$) by the median across the n categories. The median as a statistical measure is preferred to the average since it is more robust to outliers. In comparative analysis, where control samples are available, data can be standardized by dividing the relative response of each element of a sample, x_i, by the relative response of the corresponding element in the control (c) sample, x_i^c, or the median response in the control sample [64]. A similar approach is used for microarray data [89]. After normalization, data are typically transformed into logarithmic form (the choice of the base is irrelevant). Transformation by logarithm adds symmetry to positive and negative changes within the dataset as well as minimizes the impact of outlier entries.

A different approach to normalize the dataset is the use of *unit vector normalization*. This method was used by Scholz et al. [66] to normalize data from direct infusion Q-TOF MS fingerprinting. Unit vector normalization is done by scaling each sample vector to the unit vector norm [64]. Geometrically interpreted, this means that the length of each sample vector is scaled to one, and each sample can be interpreted as a projection onto a hypersphere. Here vectors that have similar direction are found close to each other, although the un-scaled vectors may differ in intensity.

Z-score transformation provides an additional standardizing tool across a wide range of experiments and separate measurements. By definition, Z-score transformation of a sample vector results in a mean of zero and a standard deviation of one. Comparisons across samples or experiments are performed on the transformed data, and changes in gene expression or metabolite levels are expressed as differences between Z scores (Z ratios). Z-score transformation allows the comparison of, for instance, microarray data independent of the original hybridization intensities [12]. Normalized data can be used in the calculation of significant changes in transcript/metabolite abundance between different samples and conditions. Comparative analysis of data originating from different instruments/ measurements is thus possible.

When approaching the experimental data, often the assumption is made that biological data follow a parametric distribution (e.g., normal distribution) and that skewing in the data distribution is solely of technical origin. However, several studies have shown that metabolic data can follow a non-parametric distribution [19, 38, 42, 85]. Normalization of such data can result in the loss of biologically significant information. In order to avoid this, bootstrapping tools can be used for the elaboration of non-normalized data showing skewed parametrics. Non-parametric methods should otherwise be implemented for the analysis of bimodal or non-parametric data, like small samples, for which population parameters are not estimable (Section 6.3).

Several statistical methods, known as goodness of fit tests, exist to test the nature of the distribution of experimental data (e.g., normal, log-normal, Weibull, exponential). Examples of such tests are the chi-square goodness-of-fit test, which can be applied to discrete distributions such as the binomial and the Poisson, the Kolmogorov–Smirnov and Anderson–Darling tests, which are restricted to continuous distributions, and the Shapiro–Wilk test, which specifically tests whether a random sample, $x_1, x_2, ..., x_n$ comes from a normally distributed population.

6.2.2 Outliers Detection

Additional pretreatment of data prior to statistical analysis includes the detection of outliers. These are defined as atypical observations, namely data points that do not appear to follow the characteristic distribution of the rest of the data. The Grubbs' test or the kurtosis test [47] can identify such outliers in order to be removed from the dataset. In addition to the problem of outliers, the interpretation of intrinsically small values may be problematic if they are not well above background noise. While a small change in intensity might have high biological significance, background noise may render these small values useless [56].

6.2.3 Missing Values

Similar to outliers, a common phenomenon in metabolomics measurements is the occurrence of missing values in the data matrix. A certain metabolite may be present in low concentration under the detection limit in most samples while being detectable only under a certain condition, e.g., a certain genetic alteration. Deciding a priori to entirely delete such a metabolite in the dataset will obviously cause a significant loss of potentially relevant information. Since many multivariate methods require a complete data matrix without absentees, estimation of missing values might be necessary. In result, methods of missing value imputation can significantly improve clustering and interpretation of large datasets. In a comparative study of a number of advanced imputation methods on microarray datasets, Tuikkala et al. [80] concluded that, regardless of the evaluation approach, imputation always gave a better reproduction of the original gene clusters or biological interpretation, than ignoring missing values. A simple method for missing data imputation is replacement of a missing entry by the mean of the values of the whole data row in the data matrix. More complex is the use of a k-nearest neighbors algorithm. However, such linear methods can have limitations for non-linear data structures. Non-linear pattern of data distribution was shown to occur in instances such as temporal response to cold stress [38] or as part of natural variation [19]. Thus, a different approach was proposed by Scholz et al. [65]. The authors compared different linear and non-linear missing values algorithms (Fig. 6.1) to test their efficiency in estimating missing values in an artificial non-linear dataset. The authors suggested an inverse non-linear principle component analysis (PCA[2]) algorithm to extract the non-linear component from a highly incomplete dataset (artificially produced). The algorithm performed far better for estimation of missing values than the other methods used, except for the self-organizing map (SOM) algorithm (Fig. 6.1). The latter was also successfully used in the analysis of the relationship between biomass and metabolites profile in an *Arabidopsis* RIL population with 6% of the data missing [49]. In the analysis of the non-linear dataset by Scholz et al. [65], linear approaches such as PCA

---

[2] See Section 6.3.2 for an introduction to PCA.

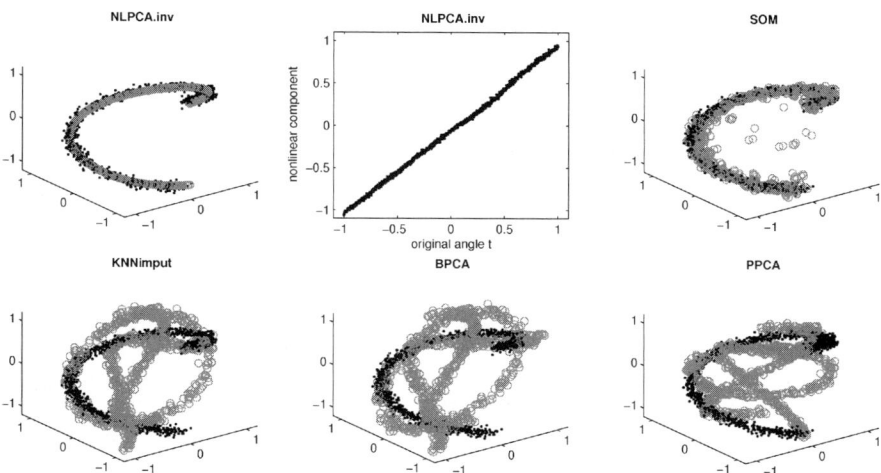

Fig. 6.1 Test of different missing value algorithms on artificial data with non-linear structure. A total of 1,000 three-dimensional samples were generated (*black dots*). Consecutively, each of the three-dimensional samples was removed and then re-estimated by a missing value algorithm (*grey circles*). The inverse NLPCA, followed by SOM, is able to extract the non-linear component from highly incomplete dataset (see text), and hence it can give a very good estimation of the missing values. In contrast, linear approaches BPCA and PPCA, as well as the k-nearest neighbor-based approach KNNimpute, fail with this non-linear dataset. (Reprinted with permission from [65])

and k-nearest neighbor algorithm (KNN) displayed the lowest predictive efficiency. When the different algorithms were tested against a real experimental dataset, different percentages of values were randomly removed and regarded as missing for the estimation test.

In this case, the linear methods were shown to be highly comparable to the non-linear methods and, to a certain degree, even with higher estimation efficiency. Nonetheless, in a successive analysis, it was shown that the performance of each approach fairly depends on the linearity of the dataset. The authors conclude that in the analysis of datasets with a limited number of variables (metabolic profiles) and which are non-linearly structured, non-linear techniques can improve the missing value estimation performances. In larger datasets (gene expression matrices), best missing data estimations were achieved by the Bayesan and probabilistic principle component analysis, BPCA and PPCA, respectively.

6.3 The Analysis of the Dataset

6.3.1 Correlation Analysis

Both transcripts and metabolites are usually characterized by high natural variation. Correlation analysis rather than average-based statistics can provide a more reliable and less noise-dependent statistical analysis. Indeed, like other post-genomic techniques, the data produced by metabolic profiling is usually in the form of ratios, while measuring absolute concentrations is limited to targeted analysis. Furthermore, the correlation-based approach has also a biological reasoning behind: the remarkable and organized biological variability in metabolite abundance is suggested to reflect the intrinsic flexibility of metabolic networks [72, 73, 81, 86]. In this regard, a perturbation which propagates through the metabolic net is expressed as measurable pattern of correlations among metabolites [50]. In other words, metabolites are produced from other metabolites, and this creates a network of interdependencies based on the stoichiometry of the metabolic pathways involved.

It should be noted however that metabolic co-regulation does not depend solely on sharing a biochemical pathway but rather on being part of all reactions and regulatory interactions present in a system. In this regard, several studies across different biological systems have shown the occurrence of weak, but widespread, correlations between metabolites on neighboring biochemical pathways, while few strong correlations characterized the pairs of metabolites – some of which are distant in biochemical steps [18, 48, 58]. A recent co-response analysis of metabolite and transcript data [16, 63] suggests a novel regulatory mechanism for the production of GABA. The GABA shunt has been classically defined as a bypass of the TCA cycle. Intriguingly, by querying the Comprehensive Systems-Biology Database (http://csbdb.mpimp-golm.mpg.de/) for the response pattern of genes of the GABA shunt and of the TCA cycle, the authors found a highly specific and positive co-response, suggesting that these two pathways act in a concerted rather than alternate manner [16]. However, when analyzing the metabolite profiles of a tomato introgression lines collection for the metabolite:metabolite associations, GABA was shown to cluster deeply within the amino acid group, suggesting the occurrence of an N-metabolism driven co-regulation. It is notable that, within this class of N-rich metabolites, Glu – a direct biochemical precursor of GABA – is located on a statistical scale more distant from GABA than other amino acids biochemically less related [16]. Finally, biological systems can differ in their set of co-responding metabolites further complicating the understanding of general regulatory mechanisms at the basis of metabolites levels. For example, in a comparative analysis of different tissue types in three species, *Arabidopsis thaliana*, *Solanum tuberosum*, and *Nicotiana tabacum*, it was found that the extent of correlation among metabolite pairs changed in magnitude (or even in sign, e.g., Glc:putrescine and Glc:Gly) between different organs, reflecting different regulatory networks [50]. While theoretical and computational analysis suggests that correlation is shaped by a combination of stoichiometric and kinetic elements [73], the examples reported suggest that the factors accountable for metabolic co-variation are yet to be fully understood.

Regardless of the cause of the correlation between a set of parameters, let us briefly address the basics of correlation analysis. Correlation and regression are often confused. While the algebraic formulas are quite similar, the structure of the data and the purpose of the investigation may make one or the other of the techniques inappropriate. In regression, the aim is to describe the dependence of a variable Y on an *independent* variable X. We employ regression analysis to explain Y (and its variation) in terms of X. On the other hand, in correlation analyses, we aim to determine the co-variance of two variables; in other words, we test if two variables are interdependent as an effect of a common cause. For example, we might want to understand the relationship between different sugars, as their content is affected by developmental stages. The structure of the data can be misleading. We might want to explain the changes in sucrose content as a function of tomato fruit weight by applying regression analysis. However, none of the variables was deliberately chosen; thus, none is independent. Therefore, either we will need to implement a modification to the commonly used regression analysis (defined as Model II, Sokal and Rohlf [71]) or, if it is the degree of association between X and Y that is of interest, we should employ a correlation analysis. Correlation coefficients are dependent on the structure and size of the dataset under analysis. In correlation analysis, rank-order correlation like Spearman or Kendall test statistics should be used for non-linear relationships. These non-parametric methods for correlation analysis are less sensitive to outliers in the dataset. Furthermore, they are appropriate also for experiments with limited number of data points, where a normal distribution cannot be estimated and thus parametric methods like Pearson correlation analysis, which assumes normal distribution of the data, cannot be implemented. Nonetheless, among the many correlation coefficients available in statistics, the most common in use is the product–moment correlation coefficient, the Pearson correlation, which assumes normal distribution of the data.

Once measured, an association between a pair of variables (e.g., metabolite:metabolite level, metabolite:transcript level) must be proved as significant. A test of significance for correlation coefficients determines whether a sample correlation coefficient could have come from a population with a parametric correlation coefficient of zero. The null hypothesis is tested as a t-test with $n-2$ degrees of freedom:

$$Ts = r\left[\frac{n-2}{1-r^2}\right]^{0.5} \tag{6.1}$$

where r is the correlation coefficient and n is the sample size. In a multiple comparison approach, as is the case for the analysis of metabolite:metabolite relationships in metabolic profiling studies, the recurring inference of "false" correlation can be avoided by P-value correction (e.g., Bonferroni correction).

Sometimes we are interested to discern between correlation coefficients in several samples. A relevant example would be to test the degree of homogeneity of correlation between metabolites in different mutants or transgenic lines as affected by changes in environmental conditions. A test of homogeneity among two or more correlation coefficients may help us to determine whether these can be considered samples from a population exhibiting a common correlation among variables. Without entering into details, we suggest the interested reader to consult biometry [71]. In correlation analysis of extensive datasets and multiple tests, we might incur into what are called non-sense correlations. In other words, often a significant correlation can be confused with causation. This means, significant correlation can lead to terribly false conclusions. One variable can be the entire or partial cause of another; in other instances, both variables have a common cause(s); finally, the correlation might result as a complex interaction of both direct and indirect common causes. The distinction between the different types of correlation is particularly acute in metabolic data, where metabolites are converted into other metabolites or can regulate the level of other, possibly distant – on a scale of biochemical steps – metabolites. A useful approach to identify common causes of relationships between variables would be to measure partial correlation coefficients. In this approach, we measure the correlation between any pair of variables when others have been held constant. Via this approach, we can test the possibility of common causes of a series of variable by excluding the one kept constant.

The described abundance-based correlation analysis is commonly used in network reconstruction from gene expression data [40] and metabolite profiles data [41]. However, it might not be optimal in the analysis of metabolite profile data for the reasons we already mentioned: the spike character often observed in metabolite abundance data and consequent departure from normal Gaussian distribution; the intrinsic redundant nature of metabolic profiling datasets; and "false-positive" scores for significant correlation. Other more complex forms of correlation analysis have been thus developed. *Canonical correlation analysis* (CCA) is a multivariate statistical model that facilitates the study of interrelationships among sets of multiple dependent variables and multiple independent variables (for in depth covering of CCA, the reader is referred to Hair et al. [24]). CCA calculates the highest possible correlation between linear combinations of each set of variables. The correlation thus found is called canonical correlation and the corresponding linear combination, canonical variate. Another objective of CCA is explaining the nature of the relationships existing between the variables, by measuring the relative contribution of each to the canonical function. Backdrops of CCA are the sensitivity to sample size (small dataset will not represent the correlations well; large size could create inaccurate significant instances); and the assumption of linearity between any two variables; normal distribution of the data is not compulsory, but highly recommended. The power of CCA in analyzing multivariate datasets and the relationships among variables was shown by Meyer et al. [49] to test the hypothesis that changes in biomass are expressed in – and can be predicted from – the metabolite composition. Using *Arabidopsis thaliana* RIL populations, the authors conducted a parallel analysis of biomass and metabolite profiles. By implementing CCA, the authors show that major global changes in metabolism are the result of variation in growth rather than vice versa. By comparing the data matrices, the authors found that the compound changes of a combination of the levels of a large number of metabolites, rather than few individual metabolites, showed to be highly associated with changes in biomass. Variation in growth, thus, coincides with characteristic combinatorial changes of metabolite levels, whereas individual metabolites may fluctuate largely independently of alterations in growth. In fact when pair-wise correlation was applied, it yielded a mere 7% explanation of the total variance observed in biomass. By further analyzing the metabolites highly ranked in CCA, the authors found that central-metabolism-derived intermediates were significantly represented, specifically those of the hexose phosphate pool.

A different computational protocol, based on complex correlations, was recently developed to cope with aforementioned pitfalls in metabolite-abundance-based correlation analysis [20]. Here, associations between metabolites of an *Arabidopsis* RIL population were not inferred based on correlation matrices between abundance data, but rather between metabolites' Quantitative Trait Locus (QTL) profiles (vectors of *P*-values associated with markers along the genome for each mass). In principle, once associations between metabolites are drawn, the most relevant relationships can be isolated by second-order correlations. These are defined as correlation between metabolite pairs independent of covariance with any other pair. This method was successfully implemented by Keurentjes et al. [41] to reconstruct glucosinolate biosynthetic pathway. The resulting correlation network was found to reasonably represent the known pathway for glucosinolate formation, validating the method used.

Correlation analysis is thus an efficient method to extract biological meaningful information from the complex interactions between multiple molecular compounds and mechanisms stretched at different levels and which give rise to a certain cellular response.

6.3.2 Principle and Independent Component Analyses

Principle component analysis (PCA) and independent component analysis (ICA) are unsupervised ("blind") methods to analyze data, unbiased from the experimental target knowledge [88]. They are used to extract biologically relevant information from multidimensional complex datasets and to reduce data dimensionality. A high-throughput experiment can consist of a number of conditions (e.g., genotypes, developmental stages, treatments) and a number of variables (e.g., metabolites, genes) with hundreds to thousands of entries. The analysis of the resulting data matrix necessitates algorithms aimed at expressing its multidimensional nature by a set of meaningful new variables (components). Generally speaking, components are sought that best reflect a specific aspect of the study, e.g., genotype- or development-induced changes within the dataset. The data is then often visualized in two dimensions by projection of the original data into two selected components. Eventually, original variables can be identified which highly contribute to each component. They can be detected by ranking according to the respective weights in each component (loadings, eigenvalues).

The classic multivariate analysis used is principle component analysis (PCA). The data containing n variables are expressed with n new variables (PCs). Geometrically, the n-multidimensional data are transformed into in a new coordinate system in a way that the highest variability (variance) is displayed by the first axis (1st PC). Along the following axes (PCs) of increasing order, the variance successively diminishes. Often, most of the variability in the dataset is captured by the first two or three components, i.e., the data can be visualized without much loss of information.

PCA was successfully used by Taylor et al. [77], to discriminate between parent lines and progeny in an *Arabidopsis* ecotypes Col0*C24 crossing. The scores of the original variables enabled the detection of variables which contributed to a significant section of the variance within the dataset. These corresponded to the metabolites citrate and malate on principle component 1 and the sugars glucose and fructose on principle component 2. These results were in line with the fact that the genetic differences in the crossings were the maternally inherited mitochondria and chloroplast, key sites for the production of the above mentioned metabolites. The output of PCA was also used to look for noisy metabolites, metabolites with great variance among samples, or those in which outliers were present. PCA analysis was also shown to efficiently discriminate between developmental stages during seed maturation in *Arabidopsis* [15]. The analysis, in combination with ANOVA, was employed to determine what metabolites significantly contributed to the metabolic shift measured during the developmental process and identify a previously not reported metabolic switch [23] at the transition between late maturation and desiccation [15]. This event was characterized by the accumulation of specific free metabolic intermediates, which decreased during the very early stages of germination and are presumably involved in the reactivation of the metabolic processes [15].

Similar to PCA, independent component analysis (ICA) is an unsupervised method to analyze data [88]. In recent years, it has become increasingly popular as an alternative to PCA. However, in ICA, different components represent non-overlapping information, thought to be caused by unrelated processes or causes, thus termed independent components. ICA differs from other factor analysis in that it considers the non-Gaussianity (or super-Gaussianity) of the biological data, thus providing a more flexible model to represent them. In biological systems, data often behave in a non-Gaussian manner. For instance, variables that are characterized by more values close to zero and/or very large ones are said to have a super-Gaussian distribution. The probability density thus is peaked at zero and has heavy tails. Scholz et al. [66] proposed a PCA preprocessing coupled to ICA to elaborate data arising from metabolic fingerprinting. This PCA processing aims at reducing the numbers of variables, since ICA is best suited to low number of variables against high number of samples. On the given experimental dataset, the authors have shown advantages of ICA in the detection of biological features as well as technical factors, not detected by a comparative PCA analysis. In a different study [62], ICA was used to analyze co-regulation of genes in endometrial cancer cells. The analysis employing ICA showed increased biological relevance and robustness as compared to other statistical approaches, filtering out elements of noise and unrelated patterns. ICA can be found as freely available web-based tool at MetaGeneAlyse (http://metagenealyse.mpimp-golm.mpg.de).

6.3.3 Clustering Techniques

Clustering techniques allow classification of objects into different groups according to similarity and visualization in dendrograms. A dataset is partitioned into subsets (clusters), typically in an iterative procedure. Similarity between data subsets is defined according to some distance measure. Clustering techniques can reveal co-response of metabolites/genes in a given biological process or to an environmental signal. Clustering can also be used to distinguish between biological defined entities, e.g., genotypes/environmental conditions. Statistical clustering approaches can be supervised or unsupervised [25]. Un-supervised procedures define homogeneous clusters from a given collection of objects or statistical units. The cluster structure is then interrogated for biological/experimental significance. On the contrary, in supervised classification, the groups are a priori inferred. The choice of a specific algorithm for the analysis can be a cumbersome task given the increasing number of alternatives, tuning parameters, threshold values, etc. It is important to note that results can vary based on the chosen method [14, 56]. Consequently, decision making on the statistical method to be used in the elaboration of a given dataset is crucial. Datta and Datta [13] proposed a protocol to evaluate the efficiency of a clustering method on a given dataset in producing biologically meaningful clusters. Two measures are proposed by the authors. The first is a biological homogeneity index (BHI), which measures the biological homogeneity of the clusters on a given set of data. The second performance measure is defined as biological stability index (BSI), which measures the reproducibility of the clustering output on a specific dataset through a number of test repetitions. By comparing ten different clustering algorithms, the authors conclude that the proposed measures are useful in identifying the optimal clustering algorithm, in terms of biological-based discriminative efficiency, for *each given* dataset, since no single clustering method can be generally applied to all datasets. When choosing a specific algorithm, the distribution nature of the dataset must be taken into account.

Clustering analysis is a complex task which involves data pre-processing (see Section 6.2), choosing of an appropriate measure of similarity, selection of a clustering algorithm, and finally, careful interpretation of the results: First, *data pre-processing* (the way of scaling or weighting, see Section 6.2.1) is influential on the final results and interpretation of the analysis. Of similar influence is the choice of the *similarity measure* which defines the inter-observation distances. Among the distance functions are the Manhattan metric or the squared distance functions. By choosing to cluster categories (e.g., genotypes or treatments) instead of observations, a measure of association (e.g., correlation coefficients) between every pair of category should be used. Furthermore, there are numerous different *clustering algorithms* available, which fall under hierarchical or k-means clustering (see also Chapter 7). Hierarchical algorithms find successive clusters using previously established clusters. Agglomerative hierarchical clustering starts with all data elements separately. Clusters are then obtained by successively merging existing elements/clusters. In divisive hierarchical clustering, existing clusters are divided successively. In biological research concerned with grouping of entities (species, metabolite, genes), hierarchical clustering might be the method of choice, although in certain instances clustering per se of biological entities could have no clear biological meaning.

In addition to the different distance functions between the observations used in clustering, there are different measures to define the distance between clusters. Among those are the single linkage method (minimum distance between elements of two clusters), the complete linkage method (maximum distance between elements of two clusters), or the average linkage method (mean distance between elements of two clusters). The single and the complete linkage methods can yield different results when applied to the same dataset. The first tends to create fewer distinct clusters than the second, thus finding more compact and smaller clusters. The average linkage method is an intermediary option between the two (for more detailed description of the different methods see Johnson [34]). After the cluster analysis is completed, the reliability of the dendrogram can be investigated by statistical methods (see Levenstien et al. [46] or Gnanadesikan [22] and references therein).

6.4 Bridging Platforms: Correlation-Based Integration of Large Datasets of Different Origin

The metabolic network interacts with and is modeled by many other cellular processes, such as transcriptional regulation and protein–protein interaction dynamics [75]. The study of metabolic networks from transcript data provides important knowledge on regulatory metabolic processes in a wide range of systems from *Arabidopsis* seeds [61] to *Caenorhabditis elegans* [29], from which novel information can be acquired on the impact of environmental signals and genetic alteration on gene expression and metabolic adjustment [8, 26, 28, 45, 67]. Nowadays, integration of multiple matrices describing the structure and kinetics of the system under study with extensive post-genomic datasets is, however, possible only for few microbial model organisms and limited to their primary metabolism. In yeast [52], the transcriptional regulatory architecture of metabolic networks as an expression of environmentally induced perturbations was investigated. The analysis (correlation based) led to insights in the relationship between environment, metabolites, and gene expression. The grouping into subnetworks (modules) with biological significance was possible when the experimental data were integrated with structural data of the very well-characterized yeast metabolic network, using the KEGG database (see also Section 6.5.2). In this way, the authors could identify differential response of the metabolic network to environmental signals. For example, stress responses to temperature shocks and amino acid starvation induced a larger number of genes scattered on the metabolic map, while the impact of nitrogen depletion or stationary phase was of more repressing nature. Metabolic subnetwork responses to specific conditions were also identified. Namely, heat shock had specific impact on carbohydrate metabolism, while lipid and amino acid metabolism were repressed specifically under nitrogen depletion experiments and energy metabolism for hyperosmotic shocks. The authors, further, show that expression responses to certain conditions are significantly correlated, e.g., alternative carbon sources and hypoosmotic shock, nitrogen depletion and stationary phases form exemplary pairs.

Considering the fact that 50% of the genes with maximum response (induction or repression) to given conditions were shown to have no assigned molecular function, this co-response analysis can lead to the identification of novel regulatory elements and modules in response to environmental signals.

Given the importance of posttranslational and allosteric regulation in governing metabolic processes, the understanding of cellular physiology is incomplete without information on the metabolite steady-state levels and fluxes. An analytical approach based solely on gene expression data and structural information of the system, while ignoring the interdependency between enzymatic regulation, metabolite levels, and fluxes [54], is prone to speculation. Combined analysis enables to elucidate gene:gene or gene:metabolite/metabolite:gene networks and to describe different patterns of regulation, from metabolic to hierarchical, namely enzyme production/activity. Modules or co-regulated entities in response to environmental signals or genetic manipulation might then be detected. Post-genomic studies will thus increasingly focus on the integration of different level of analysis via bioinformatics tools. However, in the process of association between genes and reactions, one must keep in mind that not all genes have one-to-one relationship with their corresponding enzymes or metabolic reactions. Examples are genes that code for subunits of proteins governing one reaction or those that encode for so-called promiscuous enzymes that can catalyze several different reactions.

The major goal of combining data originating in different platforms is eventually to obtain an overall quantitative description of cellular systems, the core idea of systems biology. The limit for this to be achieved is the large parameter space due to the high number of components and interactions. Reducing dimensionality is thus a priority to get started with inter-platform analysis and isolate key points in the numerous cellular processes.

To analyze datasets of different origin, e.g., microarray data and GC-MS profiles, co-responsive pattern of change in gene expression and metabolite level can be identified by non-parametric Spearmans' rank order correlation analysis. In one of the first parallel analysis of transcript and metabolic profiles in plants, Urbanczyk-Wochniak et al. [84] have identified three classes of correlation between transcripts and metabolites in *Arabidopsis thaliana*: (a) confirmatory correlations, in line with previously identified co-responses; (b) correlation with a functional basis and can be retrospectively assigned; and (c) novel correlations unrelated to the biochemical pathway, where the gene product operates. Intriguingly, the majority of the cases fell in the third class suggesting a non-linear, non-causative, and non-trivial relation between metabolome and transcriptome in response to changing environmental and/or genetic conditions.

Similar conclusions were drawn when studying the response of *Arabidopsis* to low temperature exposures, resulting in cold acclimation. Kaplan et al. [37] performed parallel analyses of transcript and metabolite changes during cold acclimation to study the dynamics of gene–metabolite relationships. PCA revealed temporally dependent, global changes in both gene expression and metabolite profiles, in response to the stress conditions. Changes in transcript abundance reflected changes in many metabolic processes, and part of these temporally correlated with changes in metabolite levels. For other metabolic processes, however, this was not the case, suggesting that regulatory processes independent of transcript abundance represent a significant portion of the metabolic adjustments during cold acclimation. In a different study, Hirai and coworkers [30] have analyzed the time-dependent changes in the metabolome and transcriptome of *Arabidopsis* plants subjected to sulfur deprivation. In their study, around 2,000 metabolites were detected, by targeted and non-targeted analysis, and 21,500 transcripts. Among these, variables which did not show significant changes were filtered out prior to the analysis. Multivariate analysis, specifically batch-learning self-organizing map (BL-SOM), was used for the integrated analysis on the combined log-normalized dataset. This method classified metabolites and transcripts according to their time-dependent pattern of changes in accumulation and expression, building a matrix of cells in a two-dimensional lattice. Each cell (or neighboring cells) contains entities showing similar patterns of change [30]. Using this approach, the authors could show that a group of metabolites/genes regulated by the same mechanism clustered together, e.g., glucosinolates metabolism and glucosinolate biosynthesis genes and sulfotransferases or genes involved in sulfur assimilation and *O*-acetylserine, a positive regulator of these genes. The BL-SOM algorithm implements a characteristic non-linear projection from the high-dimensional space of input data onto a two-dimensional array of weight vectors [1, 43]; the reduction in dimensions is achieved by PCA analysis [3]. This strategy was applied successfully to classify complex data in diverse system; for instance, in the analysis of global characteristics of genome sequences to identify species-specific genome signatures [2].

Carrari et al. [9, 10] investigated tomato fruit development on a broad scale, measuring a total of 92 metabolites comprising sugars, sugar alcohols, organic acids, amino acids, vitamins, and a select few secondary metabolites in addition to pigments and the monosaccharide composition of the cell wall, in parallel to transcript

Table 6.1 Online freely available resources for metabolic network reconstruction

	Reconstruction tools
MetaFluxNet	http://mbel.kaist.ac.kr/mfn
MFAML	http://mbel.kaist.ac.kr/mfaml
MapMan	http://gabi.rzpd.de/projects/MapMan/
CentiBiN	http://centibin.ipk-gatersleben.de/
PATIKA	http://www.patika.org
	Pathway databases
BioSilico	http://biosilico.org
KEGG	http://kegg.com
KaPPA-View	http://kpv.kazusa.or.jp/kappa-view/
MetaCyc	http://metacyc.org
MRAD	http://capb.dbi.udel.edu/whisler
Phylosopher	http://www.genedata.com/phylosopher.php
PUMA2	http://compbio.mcs.anl.gov/puma2/cgi-bin/index.cgi
EMP	http://www.empproject.com

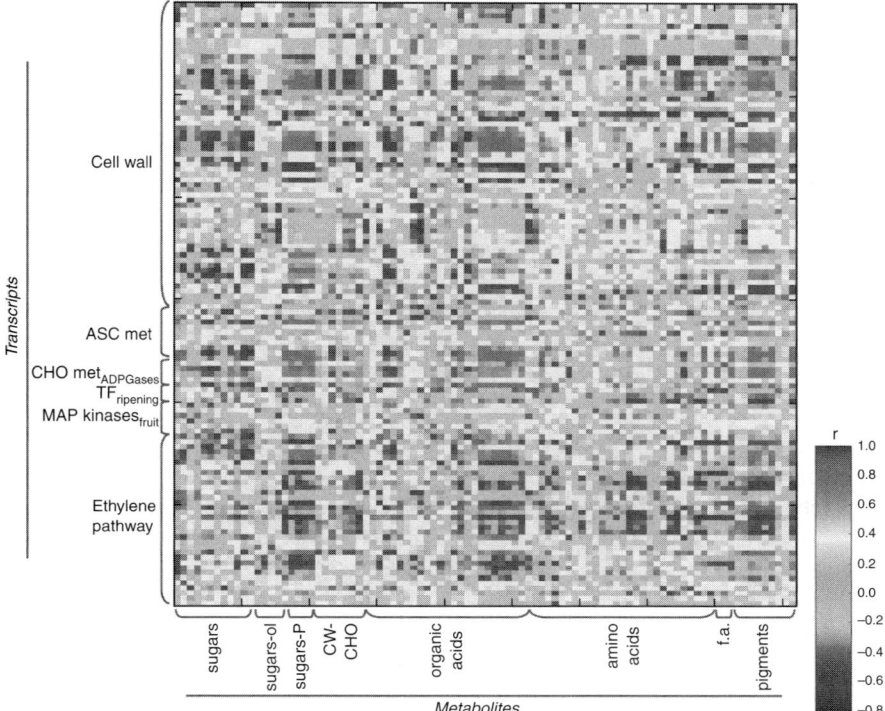

Fig. 6.2 Selected transcript–metabolite correlation visualizations. Heat map surface of selected transcript–metabolite correlations was drawn, and correlation coefficients were calculated. Each dot indicates a given r value resulted from a Spearman correlation analysis in a false color scale. Asc met, ascorbate metabolism; CHO met, carbohydrate metabolism; CW, cell wall; sugars-ol, sugar alcohols; sugars-P, sugar phosphates; f.a., fatty acids; TFs, transcription factors [63] (*see also Color Insert*)

levels. The pattern of changes in this multidimensional parametric matrix was evaluated on a temporal scale utilizing the recently developed *Solanaceous* MapMan [82]. This is one of the increasingly available online tools for bioinformatics elaboration of complex dataset (Table 6.1). The study suggested that transcript abundance was less strictly coordinated by functional group than metabolite abundance. This implies that posttranslational mechanisms govern metabolic regulation during tomato fruit ripening. Nevertheless, correlations between specific transcripts, e.g., ripening associated, and distinct metabolites classes were identified, such as TCA-cycle organic acids and sugar phosphates (Fig. 6.2). The integrated analysis identified specific metabolic pathways with characteristic behavior during tomato fruit development, suggesting a functional significance in the process of fruit ripening and which might have importance for biotechnology applications [10].

Taken together, the conclusions drawn in these case studies add to the importance of parallel and integrative analyses. Indeed, while it is not yet possible to quantify all or even the majority of the metabolites in a cellular system [17], integration of datasets originating from different platforms is being increasingly used [27, 30, 31, 32, 79].

6.5 Entering the Systems Biology Era: The Challenges

6.5.1 Integrating Flux Profiling into Network Modeling

Identifying regulatory elements and the type of regulatory events in cellular processes is challenging. A major obstacle for this kind of analysis is the restricted availability of flux data. Although metabolic networks are highly redundant, the evaluation of the flux load of various metabolic pathways can be achieved [6, 35, 57, 59, 68, 69, 70]. That said, we lack a great deal of information on the majority of the fluxes involved. In most flux analysis studies, an important assumption is that of metabolic steady state. This and other simplifying assumptions [87] can preclude its use in certain biological systems. For example, fluxes in the central metabolism of developing embryos – a closed system – of *Brassica napus* were characterized with success, and a modeling was possible [69]. However, the approach is not at all effective for the description of other tissues, for example, potato tubers [59]. For this reason, some groups attempt to develop dynamic flux modeling approaches [6, 57], requiring fewer simplifying assumptions and being based on the accurate measurement of isotope labeling in many metabolite pools at multiple time points during an experiment. A major constraint of this strategy is its mathematical complexity and the need to determine concentration levels of metabolites (see Chapters 9, 10). In addition, both flux methods are limited by our current paucity of knowledge of pathway structure. In combination, these factors represent major hurdles for a broader scale flux analysis; however, given the impressive recent progress in this area, it remains likely that it will evolve into an important component of metabolic network study in the next years.

6.5.2 Putting Things into Their Context

Scientific literature is the ultimate end of experimental information – a fact that has prompted the establishment of an increasing number of publicly available databases to collect such "hidden" information [53]. Among the best known (Table 6.1) are *KEGG* [36] and *Metacyc* [11, 90]. The first provides a frame on top of which one can overlay functional and pathway information onto rank-ordered gene lists. The second provides an interspecific reference database for metabolism and can be used to computationally predict the components of a metabolic pathway of an annotated genome. Due to the increasingly growing high-throughput experimental data hidden in published scientific literature, computational tools need to be designed to efficiently extract relevant information on molecular components (metabolites, genes, proteins), to reconstruct the system [33]. In this context, synonym extraction and terminology disambiguation are a must. Furthermore, the establishment of publicly available databases, where to collect, elaborate, and analyze experimental results, is needed.

6.6 Conclusion

An integrated approach is the key for robust conclusion drawing in systems biology [75]. This approach requires an unprecedented effort to reduce the complex multidimensional nature of the data matrices produced, in order to extract biologically relevant information on the system in study. Great attention must however be given to avoid losing important information during the process of data analysis. Success depends on an experimental design to minimize non-biological variability and missing data in large-scale "omics" measurements. Statistical and computational tools developed must cope with large experimental dataset, the implicit variance of technical or biological origin and eventually extract biologically relevant information, e.g., regulatory elements or modules. In this context, software tools are developed to store, manage, integrate, and visualize the dataset in a user friendly but comprehensive manner. Databases and literature mining are used to extract relevant facts from the scientific literature. Finally, mathematical modeling for the extrapolation and prediction of biological processes is used to test hypotheses and generate new ones.

References

1. Abe T, Sugawara H, Kanaya S, Kinouchi M, Ikemura T (2006) Self-Organizing Map (SOM) unveils and visualizes hidden sequence characteristics of a wide range of eukaryote genomes. Gene 365:27–34.
2. Abe T, Sugawara H, Kinouchi M, Kanaya S, Ikemura T (2005) Novel phylogenetic studies of genomic sequence fragments derived from uncultured microbe mixtures in environmental and clinical samples. DNA Res 12:281–290.
3. Abe T, Kanaya S, Kinouchi M, Ichiba Y, Kozuki T, Ikemura T (2003) Informatics for unveiling hidden genome signatures. Genome Res 13:693–702.
4. Alba R, Fei Z, Payton P, Liu Y, Moore SL, Debbie P, Cohn J, D'Ascenzo M, Gordon JS, Rose JK, Martin G, Tanksley SD, Bouzayen M, Jahn MM, Giovannoni J (2004) ESTs, cDNA microarrays, and gene expression profiling: tools for dissecting plant physiology and development. Plant J 39:697–714.
5. Baum G, Lev-Yadun S, Fridmann Y, Arazi T, Katsnelson H, Zik M, Fromm H (1996) Calmodulin binding to glutamate decarboxylase is required for regulation of glutamate and GABA metabolism and normal development in plants. EMBO J 15:2988–2996.
6. Baxter CJ, Liu JL, Fernie AR, Sweetlove LJ (2007) Determination of metabolic fluxes in a non-steady-state system. Phytochemistry 68:2313–2319.
7. Sumner LW, Urbanczyk-Wochniak E, Broeckling CD (2008) Metabolomics data analysis, visualization, and integration. Methods Mol Biol 406:409–436.
8. Cakir T, Patil KR, Onsan Z, Ulgen KO, Kirdar B, Nielsen J (2006) Integration of metabolome data with metabolic networks reveals reporter reactions. Mol Syst Biol 2:50.

9. Carrari F, Baxter C, Usadel B, Urbanczyk-Wochniak E, Zanor MI, Nunes-Nesi A, Nikiforova V, Centero D, Ratzka A, Pauly M, Sweetlove LJ, Fernie AR (2006) Integrated analysis of metabolite and transcript levels reveals the metabolic shifts that underlie tomato fruit development and highlight regulatory aspects of metabolic network behavior. Plant Physiol 142:1380–1396.
10. Carrari F, Fernie AR (2006) Metabolic regulation underlying tomato fruit development. J Exp Bot 57:1883–1897.
11. Caspi R, Foerster H, Fulcher CA, Hopkinson R, Ingraham J, Kaipa P, Krummenacker M, Paley S, Pick J, Rhee SY, Tissier C, Zhang P, Karp PD (2006) MetaCyc: a multiorganism database of metabolic pathways and enzymes. Nucleic Acids Res 34(Database issue):D511–6.
12. Cheadle C, Cho-Chung YS, Becker KG, Vawter MP (2003) Application of z-score transformation to Affymetrix data. Appl Bioinformatics 2:209–217.
13. Datta S, Datta S (2006) Methods for evaluating clustering algorithms for gene expression data using a reference set of functional classes. BMC Bioinformatics 7:397.
14. Datta S, Datta S (2003) Comparisons and validation of statistical clustering techniques for microarray gene expression data. Bioinformatics 19:459–466.
15. Fait A, Angelovici R, Less H, Ohad I, Urbanczyk-Wochniak E, Fernie AR, Galili G (2006) Arabidopsis seed development and germination is associated with temporally distinct metabolic switches. Plant Physiol 142:839–854.
16. Fait A, Fromm H, Walter D, Galili G, Fernie AR (2008) Highway or byway: the metabolic role of the GABA shunt in plants. Trends Plant Sci 13:14–19.
17. Fernie AR, Trethewey RN, Krotzky AJ, Willmitzer L (2004) Metabolite profiling: from diagnostics to systems biology. Nat Rev Mol Cell Biol 5:763–769.
18. Fiehn O, Weckwerth W (2003) Deciphering metabolic networks. Eur J Biochem 270:579–588.
19. Fridman E, Carrari F, Liu YS, Fernie AR, Zamir D (2004) Zooming in on a quantitative trait for tomato yield using interspecific introgressions. Science 305:1786–1789.
20. Fu J, Swertz MA, Keurentjes JJ, Jansen RC (2007) MetaNetwork: a computational protocol for the genetic study of metabolic networks. Nature Protoc 2:685–694.
21. Gibon Y, Usadel B, Blaesing OE, Kamlage B, Hoehne M, Trethewey R, Stitt M (2006) Integration of metabolite with transcript and enzyme activity profiling during diurnal cycles in Arabidopsis rosettes. Genome Biol 7:R76.
22. Gnanadesikan R (1977) Methods for Statistical Analysis of Multivariate Observations, Wiley, New York.
23. Gutierrez L, Van Wuytswinkel O, Castelain M, Bellini C (2007) Combined networks regulating seed maturation. Trends Plant Sci 12:294–300.
24. Hair JF Jr, Anderson RE, Tatham RL, Black WC (1998) Multivariate Data Analysis, 5th ed., Prentice Hall Inc, Prentice Hall: Upper Saddle River.
25. Hand DJ (2007) Principles of data mining. Drug Saf 30:621–622.
26. Harmer SL, Hogenesch JB, Straume M, Chang HS, Han B, Zhu T, Wang X, Kreps JA, Kay SA (2000) Orchestrated transcription of key pathways in Arabidopsis by the circadian clock. Science 290:2110–2113.
27. Higashi Y, Hirai MY, Fujiwara T, Naito S, Noji M, Saito K (2006) Proteomic and transcriptomic analysis of *Arabidopsis* seeds: molecular evidence for successive processing of seed proteins and its implication in the stress response to sulfur nutrition. Plant J 48:557–571.
28. Hihara Y, Kamei A, Kanehisa M, Kaplan A, Ikeuchi M (2001) DNA microarray analysis of cyanobacterial gene expression during acclimation to high light. Plant Cell 13:793–806.
29. Hill AA, Hunter CP, Tsung BT, Tucker-Kellogg G, Brown EL (2000) Genomic analysis of gene expression in *C. elegans*. Science 290:809–812.
30. Hirai MY, Klein M, Fujikawa Y, Yano M, Goodenowe DB, Yamazaki Y, Kanaya S, Nakamura Y, Kitayama M, Suzuki H, Sakurai N, Shibata D, Tokuhisa J, Reichelt M, Gershenzon J, Papenbrock J, Saito K (2005) Elucidation of gene-to-gene and metabolite-to-gene net-

works in *Arabidopsis* by integration of metabolomics and transcriptomics. J Biol Chem 280: 25590–22595.
31. Hirai MY, Saito K (2004) Post-genomics approaches for the elucidation of plant adaptive mechanisms to sulphur deficiency. J Exp Bot 55:1871–1879.
32. Hirai MY, Yano M, Goodenowe DB, Kanaya S, Kimura T, Awazuhara M, Arita M, Fujiwara T, Saito K (2004) Integration of transcriptomics and metabolomics for understanding of global responses to nutritional stresses in *Arabidopsis thaliana*. Proc Natl Acad Sci U S A 101:10205–10210.
33. Jensen LJ, Saric J, Bork P (2006) Literature mining for the biologist: from information retrieval to biological discovery. Nat Rev Genet 7:119–29.
34. Johnson SC (1967) Hierarchical clustering schemes. Psychometrika 2:241–254.
35. Junker BH, Lonien J, Heady LE, Rogers A, Schwender J (2007) Parallel determination of enzyme activities and in vivo fluxes in *Brassica napus* embryos grown on organic or inorganic nitrogen source. Phytochemistry 68:2232–2242.
36. Kanehisa M, Goto S, Hattori M, Aoki-Kinoshita KF, Itoh M, Kawashima S, Katayama T, Araki M, Hirakawa M (2006) From genomics to chemical genomics: new developments in KEGG. Nucleic Acids Res. 34(Database issue):D354–7.
37. Kaplan F, Kopka J, Sung DY, Zhao W, Popp M, Porat R, Guy CL (2007) Transcript and metabolite profiling during cold acclimation of *Arabidopsis* reveals an intricate relationship of cold-regulated gene expression with modifications in metabolite content. Plant J. 50: 967–981.
38. Kaplan F, Kopka J, Haskell DW, Zhao W, Schiller KC, Gatzke N, Sung DY, Guy CL (2004) Exploring the temperature-stress metabolome of *Arabidopsis*. Plant Physiol 136:4159–4168.
39. Kell DB (2004) Metabolomics and systems biology: making sense of the soup. Curr Opin Microbiol 7:296–307.
40. Keurentjes JJ, Fu J, Terpstra IR, Garcia JM, van den Ackerveken G, Snoek LB, Peeters AJ, Vreugdenhil D, Koornneef M, Jansen RC (2007) Regulatory network construction in *Arabidopsis* by using genome-wide gene expression quantitative trait loci. Proc Natl Acad Sci U S A 104:1708–1713.
41. Keurentjes JJ, Fu J, de Vos CH, Lommen A, Hall RD, Bino RJ, van der Plas LH, Jansen RC, Vreugdenhil D, Koornneef M (2006) The genetics of plant metabolism. Nature Genet 38:842–849.
42. Kliebenstein DJ, Gershenzon J, Mitchell-Olds T (2001) Comparative quantitative trait loci mapping of aliphatic, indolic and benzylic glucosinolate production in *Arabidopsis thaliana* leaves and seeds. Genetics 159:359–370.
43. Kohonen T (2006) Self-organizing neural projections. Neural Netw 19:723–733.
44. Kümmel A, Panke S, Heinemann M (2006) Putative regulatory sites unraveled by network-embedded thermodynamic analysis of metabolome data. Mol Syst Biol 2:2006.0034.
45. Liu F, Vantoai T, Moy LP, Bock G, Linford LD, Quackenbush J (2005) Global transcription profiling reveals comprehensive insights into hypoxic response in Arabidopsis. Plant Physiol 137:1115–1129.
46. Levenstien MA, Yang YN, Ott J (2003) Statistical significance for hierarchical clustering in genetic association and microarray expression studies. BMC Bioinformatics 4:62.
47. Livesey JH (2007) Kurtosis provides a good omnibus test for outliers in small samples. Clin Biochem 40:1032–1046.
48. Martins AM, Camacho D, Shuman J, Sha W, Mendes P, Shulaev V (2004) A systems biology study of two distinct growth phases of Saccharomyces cerevisiae cultures. Current Genomics 5:649–663.
49. Meyer RC, Steinfath M, Lisec J, Becher M, Witucka-Wall H, Torjek O, Fiehn O, Eckardt A, Willmitzer L, Selbig J, Altmann T (2007) The metabolic signature related to high plant growth rate in *Arabidopsis thaliana*. Proc Natl Acad Sci U S A 104:4759–4764.
50. Morgenthal K, Weckwerth W, Steuer R (2006) Metabolomic networks in plants: Transitions from pattern recognition to biological interpretation. Biosystems 83:108–117.

51. Morgenthal K, Wienkoop S, Wolschin F, Weckwerth W (2007) Integrative profiling of metabolites and proteins: improving pattern recognition and biomarker selection for systems level approaches. Methods Mol Biol 358:57–75.
52. Nacher JC, Schwartz JM, Kanehisa M, Akutsu T (2006) Identification of metabolic units induced by environmental signals. Bioinformatics 22:e375–e383.
53. Ng A, Bursteinas B, Gao Q, Mollison E, Zvelebil M (2006) Resources for integrative systems biology: from data through databases to networks and dynamic system models. Brief Bioinform 7:318–330.
54. Nielsen J (2003) It is all about metabolic fluxes. J Bacteriol 185:7031–7035.
55. Quackenbush J (2007) Extracting biology from high-dimensional biological data. J Exp Biol 210:1507–1517.
56. Quackenbush J (2002) Microarray data normalization and transformation. Nature Genet 32 Suppl:496–501.
57. Ratcliffe RG, Shachar-Hill Y (2006) Measuring multiple fluxes through plant metabolic networks. Plant J 45:490–511.
58. Roessner U, Willmitzer L, Fernie AR (2001) High-resolution metabolic phenotyping of genetically and environmentally diverse potato tuber systems. Identification of phenocopies. Plant Physiol 127:749–764.
59. Roessner-Tunali U, Liu J, Leisse A, Balbo I, Perez-Melis A, Willmitzer L, Fernie AR (2004) Kinetics of labelling of organic and amino acids in potato tubers by gas chromatography-mass spectrometry following incubation in (13)C labelled isotopes. Plant J 39: 668–679.
60. Roessner-Tunali U, Hegemann B, Lytovchenko A, Carrari F, Bruedigam C, Granot D, Fernie AR (2003) Metabolic profiling of transgenic tomato plants overexpressing hexokinase reveals that the influence of hexose phosphorylation diminishes during fruit development. Plant Physiol 133:84–99.
61. Ruuska SA, Girke T, Benning C, Ohlrogge JB (2002) Contrapuntal networks of gene expression during Arabidopsis seed filling. Plant Cell 14:1191–1206.
62. Saidi SA, Holland CM, Kreil DP, MacKay DJC, Charnock-Jones DS, Print CG, Smith SK (2004) Independent component analysis of microarray data in the study of endometrial cancer. ONCOGENE 23: 6677–6683.
63. Schauer N, Semel Y, Roessner U, Gur A, Balbo I, Carrari F, Pleban T, Perez-Melis A, Bruedigam C, Kopka J, Willmitzer L, Zamir D, Fernie AR (2006) Comprehensive metabolic profiling and phenotyping of interspecific introgression lines for tomato improvement. Nature Biotechnol 24:447–454.
64. Scholz M, Selbig J (2007) Visualization and analysis of molecular data. Methods Mol Biol 358:87–104.
65. Scholz M, Kaplan F, Guy CL, Kopka J, Selbig J (2005) Non-linear PCA: a missing data approach. Bioinformatics. 21:3887–3895.
66. Scholz M, Gatzek S, Sterling A, Fiehn O, Selbig J (2004) Metabolite fingerprinting: detecting biological features by independent component analysis. Bioinformatics 20:2447–2454.
67. Schramm G, Zapatka M, Eils R, Konig R (2007) Using gene expression data and network topology to detect substantial pathways, clusters and switches during oxygen deprivation of Escherichia coli. BMC Bioinformatics. 8:149.
68. Schwender J, Ohlrogge J, Shachar-Hill Y (2004) Understanding flux in plant metabolic networks. Curr Opin Plant Biol 7:309–317.
69. Schwender J, Ohlrogge JB, Shachar-Hill Y (2003) A flux model of glycolysis and the oxidative pentosephosphate pathway in developing *Brassica napus* embryos. J Biol Chem 278:29442–29453.
70. Schwender J, Shachar-Hill Y, Ohlrogge JB (2006) Mitochondrial metabolism in developing embryos of *Brassica napus*. J Biol Chem 281:34040–34047.
71. Sokal RR, Rohlf FJ (1995) Biometry: the principles and practice of statistics in biological research. 3rd ed., W.H. Freeman, New York.

72. Steuer R, Morgenthal K, Weckwerth W, Selbig J (2007) A gentle guide to the analysis of metabolomic data. Methods Mol Biol 358:105–126.
73. Steuer R, Kurths J, Fiehn O, Weckwerth W (2003) Observing and interpreting correlations in metabolomic networks. Bioinformatics 19:1019–1026.
74. Studart-Guimaraes C, Fait A, Nunes-Nesi A, Carrari F, Usadel B, Fernie AR (2007) Reduced expression of succinyl-coenzyme A ligase can be compensated for by up-regulation of the gamma-aminobutyrate shunt in illuminated tomato leaves. Plant Physiology 145:626–639.
75. Sweetlove LJ, Fernie AR (2005) Regulation of metabolic networks: understanding metabolic complexity in the systems biology era. New Phytol 168:9–24.
76. Sweetlove LJ, Last RL, Fernie AR (2003) Predictive metabolic engineering: a goal for systems biology. Plant Physiol 132:420–425.
77. Taylor J, King RD, Altmann T, Fiehn O (2002) Application of metabolomics to plant genotype discrimination using statistics and machine learning. Bioinformatics 18 Suppl 2:S241–S248.
78. Thomas CE, Ganji G (2006) Integration of genomic and metabonomic data in systems biology – are we 'there' yet? Curr Opin Drug Discov Devel 9:92–100.
79. Tohge T, Nishiyama Y, Hirai MY, Yano M, Nakajima J, Awazuhara M, Inoue E, Takahashi H, Goodenowe DB, Kitayama M, Noji M, Yamazaki M, Saito K (2005) Functional genomics by integrated analysis of metabolome and transcriptome of Arabidopsis plants over-expressing an MYB transcription factor. Plant J 42:218–235.
80. Tuikkala J, Elo LE, Nevalainen OS, Aittokallio T (2008) Missing value imputation improves clustering and interpretation of gene expression microarray data. BMC Bioninformatics 9:202.
81. Urbanczyk-Wochniak E, Willmitzer L, Fernie AR (2007) Integrating profiling data: using linear correlation to reveal coregulation of transcript and metabolites. Methods Mol Biol 358:77–85.
82. Urbanczyk-Wochniak E, Usadel B, Thimm O, Nunes-Nesi A, Carrari F, Davy M, Blasing O, Kowalczyk M, Weicht D, Polinceusz A, Meyer S, Stitt M, Fernie AR (2006) Conversion of MapMan to allow the analysis of transcript data from Solanaceous species: effects of genetic and environmental alterations in energy metabolism in the leaf. Plant Mol Biol 60:773–792.
83. Urbanczyk-Wochniak E, Baxter C, Kolbe A, Kopka J, Sweetlove LJ, Fernie AR (2005) Profiling of diurnal patterns of metabolite and transcript abundance in potato (*Solanum tuberosum*) leaves. Planta 221:891–903.
84. Urbanczyk-Wochniak E, Luedemann A, Kopka J, Selbig J, Roessner-Tunali U, Willmitzer L, Fernie AR (2003) Parallel analysis of transcript and metabolic profiles: a new approach in systems biology. EMBO Rep 4:989–993.
85. Weckwerth W, Loureiro ME, Wenzel K, Fiehn O (2004) Differential metabolic networks unravel the effects of silent plant phenotypes. Proc Natl Acad Sci U S A 101:7809–7814.
86. Weckwerth W (2003) Metabolomics in systems biology. Annu Rev Plant Biol 54:669–689.
87. Wiechert W (2002) An introduction to ^{13}C metabolic flux analysis. Genet Eng (N Y) 24:215–238.
88. Yamanishi Y, Itoh M, Kanehisa M (2002) Extraction of organism groups from phylogenetic profiles using independent component analysis. Genome Inform 13:61–70.
89. Yang YH, Dudoit S, Luu P, Lin DM, Peng V, Ngai J, Speed TP (2002) Normalization for cDNA microarray data: a robust composite method addressing single and multiple slide systematic variation. Nucleic Acids Res 30:e15.
90. Zhang P, Foerster H, Tissier CP, Mueller L, Paley S, Karp PD, Rhee SY (2005) MetaCyc and AraCyc. Metabolic pathway databases for plant research. Plant Physiol 138:27–37.

Chapter 7
Topology of Plant Metabolic Networks

Eva Grafahrend-Belau, Björn H. Junker, Christian Klukas, Dirk Koschützki, Falk Schreiber, and Henning Schwöbbermeyer

Metabolic networks can be modeled as graphs, i.e., mathematical structures consisting of vertices (representing objects such as metabolites) and edges/hyper-edges (representing the connection between objects such as reactions). An example of a very simple metabolic network is shown in Fig. 7.1. Often the term *network* refers to an informal concept describing a structure composed of objects and connections, whereas the term graph refers to an abstract mathematical structure formed by a set of vertices and a set of edges. For simplicity, we will consider both terms equivalent in the following.

In this chapter, networks/graphs, important properties, and initial findings in metabolic networks based on graph/network analysis methods will be discussed. Section 7.1 gives an overview of graphs, graph types, and simple graph properties. Section 7.2 deals with advanced network properties such as the bow-tie structure of metabolic networks. In Section 7.3, centrality measures in networks are studied. Section 7.4 presents network motifs and their application to metabolic networks. In Section 7.5, clustering methods for networks are discussed, and Section 7.6 deals with graph layout and network visualization.

7.1 Introduction

7.1.1 Graph Notation and Graph Types

A metabolic network can be seen as a hyper-graph, see Fig. 7.1. A *hyper-graph* $G = (V, E)$ is a special graph consisting of a set of *vertices* V and a set of *hyper-edges* E; each hyper-edge connects several vertices. The vertices represent the substances, and the hyper-edges represent the reactions. A hyper-edge connects all substances

---

F. Schreiber (✉)
Leibniz Institute of Plant Genetics and Crop Plant Research (IPK) Gatersleben,
Corrensstr. 3, D-06466 Gatersleben, Germany
e-mail: schreibe@ipk-gatersleben.de

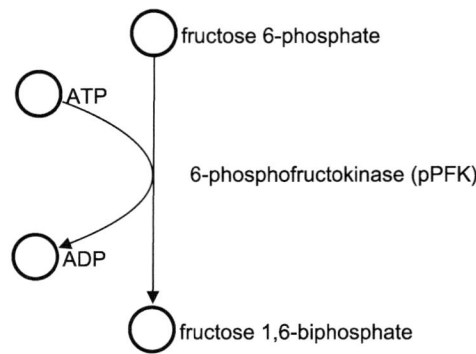

Fig. 7.1 A metabolic network where an edge connects more than two elements. This so-called hyper-edge is labeled with the name of the enzyme catalyzing the reaction

of a reaction, is directed from reactants to products, and is labeled with the enzymes that catalyze the reaction.

Hyper-graphs are not commonly used in graph theory and therefore other graphs are preferred to model metabolic networks. Let us first consider general graphs. A *(directed) graph* $G = (V, E)$ consists of a set of *vertices* V and a set of *edges* E, where each edge is assigned to two (not necessarily disjunct) vertices. A directed edge e connecting the vertex u with the vertex v is denoted by $e = (u, v)$. A graph $G = (V, E)$ is called *bipartite* if there is a partition of its vertex set $V = S \cup T$ such that each edge in E has exactly one vertex in S and one vertex in T.

Hyper-graphs mentioned before can be represented by bipartite graphs. Usually metabolic networks are modeled by bipartite graphs, see Fig. 7.2 which shows a bipartite graph representation of the hyper-graph in Fig. 7.1. Additionally to the vertices representing substances, there are vertices representing reactions (which can be labeled with the enzyme names). Edges are binary relations connecting the substances of a reaction with the corresponding reaction vertex.

We model a metabolic network by a directed bipartite graph $G = (V_1 \cup V_2, E)$. The vertices $v \in V_1$ represent the substances, and the vertices $v \in V_2$ represent the reactions with the enzymes. Edges connect substances (vertices of V_1) with the

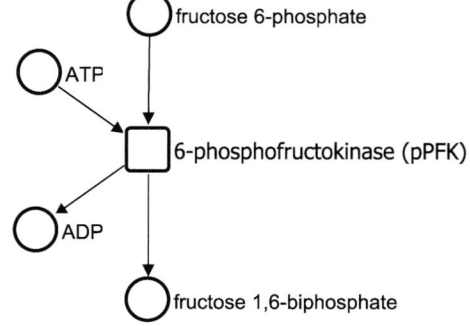

Fig. 7.2 A representation of the hyper-graph in Fig. 7.1 as directed bipartite graph. The two different sets of vertices are represented by *circles* and *squares*

corresponding reactions/enzymes (vertices of V_2). An example of this modeling of metabolic networks is shown in Fig. 7.2.

There are other networks closely related to metabolic networks which will be used in this chapter. In the following, two are particularly important:

- *Metabolite network* is a network which consists only of substances (metabolites). The reactions/enzymes are removed from this network and two substances are connected if they are both connected to the same reaction/enzyme in the metabolic network.
- *Enzyme network* is a network which consists only of the enzymes catalyzing the reactions. The substances are removed from this network and two reactions/enzymes are connected if they are both connected to the same substance in the metabolic network.

For these two types of networks, a corresponding metabolic network cannot always be directly derived. For example, the metabolites in a metabolite network are not necessarily connected according to the reactions of a metabolic network, but these connections can be established by correlation analysis of metabolite profiles [64].

To use graphs in a computer program, they have to be represented in the computer. Two representations are common: adjacency matrix and adjacency list. The choice between them depends on the operations which should be applied on the graph and whether the graph is dense or sparse, i.e., contains many or few edges in relation to the number of vertices.

A graph $G = (V, E)$ with n vertices can be represented by a $(n \times n)$ *adjacency matrix* A. The rows and columns of the matrix correspond to the vertices. A matrix-element $A_{ij} = 1$ if and only if there is an edge from vertex v_i to vertex v_j, and $A_{ij} = 0$ otherwise. A simple way to implement an adjacency matrix is an array $[1...n, 1...n]$. For the representation of biological networks, adjacency matrices are often used as their structure is simple and matrix operations can be directly applied. However, these matrices need a lot of memory, n^2 places for a network with n vertices, and several network analysis algorithms may need longer computation time than with other representations. A different representation of graphs is an adjacency list. A graph $G = (V, E)$ with n vertices can be represented by n connected lists. For each vertex $v \in V$, a list L_v contains all edges incident to this vertex (and therefore all vertices adjacent to it). A simple way to implement an adjacency list is an array $[1...n]$ of lists.

7.1.2 Definition of Network Properties

In the following, some terms are introduced which will be used in the reminder of this chapter. Given a graph $G = (V, E)$, a *walk* is a sequence $(v_0, e_1, v_1, e_2, v_2, ..., v_{k-1}, e_k, v_k)$ of vertices and edges with $e_i = (v_{i-1}, v_i)$. Often the vertices of the walk are omitted and it is denoted by a sequence $(e_1, e_2, ..., e_k)$ of edges. Such a walk $v_0 \to v_k$ *connects* vertex v_0 with vertex v_k. A walk is called a *path* if all edges of the walk are distinct. The *length* of a walk or path is given by its number of edges. A path between two vertices is a *shortest path* if no other path with shorter length exists between these vertices. There may be several shortest paths between two vertices. The distance between two vertices is the length of a shortest path between them.

If there exist paths $v_i \rightarrow v_j$ and $v_j \rightarrow v_i$ between two vertices v_i and v_j, these vertices are called *strongly connected*. If any pair of different vertices of the graph is strongly connected, the graph is *strongly connected*. If the direction of edges can be neglected (i.e., edges can be reversed if necessary) and there exists a path between two vertices, these vertices are called *connected*.

Note that in undirected graphs, the term *connected* is used if an (undirected) path exists between two vertices (or all pairs of vertices in which case the graph would be called connected).

A *subgraph* $G' = (V', E')$ of the graph $G = (V, E)$ is a graph where V' is a subset of V and E' is a subset of $V' \times V'$, that is the subset of E which contains only edges with vertices in V'. A subgraph G' is called an *induced subgraph* of G if G' is a subgraph of graph G and the edge set E' contains all edges of E which connect vertices of V'.

If two graphs contain the same number of vertices connected in the same way, these graphs are considered as the same or *isomorphic* graphs. Formally, two graphs $G_1 = (V_1, E_1)$ and $G_2 = (V_2, E_2)$ are isomorphic, if there exists a bijective mapping between the vertices in V_1 and V_2 with the property that there is an edge between any two vertices $u, v \in V_1$ if and only if there is an edge between the two corresponding vertices in the other graph, see Fig. 7.3. Such a bijection is called an *isomorphism*.

In the following, three relatively robust measures of network topology will be discussed: degree distribution, average path length, and clustering coefficient. Robust means that small changes of the networks such as the removal of some vertices only result in small changes of the measures.

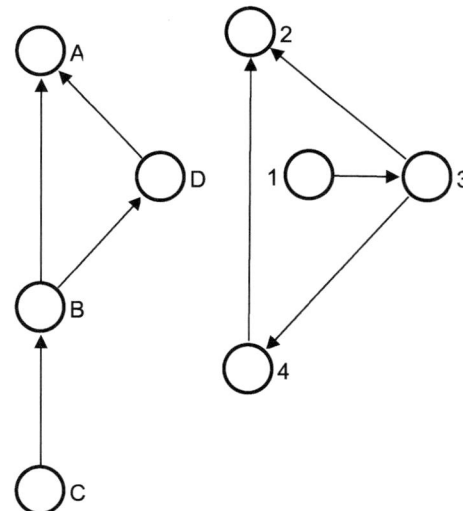

Fig. 7.3 Two isomorphic graphs

7 Topology of Plant Metabolic Networks

The degree $d(v)$ of a vertex v is the number of its neighbors. In directed graphs, we distinguish between the in-degree of a vertex v, that is, the number of neighbors with the edge pointing to v, and the out-degree of v, that is, the number of neighbors with the edge pointing away from v. The degree of a vertex v is the sum of its in-degree and its out-degree. An undirected graph with $|V|$ vertices and $|E|$ edges is characterized by an average degree

$$\bar{d} = \frac{2|E|}{|V|} \quad (7.1)$$

The *degree* distribution $P(d)$ gives the probability that a chosen vertex has exactly a degree of d, that is, that the vertex is incident to d edges. Let $X(d)$ be the number of vertices with degree d. The *degree distribution* $P(d)$ of a graph $G = (V, E)$ is

$$P(d) = \frac{X(d)}{|V|} \quad (7.2)$$

For a strongly connected graph, the *average path length* (or *characteristic path length*) is the average of the length of the shortest paths between each pair of vertices in the graph. If the graph is not strongly connected, then pairs of vertices without a path between them have to be excluded from the computation of the average path length. The average path length is a measure of how far in average an arbitrary vertex is from another vertex.

The *clustering coefficient* is a measure of the connectivity within a graph. It can be defined locally (to measure the "clustering" of a vertices) or globally (to measure the clustering of the whole graph). The *local clustering coefficient* $c(v)$ of a vertex v is defined as the number of connections that exist among the neighbors of v, divided by the number of connections that could exist if all neighbors of v were completely connected. The local clustering coefficient

$$c(v) = \frac{|E(v)|}{|E_{all}(v)|} \quad (7.3)$$

where $|E(v)|$ is the number of edges in the neighborhood of vertex v and $|E_{all}(v)|$ is the total number of possible edges in the neighborhood of v. The *(global) clustering coefficient* c is the average of the local clustering coefficient $c(v)$ of each vertex $v \in V$ in the graph.

7.2 Special Properties of Metabolic Networks

Even though metabolic networks share many common properties with biological, technical, and social networks, they also have some unique properties. The global and local properties of metabolic networks strongly depend on the way that these networks have been prepared, a fact that is unfortunately often neglected.

When a metabolic network is modeled as a bipartite graph (see Fig. 7.2), in which one kind of vertices represents the enzymes and the other kind of vertices the metabolites, we have to keep in mind that this is already a fundamental simplification of the real-world situation: the stoichiometric constraints are lost. If we look at Fig. 7.2, we could in principle make ADP from fructose 6-phosphate. Stoichiometric constraints are therefore often considered by modeling metabolic networks as Petri nets [62], which have to obey a "firing rule", which means that only when all substrates are present a reaction can take place by transforming the set of substrates into the set of products.

In many studies, metabolic networks are simplified to *metabolite* networks or *enzyme* networks (see Section 7.1.1 for a definition). By this transformation, especially the global network properties will change dramatically, as will be outlined in the next two sections. Although most of the results have been obtained from bacteria or single-celled eukaryotes, the observations are in principle transferable to plant metabolic networks.

7.2.1 Local Properties

Just like many other large biological and non-biological networks, it has been shown that metabolic networks exhibit properties that are typical for small-world networks. Jeong and coworkers described for the first time that the degree distribution of metabolic networks follows a power law [49], a characteristic that has been confirmed many times since then, e.g., in [70]. This means that there are many metabolites that are connected only to one or two others, while there are a few metabolites ("hubs") that are connected to hundreds. It has been proposed that this network structure is responsible for a high robustness against failure, e.g., a mutation, a feature that is clearly an evolutionary advantage [60].

The clustering coefficient for the metabolite network of *E. coli* was found to be significantly higher than could be expected for random networks with preserved degree distribution [111]. These results have been confirmed by analyzing the metabolite network of *Saccharomyces cerevisiae* [119]. Interestingly, by taking the bipartite metabolic network instead of the metabolite graph, and by choosing an appropriate random rewiring scheme to generate the random networks for comparison, it was concluded that the metabolic network of *S. cerevisiae* shows no significant clustering beyond the trivial clustering imposed by the construction of the network – namely a projection that was constructed from a bipartite graph [119]. Thus, depending on the preparation of the network, entirely different conclusions may be drawn on the degree of clustering in metabolic networks.

7.2.2 Global Properties

The global structure of metabolic networks has been described as a bow-tie structure. This analogy to the elegant piece of clothing becomes apparent from Fig. 7.4.

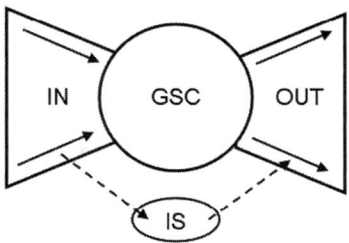

Fig. 7.4 Bow-tie structure of metabolic networks. See text for details. GSC: giant strong component; IS: isolated components

In a directed metabolic network, a subgraph can be found in which all nodes are reachable from all other ones. This *giant strong component* (GSC) is connected to an in-component, which contains nodes from which it is possible to reach the GSC, but it is impossible to go back to these nodes from the GSC. On the other side, the GSC is connected to the out-component, which contains nodes that can be reached from the GSC, but from which it is impossible to go back to the GSC. From a biological point of view, the GSC is the central metabolic network which contains all sorts of anabolic and catabolic pathways, but from which it is always possible to get back to the central part. The in-component contains specialized pathways for the usage of different substrates. The out-component yields metabolic products, e.g., secondary metabolites. The isolated components are subgraphs similar to the GSC, as they connect in- with out-metabolites, but these subgraphs are significantly smaller than the GSC.

There has been a long debate about the average path length of metabolic networks (APL; see Section 7.1.2 for a definition). In the first study, Jeong and coworkers [49] started with the metabolic network of *E. coli*, and generated a metabolite network from this by the simple algorithm described in Section 7.1.1. They calculated an APL of 3.2, which means that any metabolite can be reached from any other one by an average, a bit more than three reaction steps. However, the algorithm will have the effect that for any pair of metabolites A and B, in which A is substrate of a reaction that produces ATP, and the other one is product of a reaction that uses ATP, the network will suggest that B can be made from A in two reaction steps, which is clearly not possible in most cases. Thus, all highly connected metabolites, such as water, phosphate, ATP, and all other co-factors, will render the network seemingly small.

Recognizing this problem, Wagner and Fell [111] took an undirected metabolite network and removed the metabolites ATP, ADP, NAD, NADP, NADH, NADPH, carbon dioxide, ammonia, sulfate, thioredoxin, phosphate, and pyrophosphate. Nevertheless, this resulted in an only slightly higher APL of 3.8, probably because there are still many other highly connected metabolites such as glutamate that serves as a donor of an amino group in many transaminase reactions.

Ma and Zeng [70] defined one or more main reactant pair(s) in every reaction and removed all other metabolites, which they termed "current" metabolites (elsewhere also termed "currency" metabolites). As a result, an APL of 8.2 was obtained, which is most probably a much more realistic number for the average number of steps it takes to convert any metabolite into any other.

Arita [2] introduced the concept of carbon atomics traces. For each reaction, the fate of individual carbon atoms are traced and metabolites are connected if at least one carbon atom is transferred from the substrate to the product (see also Section 7.3.2). With an APL of 8.4, this strategy yielded a value close to that of Ma and Zeng [70]. Arita concluded correctly in the title of his publication that "the metabolic world of *Escherichia coli* is not small."

In conclusion, it can be argued that the average path length of a metabolic network depends more on the way the network was prepared than on the network itself.

7.3 Centralities in Metabolic Networks

7.3.1 Basic Concepts

Centrality analysis of metabolic networks is easily explained by an example: the election of a class spokesperson in school. Every pupil has a vote and may issue his vote for himself or for another pupil. Clearly the pupil with the most votes becomes the class spokesperson, and the pupil receiving the second most votes becomes the substitute.

The election can be modeled as a directed graph. Every pupil is represented by a labeled vertex, and votes are modeled as directed edges from a pupil giving his vote to the pupil voted for. Fig. 7.5 shows an example. In this example, Uta received the most votes (6) and is therefore the class spokesperson and Klaus is her substitute.

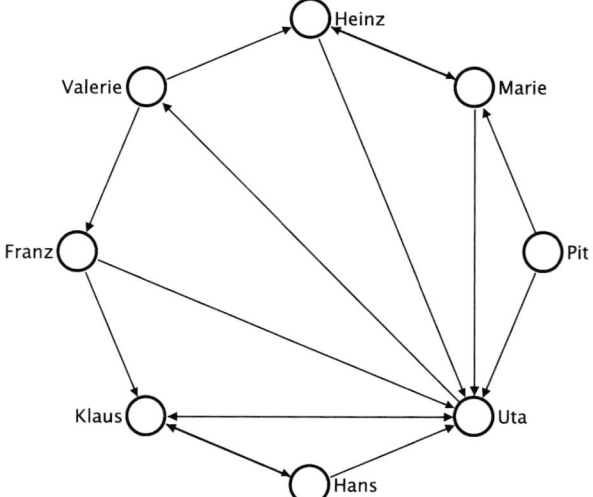

Fig. 7.5 A directed network modeling the election of a class spokesperson. In this example, every pupil has two votes and votes have to be given to other pupils. By counting the number of votes, a ranking of the pupils is established and Uta is elected as the spokesperson. The centrality applied in this example is called the *in-degree centrality*

7 Topology of Plant Metabolic Networks

Obviously within cells no votes for the most prominent metabolites are casted. Nevertheless a ranking of enzymes, metabolites, or other network elements based on the network structure is of interest. The prioritization of candidate target genes during drug discovery is a plausible example [42].

The general idea of ranking vertices based on the structure of the underlying network is known from sociology since about 1930 [81] and is termed *centrality analysis*. A *centrality* is a function which assigns every vertex of a given network a real value. By convention, vertices with higher centrality value are more important than others. A formal definition of a centrality as a function from the set of the vertices (or edges) to the set of the reals is given in a recent review [63].

More than 20 different centrality measures are known [63]. Some of them are simple, for example, the *degree centrality* simply counts the number of edges connected to a vertex. Others are more complicated, for example, *PageRank* or *eigenvector centrality*, which uses feedback as the underlying concept. By this idea, the centrality value of a vertex is more or less dependent on the centrality values of adjacent vertices. For the analysis of metabolic (and other biological) networks, shortest-path based centralities were applied in several cases.

The following three centralities using information about shortest paths inside a network are applied most:

- *Eccentricity centrality* assigns to the vertex under analysis the reciprocal of the longest distance from the vertex of interest to all other vertices.
- *Closeness centrality* assigns to the vertex under analysis the reciprocal of the sum of distances from the vertex of interest to all other vertices.
- *Shortest-path betweenness centrality* assigns to the vertex under analysis the ratio of the number of shortest paths between all vertex pairs which use the vertex of interest as an inner vertex on the shortest paths.

Even in the simple network in Fig. 7.6, the three shortest-path based centralities produce different rankings, see Table 7.1.

The selection of a centrality for the analysis of a network is based on two aspects: the process modeled by the network and the restrictions which some centralities enforce on the network. Every network somehow models a process. In the case of the election shown above, the process is the cast of a vote. In a transport network, for example, a map of a road system, the process is the transport of individuals or

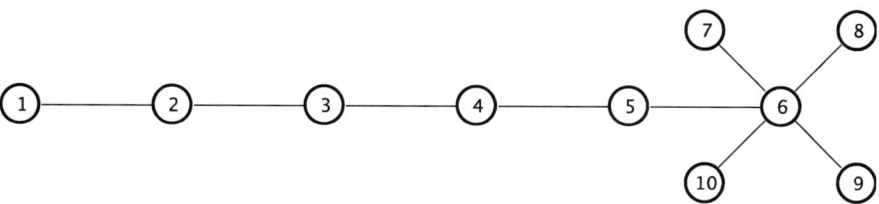

Fig. 7.6 An example graph to explain the three different shortest-path based centralities

Table 7.1 Centrality values for the network in Fig. 7.6 for the three different shortest-path (SP)-based centralities

Vertex	Eccentricity	Closeness	SP Betweenness
1	0.167	0.026	0
2	0.200	0.032	8
3	0.250	0.040	14
4	0.333	0.048	18
5	0.250	0.053	20
6	0.200	0.053	26
7	0.167	0.037	0
8	0.167	0.037	0
9	0.167	0.037	0
10	0.167	0.037	0

goods. Depending on the process and on the notion which vertices are important inside the modeled process, a centrality has to be selected. As a counter example take again the election and the shortest-path betweenness. The ranking produced by the shortest-path betweenness is different from the ranking produced by the in-degree centrality (which counts how many vote or incoming edges a vertex receives) and does not reflect the result of the election.

Due to restriction given by the formal definition of some of the centrality measures, not every centrality can be applied to a given network. For example, the degree centrality can be applied to any network. On the other hand, the centralities eccentricity and closeness can only be applied to strongly connected, or in the case of undirected networks to connected, networks. Those restrictions have to be obeyed if a centrality is applied for the analysis of a given network. Otherwise the computed results might lead to wrong results.

7.3.2 Metabolite Ranking

In several publications, rankings of metabolites or enzymes based on different centrality measures are given. In general, the produced rankings are incomparable between different studies due to huge differences in the used data sources and in the representation of biochemical reactions as metabolic or metabolite networks.

Jeong et al. [49] analyzed 43 metabolic networks from all three domains of life. The networks were represented as directed bipartite networks, and the metabolites occurring in all networks were ranked according to in- and out-degree. Without the removal of currency metabolites, see Section 7.2.2, these currency metabolites, e.g., H_2O, ADP, inorganic phosphor, and ADP, ranked highest.

The intermediate metabolism for energy generation and small building block synthesis of *E. coli* under a fixed growth condition was analyzed by Wagner and Fell [111]. The network was represented as an undirected metabolite network, and several metabolites were deleted (CO_2, NH_3, SO_4, thioredoxin, organic phosphate, and pyrophosphate) prior to the analysis. Metabolites were ranked on degree and on the so-called importance number, which is equivalent to the closeness centrality. The provided ranking of metabolites contains mainly metabolites that are the common biosynthetic source of all cell materials. The metabolites NAP, ATP, and their derivates were removed from the listing, otherwise these were ranked highest.

Based on data from the KEGG-LIGAND database, Ma and Zeng [70] reconstructed the metabolite networks of 80 organisms. These networks were created on the basis of a manually curated list of reactions. Within each reaction occurring, current metabolites were removed. The metabolites in each network were ranked according to the degree centrality. An evaluation of the frequency of the metabolites appearing in the top 20 position according to the degree centrality resulted in a ranking of metabolites which mainly shows intermediates of the glycolysis, the pentosephosphate pathway and acetyl-CoA, the linking metabolite between glycolysis, citric acid cycle, and the fatty acid synthesis pathway. Additionally, the amino acids glutamate and aspartate were ranked high, mirroring their importance as precursors for other amino acids.

In another paper, Ma and Zeng [69] ranked metabolites occurring in *E. coli* according to three variants of the closeness centrality. Among all the three variants (input closeness, output closeness, and overall closeness), pyruvate is the most central metabolite. Eight of the ten top-ranked metabolites occur either in the glycolysis or the citrate acid cycle.

Arita ranked metabolites occurring in the small-molecule metabolism of *E. coli* according to their degree [2]. In the resulting ranking metabolites like CO_2, pyruvate, and acetyl CoA, which were identified as important in earlier studies [49], were ranked highest. In contrast to other approaches, Arita used carbon atomic traces instead of the reaction equation to construct a metabolite network. In his approach, for each reaction, the fate of individual carbon atoms is traced, and the metabolites are connected if at least one carbon atom is transferred from the substrate to the product metabolite.

Wuchty and Stadler [118] applied three similar centralities (eccentricity, closeness, and centroid value) to the metabolic network of *E. coli*. According to them, the central metabolites are the crossroads of the networks which are believed to be evolutionary oldest and a centrality should reflect both the age and the importance of metabolites to the organism. All three centralities resulted in a similar ranking, and currency metabolites such as inorganic phosphate, ADP, and the metabolite with the abbreviation HEXT are ranked highest.

Ranking enzymes to find potential drug targets is another application of centrality analysis of metabolic networks [42]. Several authors applied centrality-like methods to correlate the computed ranks of enzymes with information about the viability of the respective genes [68, 90].

7.3.3 Tools

Numerous software systems are available for the analysis of (biological) networks. Cytoscape [101], Osprey [13], and VisANT [43] are the three which are often cited.

Currently, none of theses supports the centrality analysis of biological networks. Pajek [6], a software system for the analysis of large networks, is able to compute degree, closeness, and shortest-path betweenness centrality. Two tools which allow the computation of a larger number of centralities are Visone [12] and CentiBiN [51]. The latter, CentiBiN, focuses on the computation and exploration of centralities in biological networks.

7.4 Motifs in Metabolic Networks

7.4.1 Basic Concepts

The increasing availability of complex networks from biological and technical domains has boosted the development of network analysis methods to obtain a better understanding of the structure and function of these networks. The analysis methods deal either with the global and large-scale organization of networks or with properties of individual vertices. A concept that lies between local and global network structure with a focus on biological networks has been introduced with *network motifs*. Motifs of a network are particular subgraphs representing patterns of local interconnections between network elements [67, 79, 102]. They have been originally introduced as patterns which are statistically over-represented, compared to random networks. For biological networks, whose structure has been shaped during evolution due to functional constraints, it has been supposed that a positive selection for these interaction patterns based on their functional or structural properties has caused their overabundance. Accordingly, network motifs are regarded as basic building blocks and design patterns of complex networks.

Formally a network motif is defined as a small connected graph G, and the size is usually given by the number of vertices. A match of a motif G within an analyzed target graph/network G_t is a graph G_m which is isomorphic to the motif G and is a subgraph of G_t. See Fig. 7.7 for an example of a motif match within a graph. Note that a match does not have to be an induced subgraph of G_t. The frequency of a motif within an analyzed network is typically given by the number of all matches. These matches can partly overlap, i.e., for a pair of overlapping matches, some of the graph elements (vertices/edges) are shared and some elements are unique for each match. The analyzed networks are usually directed, simple and loop free, that is, there are no multiple edges in the same direction between two vertices and there are no edges from a vertex to itself. Less commonly studied in network motif analysis are undirected networks as well as mixed networks that contain both directed and undirected edges.

There are several motifs that have been detected, and the properties of some motifs have been studied in more detail, e.g., the feed-forward loop motif [46, 112] and the bi-fan motif [45]. The feed-forward loop motif was shown to act as a persistence detector by filtering out noise within the process of gene regulation [72, 73]. The structure of some well-studied motifs is illustrated in Fig. 7.8.

7 Topology of Plant Metabolic Networks

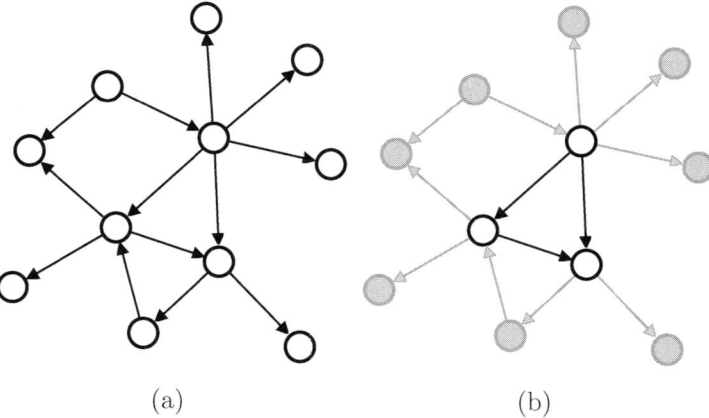

Fig. 7.7 Illustration of a motif match. In (**a**), a target graph is shown, and in (**b**), a match of the feed-forward loop motif (see Fig. 7.8 (a)) within the target graph is *highlighted* (all graph elements that are part of the match are displayed *black*, the remaining elements are displayed *grey*)

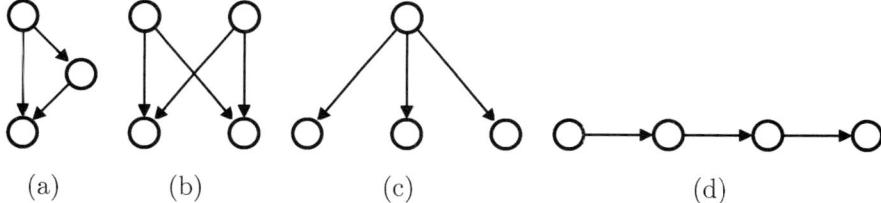

Fig. 7.8 Example of some well-studied network motifs: (**a**) feed-forward loop motif, (**b**) bi-fan motif, (**c**) single-input motif, and (**d**) regulatory chain motif

Besides analyzing the functional properties of individual motifs, numerous studies have been carried out that dealt with various different aspects on network motifs. In the protein–protein interaction (PPI) network of *S. cerevisiae*, proteins that are constituents of particular motifs are more evolutionary conserved than proteins which do not occur within such motifs [117]. Conserved network motifs have been used for the prediction of potential interaction partners of proteins in the PPI network of *S. cerevisiae* [1].

The frequency of motifs, or rather to say the statistical significance of a motif based on its frequency, characterizes the local structure of networks. This property has been used by many approaches to compare different networks. Profiles based on the statistical significance or rather the frequency of motifs have been used to classify networks from different domains into distinct superfamilies [78]. A network distance measure based on the frequency distribution of a set of motifs has been applied for the selection of suitable network generation models that best reflect the structure of PPI networks [88]. For a similar task, the frequency of particular motifs has been used for the application of discriminative classification techniques adapted from machine learning for the selection of a model of network evolution which best resembles the structure of the PPI network of fruit fly (*Drosophila melanogaster*) [77].

Frequently motif matches overlap with each other and only a minor part is isolated. Higher order structures built by overlapping matches have been characterized by different concepts like motif clusters [21], motif generalizations [59], and network themes [120], and it is assumed that different levels of network organization exist.

7.4.2 Motif Statistics

Network motifs have been originally introduced as particular patterns of interconnections that are statistically over-represented [79]. Calculation of the statistical significance of network motifs is usually done by comparing the frequency of a motif in the analyzed target network to the frequency distribution of this motif in random networks. It is assumed that random networks are free of any selection for particular interaction patterns and therefore are suitable as null model networks. The over-representation of particular subgraphs indicates for a selection of these interaction patterns, which could be caused by functional properties of the motif or design principles that shaped the network [3, 79].

A popular algorithm for the generation of random networks is based on local rewiring of connections. The algorithm replaces two edges (v_1, v_2) and (v_3, v_4) by the edges (v_1, v_4) and (v_3, v_2), provided that none of these edges already exist, see Fig. 7.9. Starting with the target network, this rewiring step is applied a great number of times to generate a properly randomized version of the target network. An important property of this algorithm is the conservation of the degree of all vertices in the randomized versions of the target network. By conserving the degree distribution, the random networks are assumed to have a similar over-all structure compared

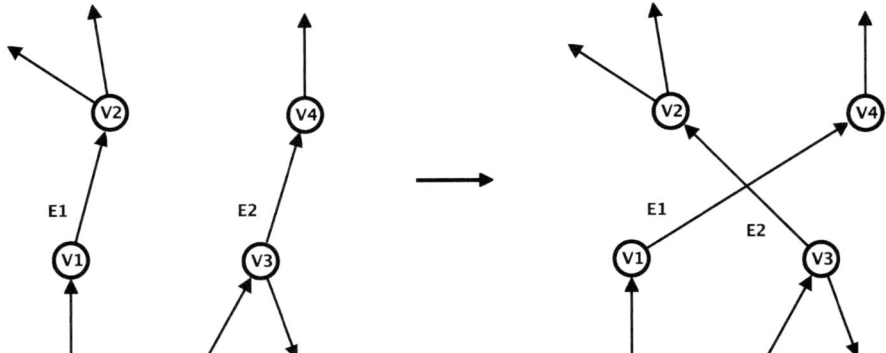

Fig. 7.9 Example of a rewiring step of the randomization algorithm. Two edges (*v*1, *v*2) and (*v*3, *v*4) are reconnected in such a way that *v*1 becomes connected to *v*4 and *v*3 to *v*2

to the target network but are locally randomized, and therefore are supposed to represent suitable null model networks for network motif detection.

The statistical significance of a motif is given by the P-value and the Z-score, which are calculated on the basis of the frequency of the motif in the target network and the frequency distribution of a sufficiently large set of random networks. Equation 7.4 shows the calculation of the Z-score for a motif m_i. F_i^{real} is the frequency of motif m_i in the target network. $\overline{F_i^{\text{rand}}}$ and σ_i^{rand} are the mean frequency and the standard deviation of motif in random networks, respectively. For a sound statistics, at least 1000 random networks should be considered [79]. Motifs of a network which are statistically significant under-represented compared to random networks are termed anti-motifs.

$$Z_i = \frac{F_i^{\text{real}} - \overline{F_i^{\text{rand}}}}{\sigma_i^{\text{rand}}} \qquad (7.4)$$

The motif significance profile SP is a vector of Z-scores of a set of motifs $\{m_1, \ldots, m_n\}$ that is normalized to the length one [78]; see Equation 7.5 for the calculation of the normalized Z-score for a motif m_i. Motif significance profiles are used for a comparison of networks independent of their size on the basis of motifs. Typically all motifs of a particular size are used as motif sets for the profile, e.g., the significance profile of all 13 (directed) motifs of size three is called the triad significance profile (TSP). TSPs have been used for the classification of various networks from different domains into distinct "superfamilies" [78] and have been applied in a study described in the following section.

$$SP_i = \frac{Z_i}{\sqrt{\sum_{j=1}^{n} Z_j^2}} \qquad (7.5)$$

7.4.3 Motifs in Metabolic Networks

The expression dynamics of genes encoding metabolic enzymes was investigated in a metabolic network of *S. cerevisiae*. The network for the study of network motifs was constructed by modeling metabolites as vertices and the conversion of metabolites through enzymatic reactions as edges. Edges are directed from the educts to the products of a reaction catalyzed by enzymes that are encoded by the studied genes. The 14 most highly connected metabolites (e.g., ATP, CO_2, H, NADP, NH_3, P_i) were excluded from the analysis, see also the discussion in Section 7.2.2. The expression distance between two genes was defined as one minus the correlation coefficient. Coexpression of the genes of all motifs with two edges and some motifs with three edges was analyzed. The matches of these motifs were detected, and the mean expression distance for each motif was calculated on the basis of the gene pairs of the corresponding matches. The mean expression distance was used to order the motifs, see Fig. 7.10. It was discovered that this ordering is in accordance with biological knowledge, e.g., a high level of coexpression of genes that follow each other and are part of a pathway (M1) and of genes that catalyze the same reaction (M4). In the latter case, these genes include homologous genes which frequently result from gene duplication. On the other hand, a low level of co-expression of genes building a futile cycle (M5) was detected as well as for genes that lead to the same metabolite in convergent branches (M3). The higher co-expression distance of motif M3 compared to motif M2 implies that co-regulation is stronger on divergent metabolic pathway than on convergent pathways, i.e., a preference for reactions that use the same metabolic precursor. The co-expression of genes in motifs with three edges supports the results obtained for motifs with two edges that co-regulation in divergent branches prevails co-regulation in convergent branches.

In another study, network motifs have been used for a comparative analysis of the local structure of metabolic networks [28]. These networks have been derived from 43 organisms covering the three domains of life (Eukaryota, Archaea, Bacteria). The metabolic networks model the connections of metabolites through reactions, i.e., metabolites are represented as vertices and directed edges are inserted from each educt to each product of a reaction. The triad significance profile (TSP) that comprises all 13 motifs of size three has been calculated for each network. All TSPs showed a similar distribution: the feed-forward loop (FFL) motif, a derivative of the FFL motif and the motif with three mutual edges between the three vertices are over-represented; three other patterns are anti-motifs that are significantly under-represented. The correlation coefficient between the TSPs of all 43 metabolic networks is 0.78 which implies a high similarity of the local organization of the networks. The comparison of these TSPs to TSPs of neuronal and transcriptional networks showed clear differences, indicating that these networks have a different local organization than metabolic networks.

Even though the overall correlation of the triad significance profile of the metabolic networks is relatively high, a more precise examination of the TSPs revealed that the taxonomy of the organisms is reflected by the correlation coefficients of the TSPs. For example, when the bacteria are divided into evolutionary subgroups, the correlation coefficient for the TSPs of the subgroups becomes much higher. Since this property also holds for the group of Eukaryota and Archaea, the

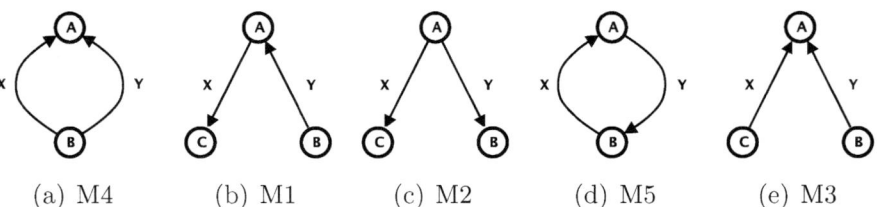

Fig. 7.10 Ordering of motifs according to mean co-expression distance of the gene pairs (X and Y) of the matches. Co-expression distance increases from (**a**) to (**e**)

grouping of subgroups based on the correlation of TSPs is in accordance with the taxonomical grouping of the organisms. Therefore, the taxonomy of the organism is reflected in the local organization of their metabolic networks measured by the triad significance profiles.

Furthermore, the distribution of 62 common metabolites (e.g., ATP, H, glycine) in matches of the motifs of size three has been studied in this work [28]. For each motif, the fraction of matches completely build with these common metabolites was divided by the fraction of matches completely build with 62 randomly selected metabolites to calculate the conservation ratio. This ratio was for all motifs except for one infrequent motif above 20 or even considerably higher, indicating that the common metabolites are not randomly distributed in metabolic networks but the topology of the network affects the distribution of these metabolites. Furthermore, it was discovered that more cohesive motifs (i.e., motifs which form triangles or which have mutual edges) have a higher fraction of matches built by common metabolites.

In a study on structural comparison of metabolic networks of single cell organisms, the same over-represented network motifs have been found in the eukaryote *S. cerevisiae* and six bacterial species [122]. In contrast, no over-represented motifs have been detected in the four studied archaeal species.

The motifs found in the networks of *S. cerevisiae* and the bacterial species are the feed-forward loop motif and two triangles with two and three mutual edges, respectively. These motifs are relatively dense and therefore support the finding that the metabolic networks of these species are more clustered and modular than those in archaeal species. The further studied topological properties support the conclusion based on motif analysis that the metabolic networks of the archaeal species are similar to each other but significantly different from those in *S. cerevisiae* and the bacterial species.

7.4.4 Tools

Various methods have been developed and implemented to analyze network motifs. For example, Pajek [6], a multipurpose program for the analysis and visualization of large networks includes also some motif analysis functionality, and Cytoscape [101], a software platform for analyzing and visualizing molecular interaction networks, offers with the NetMatch plug-in some possibilities for motif analysis, including consideration of vertex and/or edge labels. Beyond these applications, specialized tools have been developed for the detection and detailed analysis of network motifs.

Mfinder was the first specialized tool introduced for network motif detection [58, 75]. It is controlled from the command line and calculates the frequency of motifs along with the *P*-value and *Z*-score. A sampling method for a fast approximation of motif statistics is also included. The results are given as a text file, and the structure of detected motifs can be viewed in a separate motif dictionary.

MAVisto combines a motif-search algorithm with different views for the analysis and visualization of network motifs [74, 98]. It offers graph editor functionality for network manipulation and a force-directed layout algorithm to automatically generate readable drawings of the network which preserve the layout of motif matches if possible. Motif statistics are given by frequency values for different frequency concepts, P-value and Z-score. Furthermore, the analysis of vertex-labeled and/or edge-labeled networks is supported by MAVisto.

FANMOD [30, 114] is the newest of the three specialized tools and has a faster motif-detection algorithm compared to Mfinder and MAVisto, but in contrast to the other two, it detects only motifs that are induced subgraphs.

For motif detection, an exact method and a sampling method for a fast approximation of the motif number is available. Motif statistics is given by P-value, Z-score, and a proportional motif frequency that is relative to all motifs of a particular size. FANMOD offers a graphical user interface and the results are presented as text- or HTML files, whereas the latter include a representation of the structure of the motifs. FANMOD also supports the analysis of vertex labeled and/or edge-labeled networks.

7.5 Clustering of Metabolic Networks

7.5.1 Clustering

Metabolic networks are believed to be structured hierarchically into modules [35, 41, 71, 91]. Modularization allows to evolve, maintain, and coordinate cellular functions effectively, thus being an important feature of living systems on all levels of organization [27, 85].

To address the modularity of networks, efforts have been made to develop methods for the identification of modules using graph theory. Graph theoretical methods analyze networks based on topology, affording no prior knowledge about biological function and thus having the potential to give insight into metabolism based on unbiased structural modules [121]. By grouping vertices with respect to their functional meaning, these techniques are often referred to as network clustering techniques.

Cluster analysis comprises a range of methods for grouping objects of similar kind into respective subgroups. The analytical goal is to find disjoint subgroups (clusters) such that elements within the same cluster are similar to each other and elements in different clusters are dissimilar. By organizing heterogeneous data into homogeneous subgroups, clustering can help to discover natural groups in datasets, to identify representatives for homogeneous groups (data reduction) or to find unusual data objects (outlier detection).

Cluster analysis is used in bioinformatics and has a wide range of applications, such as data mining, machine learning, and pattern recognition. Due to its broad applicability, a large number of statistical and computational approaches are available for clustering. These techniques can be summarized according to [39] the following categories: partitioning methods, hierarchical methods, density-based methods, grid-based methods, and model-based methods. For an overview of the multivariate statistics used in cluster analysis, the reader is referred to [4, 29, 47]. The focus on this section will be on partitioning and hierarchical methods:

- *Partitioning methods* A partitioning method divides objects into a number of non-overlapping clusters such that the partitions optimize a given criterion. In most of the partitioning methods, an initial partitioning is chosen and, by using the defined criteria to judge the quality of the clusters, the cluster membership is changed iteratively until an optimal partitioning is obtained.
- *Hierarchical methods* A hierarchical method disposes a given dataset into a hierarchically ordered sequence of partitions. This hierarchy is represented by a hierarchical clustering tree (dendrogram), with the data samples being located in the leaves of the dendrogram and similar samples occurring in proximate branches. The dendrogram can be cut at any level to yield different clusterings (i.e., partitions) of the data. The procedure of a hierarchical clustering algorithm is illustrated using the dataset in Fig. 7.10. A dendrogram corresponding to the data points in Fig. 7.10 is shown in Fig. 7.12. There are two kinds of hierarchical clustering methods: agglomerative ("bottom-up") and divisive ("top-down") clustering. An agglomerative approach begins with each element in a distinct (singleton) cluster and successively merges clusters together until one cluster remains. A divisive method begins with all elements in a single cluster and performs splitting until each element is assigned to a separate cluster.

Prerequisite for any clustering is the selection of a distance measure, that is, a metric (or quasi-metric) on the feature space used to quantify the similarity of elements [47]. Given a set of elements $E = \{e_1, \ldots, e_k\}$, the distance between every pair of elements is computed based on the distance measure, i.e., a distance function $d = d(e_i, e_j) = d_{ij}$ with $d_{ij} \geq 0$, $d_{ij} = d_{ji}$, and $d_{ii} = 0$. The distance is usually computed by comparing feature vectors or single features of the elements. The $k \times k$ (for k elements) matrix $D = d_{ij}$ is called a *distance matrix*. A distance matrix corresponding to the data points in Fig. 7.11 is shown in Fig. 7.12, and a corresponding dendrogram in Fig. 7.13.

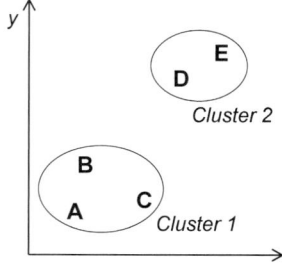

Fig. 7.11 Two-dimensional dataset falling in two clusters

Fig. 7.12 The distance matrix corresponding to the two-dimensional dataset shown in Fig. 7.11 (distance measure: Euclidean distance [48])

	A	B	C	D	E
A	0				
B	3.2	0			
C	5.1	4.4	0		
D	10.6	7.8	7.3	0	
E	14.1	11.4	10.3	3.6	0

For an overview of distance measures using quantitative as well as qualitative features, the reader is referred to [4]. Note that in graphs measures such edge weights usually do not correspond with a distance in an Euclidian geometry [23]. Therefore for network clustering, the similarity between vertices is usually defined by local or global characteristics of the graph such as degree or betweenness (Fig. 7.13).

7.5.2 Network Clustering

Network clustering deals with clustering data represented as a network. The goal of network clustering with respect to biochemical networks is to group vertices by means of their biological meaning. Network clustering techniques can be distinguished by the way natural groups (modules) are defined in a network. Network clustering is related to the field of graph partitioning, which is the study of finding the optimal partition of a graph with given constraints.

The division of a network into smaller functional units facilitates understanding the modularity and design principles of the network [71], allows to gain new information on the internal structure of the network and to validate the network

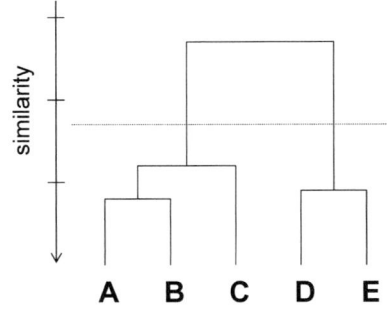

Fig. 7.13 The dendrogram corresponding to the distance matrix shown in Fig. 7.12 (clustering algorithm: UPGMA [104]). The horizontal line represents the cut which leads to the partitioning shown in Fig. 7.11

model. Besides giving insight into the functioning of the network, the decomposition may help to enhance, reconstruct, and simplify the model [27], thus supporting the process of network modeling. With respect to metabolic networks, network decomposition is also necessary for functional analysis of metabolism by pathway analysis methods such as extreme pathway analysis and elementary modes analysis as theses techniques are often hampered by the problem of combinatorial explosion [71].

Various methods of network clustering have been developed and applied to identify modules in various biological systems, including protein-interaction networks [92, 103, 105], signal transduction networks [27], metabolic networks [91, 121], and food webs [36, 65]. In the following, an overview of methods for decomposing metabolic networks is given: Ravasz et al. [91] proposed a mathematical framework to reveal the presence of hierarchical modularity and to identify modules based on the network topology.

By applying an average-linkage clustering algorithm to the topological overlap matrix of the condensed metabolite graph, the underlying modularity of the network is uncovered. Using this approach, the authors analyzed the metabolic network of E. coli, finding that the identified modules (i.e., subsets of metabolites) closely overlap with the known metabolic functions. To illustrate this approach, a small example adapted from [91] is given in Figs 7.14 and 7.15. According to [71], a potential drawback of this metabolite-graph-based method is that, from a biological viewpoint, a subset of metabolites cannot sufficiently define a module and that only local connectivity property is considered for the distance calculation.

Considering this drawback, Ma et al. [71] used a reaction/enzyme network representation of a metabolic network for decomposition and the global connectivity structure for distance calculation. Based on the bow-tie structure of metabolic networks, they proposed a network clustering technique using neighbor joining, a hierarchically agglomerative clustering method and a distance measure derived from the path length between reactions. Having decomposed the core of the giant strongest component (GSC) by this method, the decomposition of the core part is expanded to the global network by using a majority rule. Zhao et al. [121] extended this method

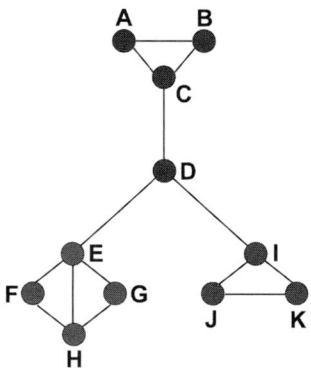

Fig. 7.14 Small hypothetical network (adapted from [91]) (*see* also color Insert)

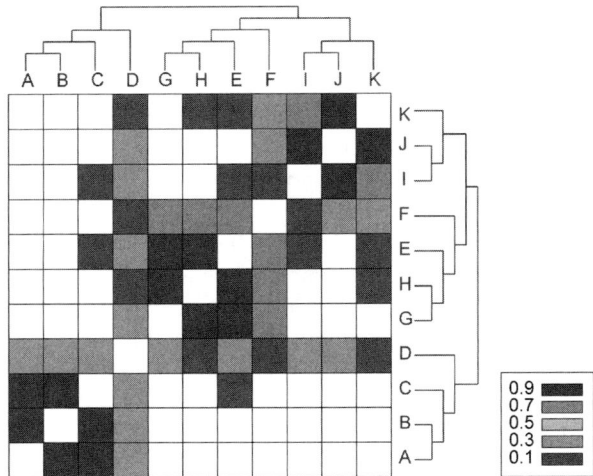

Fig. 7.15 The topological overlap matrix (adapted from [91]) corresponding to the network shown in Fig. 7.13. The topological overlap between each pair of vertices v_i and v_j is defined according to [91] as $O_T(v_i, v_j) = \frac{J_n(v_i, v_j)}{[\min(d(v_i), d(v_j))]}$, where $J_n(v_i, v_j)$ denotes the number of vertices to which both v_i and v_j are linked and $[min(d(v_i), d(v_j))]$ is the smaller of the degrees of v_i and v_j. The degree of topological overlap between the vertices is reflected by the color code. The associated hierarchical tree is obtained by applying an average linkage clustering algorithm to the topological overlap matrix. The given tree reflects three distinct modules contained in the network of Fig. 7.14 (*see also color Insert*)

by applying a distance measure computed by the Floyd algorithm and Ward's clustering, a hierarchically agglomerative clustering method.

Opposite to these topology-based network clustering techniques, Ederer et al. [27] proposed an algorithm to hierarchically divide an ordinary differential equation (ODE) model of a biological reaction network into non-overlapping functional units. After performing simulation, the distance between network compounds is computed by comparing activity vectors, and a linkage method is used to generate a dendrogram revealing the internal structure of the reaction network. Using this method, the authors analyzed two models of different biological background, with the identified modules being similar to the functional units of the considered systems.

In addition to these hierarchical agglomerative modularization techniques, a variety of other methods exit in literature: Schuster et al. [99] proposed a decomposition method based on the metabolite degree of network connectivity. By considering metabolites above a given connectivity threshold as external and removing them from the network, the metabolic network is decomposed into connected components of internal metabolites. By applying the method to the metabolic network of *Mycoplasma pneumoniae*, they showed that the obtained modules are in agreement with biochemical knowledge.

The degree-based method proposed by Gagneur et al. [35] defines modules on the basis of the connected components of the subgraph induced by the metabolites of lowest degree and their reactions. According to Ma and Zeng [71], a potential drawback of these degree-based methods is that local properties such as the connection degree may be inappropriate to reveal the global organization structure. Considering this drawback, Holme et al. [41] developed an algorithm to decompose biochemical networks into subnetworks by successively removing reactions of high betweenness centrality. In [38], a method is proposed to identify functional modules in metabolic networks by maximizing the networks's modularity using simulated annealing. An overview of recently published mathematically based top-down modularization techniques like the identification of *Correlated Reaction Sets* or *Concentration Pools* is given in [87].

Given the variety of network clustering techniques, it has to be considered that different methods predict different modules with the boundaries between modules not being sharply separated. Rather than being a limitation of the present network clustering techniques, this ambiguity is a consequence of the networks hierarchical modularity [5]. The hierarchical modularity of metabolic networks is characterized by a natural breakdown of metabolism into several large modules, which are further partitioned into smaller, but more integrated sub-modules [91]. At present, objective quantitative criteria to evaluate the decomposing quality of a network are sparse. References [37] and [121] used the modularity metric in order to identify the optimal partition of a given hierarchical tree. Due to the fact that well-defined modules occur at different levels of organization (at different levels of the hierarchical dendrogram), some authors propose to avoid looking at an absolute set of modules at a specific level, but rather to visualize the hierarchical relationship between modules of different sizes [5, 40, 91].

The identification of hierarchical modularity in biochemical networks is a key issue in network biology, and one that is likely to witness much progress in the near future [5].

7.5.3 Tools

The following websites describe some of the available computational tools for decomposing metabolic networks discussed in this section: SEPARATOR [99, 100] and HI [40, 41]. Programs for the identification of *Correlated Reaction Sets* [87] are, for example, CellNetAnalyzer [15] and Flux coupling finder [33].

7.6 Visualization of Metabolic Networks and Experimental Data

Massively parallel techniques such as metabolite profiling [32, 93] generate increasing amounts of experimental data, which offer a top-down view of the biochemistry of an organism. The data interpretation is usually limited by analysis and visualization procedures. The central task of data visualization is to bring the data into a form that shows it with reasonable precision, while at the same time being readable and understandable. Analysis of the experimental data is eased by the consideration of knowledge about reaction networks, stored in pathway databases such as KEGG [52] and BRENDA [96].

Visual exploration of metabolic networks and network-integrated data visualization are useful techniques for modern biology research.

7.6.1 Static and Dynamic Visualizations

Static visualization of metabolic networks is used not only in biochemical textbooks and on posters [8, 76, 84] but also in software systems such as BioCarta, ExPASY, KEGG [52], and MapMan [108, 110]. It is characterized by the following aspects:

1. Diagrams which are created manually long before their use.
2. A predefined view of the data (e.g., the elements shown, the level of detail), which usually cannot be changed by a user.
3. Restricted navigation (sometimes supported by links to other pictures, but the result (the new picture) either replaces the current image or is shown in an independent new view).
4. No editing of the diagram by the end-user.

In contrast to static visualization, dynamic visualization is characterized by

1. Diagrams which are created automatically by the end-user based on the up-to-date data at the time the drawing is needed.
2. A flexible view onto the data and annotation of network elements (vertex labels, links to other resources, and level of detail can be modified using different interaction techniques).
3. Navigation methods are supported and it is possible to extend existing drawings with new parts.
4. Editing is possible and the layout/graphical representations may be changed by the end-user as needed; also the structure of pathways may be modified by adding or removing elements.

Static visualizations are often of higher visual quality in comparison to computer-generated dynamic visualizations, but they are tuned to a specific use case and often not based on the most up-to-date data. They are not editable, the space for data annotation and the relative layout of network elements is fixed, and unneeded information can't be easily removed. When several small parts of a large pathway chart may be investigated or when pre-defined static visualizations of the metabolism are used (e.g., KEGG system), navigation becomes difficult. Because of these aspects, dynamic visualization is well suited for the interactive and flexible exploration of metabolic pathways and the network-integrated visualization of experimental data; it is the state-of-the-art method to present such information.

7.6.2 Interaction Times and Techniques for Dynamic Visualization Systems

Two important aspects of human–computer interaction are the aspect of time in user interactions and established interaction techniques. Both aspects are essential for the design or assessment of visualization software.

7.6.2.1 Interaction Times

Dynamic visualizations may be interactively manipulated using a wide variety of interaction techniques. Depending on the speed of the user interaction, three time-dependent levels of interaction may be differentiated [14]: roughly 0.1 s, 1 s, and 10 s. Response times of visualization systems need to be tuned to these time constants for the fluent and intuitive interaction between the user and the computer.

The time frame of 0.1 s is called *physiological moment*. Two similar images shown in less than 0.1 s after each other result in the perception of a motion. If a visualization changing user-command is processed in less than 0.1 s, the user perceives the update of the screen as a direct reaction on the user input, which eases tweaking visualization parameters.

The next level of interaction is the time frame of about 1 s, the time of an *unprepared response*. Events that occur in less than about 1 s happen too quickly for a sensible reaction of a user. By using animations of a maximum length of about 1 s, a smooth and at the same time non-disturbing interaction with a computer program is made possible. Progress information about long-lasting calculations should update at least each second.

The coarsest level of user interaction happens at roughly 10 s, a user's response time for a typical *unit task*, the minimal unit of cognitive work.

These different levels of interaction times influence the design and implementation of interaction techniques, the most common ones are introduced in the following.

7.6.2.2 Interaction Levels and Techniques

User interaction may change the parameters of a visualization at three different levels from raw data to the visualization views [14]. The first level is connected to the data transformation from raw data to the data tables (which then will be visualized). Common interaction techniques at this level are "dynamic queries" (interface elements allow the user to specify value ranges for cases to be highlighted or hidden), "direct walk" (navigation to different datasets by data linkages), "details on demand" (expansion of the visualization to show more details of an object), "attribute walk" (starting with a specific parameter value of an object under investigation, objects with the same specificity are highlighted), "brushing" (concurrent visualizations are updated accordingly as the user manipulates a subset of the visualized objects), and "direct manipulation" (enables the direct manipulation of visualization parameters within the display).

The next level of user interaction happens at the point where the data tables are mapped to visual structures. At this level, mostly domain-specific interaction techniques have been developed. Examples are "pivot tables" of spreadsheet applications and the "data-flow" technique, for the graphical specification of diagrams which determine the mapping from data to visual properties.

Finally, user interaction may influence the transformation from visual structures to data views. The following techniques (or a subset of these) are commonly used in visualization systems at this level: "direct selection" (selection of a set of objects which are highlighted for visual investigation or which are the argument of a subsequent user interaction), "camera movement" (change of observer position in 3D space), "magic lenses" (data or view transformation for objects, dependent on their x, y position; several lenses may be placed over each other, resulting in a combination of the individual transformation effects), "overview + detail" (an overview of data and a marked region are shown in one view; a second view shows the objects of the highlighted region in more detail), and "zoom" (a subset of the data is shown in more detail).

7.6.3 Pathway Layout

A metabolic pathway consists of a number of interconnected biochemical reactions. Most reactions are catalyzed by reaction-specific enzymes. During the reaction, the reactants (substrates) are transformed into products. Some enzymes require co-factors (small molecules or ions); these substances are often treated specially or even left out in the visualization of metabolic pathways. Many biochemical reactions are reversible, and thus the definition of reactants and products may vary. The preferred direction of a reaction in the cell environment is indicated by the edge direction in the graph model.

Commonly directed bipartite graphs are used as a graph model. Currently, most of the information visualization systems use differing drawing styles for pathway elements. For example, arrows often have different meanings and drawings created by different visualization systems are thus ambiguous. The SBGN (Systems Biology Graphical Notation) project works on a standard representation of pathway elements [61].

Standard graph layout approaches, such as circular, orthogonal, tree layouts, and force directed [11, 19, 20], individually applied to metabolic pathways of medium to large size, give only poor results or depending on the structure of the pathway do not work at all. Thus a number of layout algorithms especially tailored for the layout of pathways have been developed, the most notable are introduced in the following. Major improvements in the layout of metabolic pathways are achieved with these approaches by the consideration of subgraph-topologies, special layout of co-factors and enzymes, or the consideration of compartmental constraints.

Karp and Paley [54] developed a divide-and-conquer algorithm for the identification of subgraph-topologies as linear, circular, and branched (phase 1 of Karp's layout approach). The identified subgraphs are subsequently laid out using different layout approaches and placed next to each other for the complete layout.

The spacing of the graph elements depends on the number of co-substances and enzymes. Linear subgraph structures are laid out as horizontal or vertical lines or using a so-called *snake layout*, which is used in many biochemistry books. A circular layout is used for circles; for branched subgraphs a tree layout is applied. Karp's algorithm also considers co-substances and enzymes, which are processed in the last phase of the algorithm. This layout algorithm is implemented in the EcoCyc, BioCyc, and MetaCyc systems which are electronic encyclopedias for biochemical pathways of various organisms [53, 57].

Becker and Rojas developed a layout algorithm for metabolic pathways which incorporates a special force-directed layout algorithm and additional layout heuristics [7]. For this layout approach, hyper-edges are simulated by inserting dummy vertices at the front and back of an edge, which connect to reactants and products of a reaction. This algorithm starts with a search for the longest cycle in a graph. If no cycle could be found, a top-to-bottom hierarchic layout [26, 107] from the yFiles graph library [115] is used to layout the complete graph. If the longest cycle is the complete graph, a circular layout is applied. The general case is that a cyclic subgraph has been identified. The remaining vertices that have at least two connections to vertices of the cycle but not to other graph vertices are placed inside the cycle (inner components). The remaining outer components (strongly connected subgraphs) are analyzed recursively using the same approach as outlined before. After that, the identified components are collapsed into "super-vertices", which are placed and sized according to the subgraphs they represent. A customized spring-embedder (force-directed) layout [25, 89] which avoids overlapping of vertices is applied to the graph, consisting of the main cycle, the inner and outer components. During this layout, the circle subgraphs are rotated to conform to a preferred radial angle, determined by the connection to outer components. Finally, the super-vertices are replaced by the corresponding subgraphs, a process which restores the former network structure.

An advanced version of the previously described algorithm has been presented by Wegner and Kummer [113]. In contrast to the original algorithm, this approach identifies circles, beginning with the smallest ones. It is argued that these small cycles in the metabolic pathway often represent important recycling processes or shortcuts and should therefore be favored during layout. Similar to Karp et al. [55], this algorithm processes a predefined list of co-factors (e.g., ATP, NADP, etc.). In preparation of the actual layout process, co-factors and vertices which are part of more than one cycle are divided into several vertices, in a way that each split vertex has a single connection to one of the former connected vertices. In a post-processing phase, edge crossings are minimized by splitting vertices which are connected to edges crossing other edges. By lowering the number of allowed edge crossings, more vertices will be split, it is thus even possible to create planar graph drawings, which show no edge crossings at all.

As part of the BioPath project [10], which provides an electronic version of Michal's pathways poster and book [76], a layered layout algorithm has been developed [97]. In the first step of this algorithm, single reactions, reaction enzymes, and co-factors are placed taking reaction-specific information such as the ordering of co-factors and enzymes into account. Each reaction with its co-factors and enzymes forms a subgraph and is replaced by a reaction vertex. The size of this vertex is determined by the size of the drawing of the corresponding subgraph. The next step is the computation of layout constraints. Examples are horizontal and vertical constraints, which fix the ordering and relative positioning of some of the vertices (e.g., to achieve a circular layout for known metabolic cycles) and constraints for metal map preserving layouts. After that, a layered layout, extending the Sugiyama algorithm [107], is calculated. In a last step, the reaction vertices are replaced by the corresponding subgraphs, consisting of reaction enzyme and co-factor vertices. The method is able to deal with several hierarchical levels of a metabolic network.

Dogrusoz and coauthors developed a layout technique with focuses on hierarchical graphs, which is the result of the modeling of complicated biological pathways with compartmental constraints and nested drawings [22]. This layout algorithm is based on a force-directed layout approach [34]. During initialization, initial vertex and compartment sizes as well as an initial positioning are calculated. For efficiency reasons, parts of the graph that are trees are temporarily iteratively removed until no such vertex is left. The remaining part of the graph is then laid out using a spring embedder model (which models repulsive forces between all vertices, depending on the distance, attractive or repulsive forces between connected vertices). The force calculation takes varying vertex sizes, compartment locations, and relativity forces on substrates and products into account. After the spring embedder algorithm loop is finished, the former removed vertices are added back to the graph and laid out using the spring model, outlined before. The algorithm is implemented in the software system PATIKA (Pathway Analysis Tool for Integration and Knowledge Acquisition) [18].

7.6.4 Network-Integrated Data Visualization

Methods and tools which assist in the interpretation of experimental data are an important field of development in bioinformatics, and several approaches have been proposed. Examples are scatter plots of pairs of experiments [17], clustering methods with visualizations of the results [24, 31, 106], and mapping of gene expression data onto pathways and their visualization using graphical attributes (e.g., color codes) to show the level of gene expression [16, 56, 66, 80, 82, 83, 86, 108, 116].

For visualization purposes, many visual features such as object position and size, object coloring, texture, and shape of a drawing may be modified to reflect the value of quantitative, ordinal, or nominal data. Metabolite data are mostly of quantitative nature; expression data are sometimes simplified to an ordinal scale (up-regulated, unchanged, down-regulated), nominal data can be rarely found. Card [14] discusses the relative effectiveness of the mentioned visual features for visualization purposes: Object positions and size (part of diagrams and shape coding) are good techniques for the visualization of the extend of quantitative, ordinal, and nominal data. Gray-scale coloring is reported to be good for ordinal, marginal for quantitative, and poor for nominal data. For data comparison purposes, differential quantitative and ordinal data are reported to be visualized only marginally effective with a color scale, but with good effectiveness for nominal data. Object textures are marginal effective for the visualization of differential quantitative and ordinal data, and good for nominal data. Visualizations using different object shapes have poor effectiveness for quantitative and ordinal differential data, but are well suited for nominal data. Here we discuss approaches for the mapping of metabolomics data (e.g., time series data, data of different plant lines) onto metabolic networks using the techniques heatmap, shape coding, and diagram charting. Sometimes object shape or texture is changed to indicate special properties of a part of a network, but they are not commonly used for the visualization of metabolite data. To compare the mentioned visualization techniques, an example data table from a study [94] about the influence of oxygen supply on the metabolism of soybean (*Glycine max*) seeds is used.

7.6.4.1 Color Coding

Color-coding techniques, sometimes called "heatmaps", are commonly used to visualize large-scale experimental data. The simplest case is the visualization of a single value, directly measured or the result of a calculation, such as a ratio of measurements for two different environmental conditions (see left part of Fig. 7.16), or two time points.

The color of a network element or small geometric objects, placed next to graph edges or vertices, is determined by a transformation function or a discrete mapping table. In both cases, commonly two colors (or gray scales) are used to indicate the minimum and maximum value to be visualized. A third color (mostly white) is used to indicate the value zero or one, which stands for data which does not change depending on the experimental factor under investigation.

Examples for static pathway visualization systems which utilize color coding are KaPPA-View [109] and MapMan [108, 110]. Dynamic network visualization systems with support for color coding are, for example, SimWiz [95] and VANTED [50], which was used to generate the diagrams in Figs 7.16 and 7.17.

7.6.4.2 Shape Coding

For shape coding, two different approaches are imaginable: the change of network elements shapes (e.g., rectangular or circle vertex shape) and the change of network element shapes size. The use of different vertex shapes to indicate the extent of related experimental data is rare, usually the shape depends on other aspects, like the type of a network element (e.g., compound, enzyme, or gene). Systematically this visualization aspect is addressed in the SBGN (Systems Biology Graphical Notation) standard [61]. Much more common for the visualization of experimental data in context of a network is the modification of vertices size or the modification of the graph edge widths. As in the case of color coding, mapping tables or transformation functions are used to visualize single experimental values or ratios of the data from two different conditions (see right part of Fig. 7.16) or of two time points.

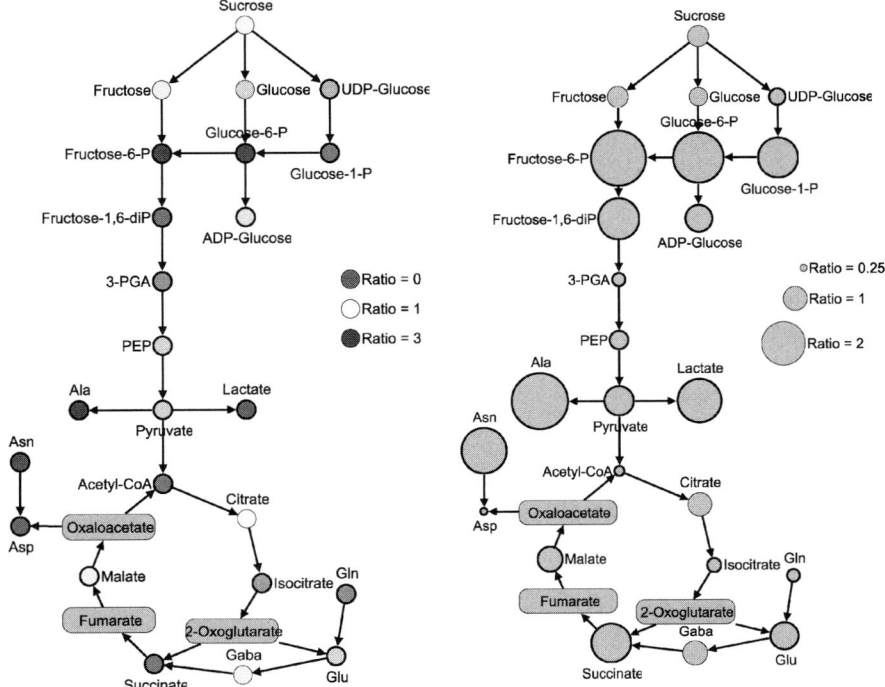

Fig. 7.16 *Left*: Color-coded visualization of the ratio of experimental data from two different environmental conditions (reduced oxygen supply condition data divided by data from normal oxygen supply condition, source of data: [94]). *Right*: Shape-coded visualization of the same data (*see* also color Insert)

7.6.4.3 Diagrams

Diagrams are widely used to visualize complex structured experimental data. The most commonly used diagram types are line or curve, column or bar, area, histogram, pie, and scatterplot. Proportional relationships at a point in time may be visualized using pie, column, or bar charts. Multiple pie charts may be used to visualize the proportional relationship over a number of time points. Trends and functional relations may be visualized using line-, curve-, area-, column- or bar charts. Ratios or other parameters may be easily compared using column- or bar charts (see Fig. 7.17).

Although diagram techniques are widely used in scientific publications and for presentations, they are not that often embedded into pathway visualizations. Among the first pathway visualization systems that supported the display of line-chart diagrams directly inside pathway drawings are the PathwayExplorer [80] and VisANT [43, 44] systems. The flexible use of line- and bar charts and the possibility to visualize the results of statistic calculations in context of dynamic pathway drawings has been pioneered in the DBE-Gravisto system [9]. VANTED, the successor of DBE-Gravisto, adds among other improvements the possibility to use (multiple) pie charts and to map and display complex structured datasets on graph vertices and edges [50].

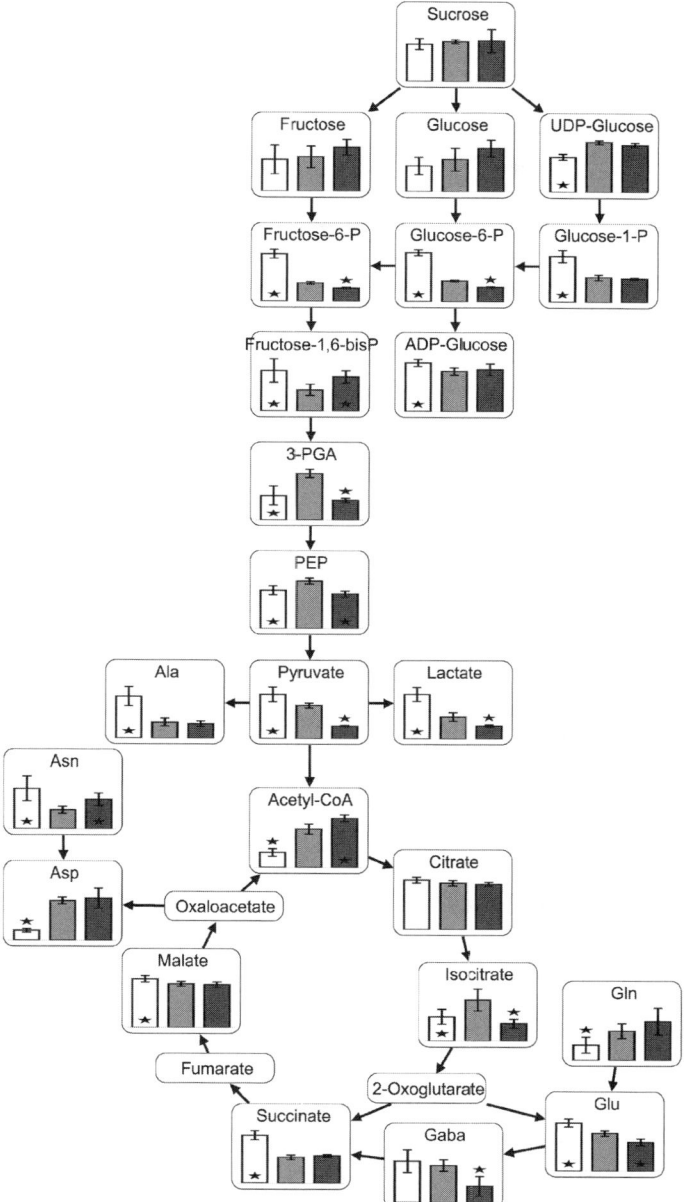

Fig. 7.17 Network-integrated visualization of data from three different environmental conditions (decreased, normal, and increased oxygen supply). Stars inside or above diagram bars indicate statistical significant differences in comparison to normal oxygen supply condition; error bars show standard deviation of replicate measurements; see Fig. 7.16 for further details

References

1. Albert I, Albert R (2004) Conserved network motifs allow protein-protein interaction prediction. Bioinformatics 20:3346–3352.
2. Arita M (2004) The metabolic world of *Escherichia coli* is not small. Proc Natl Acad Sci USA 101:1543–1547.
3. Artzy-Randrup Y, Fleishman SJ, Ben-Tal N, Stone L (2004) Comment on "Network motifs: simple building blocks of complex networks" and "Superfamilies of evolved and designed networks". Science 305:1107c.
4. Backhaus K, Erichson B, Plinke W, Weiber R (2003) Multivariate Analysis Methods. An Application-Oriented Introduction, 10th ed., Springer, Berlin.
5. Barabási AL, Oltvai ZN (2004) Network biology: understanding the cell's functional organization. Nat Rev Genet 5:101–113.
6. Batagelj V, Mrvar A (2004) Pajek – Analysis and visualization of large networks. In: Jünger M, Mutzel P (eds) Graph Drawing Software (Mathematics and Visualization). Springer, Berlin/Heidelberg, pp. 77–103.
7. Becker MY, Rojas I (2001) A graph layout algorithm for drawing metabolic pathways. Bioinformatics 17:461–467.
8. Berg JM, Tymoczko JL, Stryer L (2002) Biochemistry. W H Freeman, New York.
9. Borisjuk L, Hajirezaei MR, Klukas C, Rolletschek H, Schreiber F (2005) Integrating data from biological experiments into metabolic networks with the DBE information system. In Silico Biology 5:93–102.
10. Brandenburg FJ, Forster M, Pick A, Raitner M, Schreiber F (2004) BioPath – exploration and visualization of biochemical pathways. In: Jünger M, Mutzel P (eds) Graph Drawing Software (Mathematics and Visualization). Springer, Berlin/Heidelberg, pp. 215–236.
11. Brandenburg FJ, Jünger M, Mutzel P (1997) Algorithmen zum automatischen Zeichnen von Graphen. Informatik Spektrum 20:199–207.
12. Brandes U, Wagner D (2004) Visone – analysis and visualization of social networks. In: Jünger M, Mutzel P (eds) Graph Drawing Software (Mathematics and Visualization). Springer, Berlin/Heidelberg, pp. 321–340.
13. Breitkreutz BJ, Stark C, Tyers M (2003) Osprey: a network visualization system. Genome Biology 4: R22.
14. Card SK, Mackinlay JD, Shneiderman B (1999) Readings in Information Visualization: Using Vision to Think. Morgan Kaufmann Publ. Inc., San Francisco, CA.
15. CellNetAnalyzer http://www.mpi-magdeburg.mpg.de/projects/cna/cna.html.
16. Chung HJ, Kim M, Park CH, Kim J, Kim JH (2004) ArrayXPath: mapping and visualizing microarray gene-expression data with integrated biological pathway resources using scalable vector graphics. Nucleic Acids Res 32(Web-Server-Issue):460–464.
17. Colantuoni C, Henry G, Zeger S, Pevsner J (2002) SNOMAD (standardization and normalization of microarray data): web-accessible gene expression data analysis. Bioinformatics 18:1540–1541.
18. Demir E, Babur, O Dogrusoz U, Gursoy A, Nisanci G, Cetin-Atalay R, Ozturk M (2002) PATIKA: an integrated visual environment for collaborative construction and analysis of cellular pathways. Bioinformatics 18:996–1003.
19. Di Battista G, Eades P, Tamassia R, Tollis IG (1994) Annotated bibliography on graph drawing algorithms. Comput Geom-Theor Appl 4:235–282.
20. Di Battista G, Eades P, Tammasia R, Tollis IG (1999) Graph drawing: algorithms for the visualization of graphs. Prentice Hall, New Jersey.
21. Dobrin R, Beg QK, Barabási AL, Oltvai ZN (2004) Aggregation of topological motifs in the *Escherichia coli* transcriptional regulatory network. BMC Bioinformatics 5:10.
22. Dogrusoz U, Giral E, Cetintas A, Civril A, Demir E (2004) A compound graph layout algorithm for biological pathways. In: Pach J (ed.) Graph Drawing. Springer, Berlin/Heidelberg, pp. 442–447.

23. van Dongen SM (2000) Graph Clustering by Flow Simulation. Center for Mathematics and Computer Science, Amsterdam.
24. Dysvik B, Jonassen I (2001) J-Express: exploring gene expression data using Java. Bioinformatics 17:369–370.
25. Eades P (1984) A heuristic for graph drawing. Congr Numer 41:149–160.
26. Eades P, Sugiyama K (1990) How to draw a directed graph. J Inform Proc 13:424–437.
27. Ederer M, Sauter T, Bullinger E, Gilles ED, Allgöwer F (2003) An approach for dividing models of biological reaction networks into functional units. Simulation 79:703–716.
28. Eom YH, Lee S, Jeong H (2006) Exploring local structural organization of metabolic networks using subgraph patterns. J Theor Biol 241:823–829.
29. Everitt BS, Landau S, Leese M (2001) Cluster analysis. Oxford University Press Inc., New York.
30. FANMOD http://www.minet.uni-jena.de/~wernicke/motifs/index.html.
31. Fellenberg M, Mewes HW (1999) Interpreting clusters of gene expression profiles in terms of metabolic pathways. In: Giegerich R, Hofestädt R, Lengauer T, Mewes W, Schomburg D, Vingron M, Wingender E (eds) German Conference on Bioinformatics, Springer. pp. 185–187.
32. Fiehn O, Kopka J, Dörmann P, Altmann T, Trethewey R, Willmitzer L (2000) Metabolite profiling for plant functional genomics. Nature Biotechnol 18:1157–1161.
33. Flux Coupling Finder, http://fenske.che.psu.edu/Faculty/CMaranas/programs.html.
34. Fruchterman TMJ, Reingold EM (1991) Graph drawing by force-directed placement. Software Practice and Experience 21:1129–1164.
35. Gagneur J, Jackson DB, Casari G (2003) Hierarchical analysis of dependency in metabolic networks. Bioinformatics 19:1027–1034.
36. Girvan M, Newman MEJ (2002) Community structure in social and biological networks. Proc Natl Acad Sci USA 99:7821–7826.
37. Girvan M, Newman MEJ (2004) Finding and evaluating community structure in networks. Phys Rev E 69(026113).
38. Guimerà R, Nunes Amaral LA (2005) Functional cartography of complex metabolic networks. Nature 443:895–900.
39. Han J, Kamber M (2001) Data Mining, Concepts and Techniques. Morgan Kaufmann Publishers, USA.
40. HI www.tp.umu.se/forskning/networks/meta/.
41. Holme P, Huss M, Jeong H (2003) Subnetwork hierarchies of biochemical pathways. Bioinformatics 19:532–538.
42. Hood L, Perlmutter RM (2004) The impact of systems approaches on biological problems in drug discovery. Nature Biotechnol 22:1215–1217.
43. Hu Z, Mellor J, Wu J, Yamada T, Holloway D, DeLisi C (2005) VisANT: data-integrating visual framework for biological networks and modules. Nucl Acids Res 33(Web Server issue):W352–W3577.
44. Hu Z, Mellor JC, Wu J, DeLisi C (2004) VisANT: an online visualization and analysis tool for biological interaction data. BMC Bioinformatics 5:17.
45. Ingram PJ, Stumpf MP, Stark J (2006) Network motifs: structure does not determine function. BMC Genomics 7:108.
46. Ishihara S, Fujimoto K, Shibata T (2005) Cross talking of network motifs in gene regulation that generates temporal pulses and spatial stripes. Genes to Cells 10:1025–1038.
47. Jain AK, Dubes RC (1988) Algorithms for Clustering Data. Prentice Hall, New York.
48. Jain AK, Murty MN, Flynn PJ (1999) Data clustering: a review. ACM Computing Surveys 31:264–323.
49. Jeong H, Tombor B, Albert R, Oltvai ZN, Barabási AL (2000) The large-scale organization of metabolic networks. Nature 407:651–654.
50. Junker BH, Klukas C, Schreiber F (2006) VANTED: A system for advanced data analysis and visualization in the context of biological networks. BMC Bioinformatics 7:109.

51. Junker BH, Koschützki D, Schreiber F (2006) Exploration of biological network centralities with CentiBiN. BMC Bioinformatics 7:219.
52. Kanehisa M, Goto S, Hattori M, Aoki-Kinoshita KF, Itoh M, Kawashima S, Katayama T, Araki M, Hirakawa M (2006) From genomics to chemical genomics: new developments in KEGG. Nucleic Acids Res 34:D354–D358.
53. Karp PD, Ouzounis CA, Moore-Kochlacs C (2005) Expansion of the BioCyc collection of pathway/genome databases to 160 genomes. Nucleic Acids Research 33:6083–6089.
54. Karp PD, Paley S (1994) Automated drawing of metabolic pathways. In: Hunter L, Searls D, Shavlik J (eds) Proc. 3rd International Conference on Bioinformatics and Genome Research, AAAI Press, New Jersey, pp. 207–215.
55. Karp PD, Paley SM (1993) Representation of metabolic knowledge: pathways. In: Altman R, Brutlag D, Karp PD, Lathrop R, Searls D (eds) Proc. 2nd International Conference on Intelligent Systems for Molecular Biology. AAAI Press, Menlo Park, California, pp. 225–238.
56. Karp PD, Riley M, Paley SM, Pellegrini-Toole A, Krummenacker M (1999) EcoCyc: encyclopedia of *Escherichia coli* genes and metabolism. Nucleic Acids Res 27:55–58.
57. Karp PD, Riley M, Paley SM, Pellegrini-Toole A, Krummenacker M (2000) The EcoCyc and MetaCyc databases. Nucleic Acids Res 28:56–59.
58. Kashtan N, Itzkovitz S, Milo R, Alon U (2002) Mfinder Tool Guide. Tech. Rep. Department of Molecular Cell Biology and Computer Science & Applied Mathematics, Weizman Institute of Science, Rehovot, Israel.
59. Kashtan N, Itzkovitz S, Milo R, Alon U (2004) Topological generalizations of network motifs. Physical Review E 70:031909.
60. Kitano H (2004) Biological robustness. Nat Rev Genet 5:826–837.
61. Kitano H, Funahashi A, Matsuoka Y, Oda K (2005) Using process diagrams for the graphical representation of biological networks. Nature Biotechnol 23:961–966.
62. Koch I, Junker BH, Heiner M (2005) Application of Petri net theory for modelling and validation of the sucrose breakdown pathway in the potato tuber. Bioinformatics 21:1219–1226.
63. Koschützki D, Lehmann KA, Peeters L, Richter S, Tenfelde-Podehl D, Zlotowski O (2005) Centrality indices. In: Brandes U, Erlebach T (eds) Network Analysis: Methodological Foundations. Vol. 3418 of Lecture Notes in Computer Science (LNCS) Tutorial. Springer-Verlag, Springer Berlin/Heidelberg, pp. 16–61.
64. Kose F, Weckwerth W, Linke T, Fiehn, O (2001) Visualizing plant metabolomic correlation networks using clique-metabolite matrices. Bioinformatics 17:1198–1208.
65. Krause AE, Frank KA, Mason DM, Ulanowicz RE, Taylor WW (2003) Compartments revealed in food-web structure. Nature 426:282–285.
66. Krieger CJ, Zhang P, Müller LA, Wang A, Paley S, Arnaud M, Pick J, Rhee SY, Karp PD (2004) MetaCyc: a multiorganism data-base of metabolic pathways and enzymes. Nucleic Acids Res 32:438–442.
67. Lee TI, Rinaldi NJ, Robert F, Odom DT, Bar-Joseph Z, Gerber GK, Hannett NM, Harbison CT, Thompson CM, Simon I, Zeitlinger J, Jennings EG, Murray HL, Gordon DB, Ren B, Wyrick JJ, Tagne JB, Volkert TL, Fraenkel E, Giord DK, Young RA (2002) Transcriptional regulatory networks in *Saccharomyces cerevisiae*. Science 298:799–804.
68. Lemke N, Herédia F, Barcellos CK, dos Reis AN, Mombach JCM (2004) Essentiality and damage in metabolic networks. Bioinformatics 20(1): 115–119.
69. Ma HW, Zeng AP (2003) The connectivity structure, giant strong component and centrality of metabolic networks. Bioinformatics 19: 1423–1430.
70. Ma HW, Zeng AP (2003) Reconstruction of metabolic networks from genome data and analysis of their global structure for various organisms. Bioinformatics 19:270–277.
71. Ma HW, Zhao XM, Yuan YJ, Zeng AP (2004) Decomposition of metabolic network into functional modules based on the global connectivity structure of reaction graph. Bioinformatics 20:1870–1876.

72. Mangan S, Itzkovitz S, Zaslaver A, Alon U (2006) The incoherent feed-forward loop accelerates the response-time of the gal system of *Escherichia coli*. J Mol Biol 356: 1073–1081.
73. Mangan S, Zaslaver A, Alon U (2003) The coherent feed-forward loop serves as a sign-sensitive delay element in transcription networks. J Mol Biol 334:197–204.
74. MAVisto http://mavisto.ipk-gatersleben.de/index.html.
75. Mfinder http://www.weizmann.ac.il/mcb/UriAlon/index.html.
76. Michal G (1999) Biochemical Pathways. Spektrum Akademischer Verlag, Heidelberg.
77. Middendorf M, Ziv E, Wiggins CH (2005) Inferring network mechanisms: The *Drosophila melanogaster* protein interaction network. Proc Natl Acad Sci USA 102:3192–3197.
78. Milo R, Itzkovitz S, Kashtan N, Levitt R, Shen-Orr S, Ayzenshtat I, Sheer M, Alon U (2004) Superfamilies of evolved and designed networks. Science 303:1538–1542.
79. Milo R, Shen-Orr S, Itzkovitz S, Kashtan N, Chklovskii D, Alon U (2002) Network motifs: simple building blocks of complex networks. Science 298:824–827.
80. Mlecnik B, Scheideler M, Hackl H, Hartler J, Sanchez-Cabo F, Trajanoski Z (2005) PathwayExplorer: web service for visualizing high-throughput expression data on biological pathways. Nucleic Acids Research 33:W633–W637.
81. Moreno JL (1934) Who Shall Survive?: A New Approach to the Problem of Human Interrelations. Nervous and Mental Disease Publishing Company, Washington.
82. Müller LA, Zhang P, Rhee SY (2003) AraCyc: A biochemical pathway database for Arabidopsis. Plant Physiol 132:453–460.
83. Nakao M, Bono H, Kawashima S, Kamiya T, Sato K, Goto S, Kanehisa M (1999) Genome-scale gene expression analysis and pathway reconstruction in KEGG. Genome Informatics 10:94–103.
84. Nicholson, DE (1997) Metabolic Pathways Map (Poster). Sigma Chemical Co., St. Louis, St. Louis, MO.
85. Oltvai ZN, Barabási AL (2002) Systems biology: life's complexity pyramid. Science 298:763–764.
86. Pan D, Sun N, Cheung KH, Guan Z, Ma L, Holford M, Deng X, Zhao H (2003) PathMAPA: a tool for displaying gene expression and performing statistical tests on metabolic pathways at multiple levels for Arabidopsis. BMC Bioinformatics 4:56.
87. Papin JA, Reed JL, Palsson BO (2004) Hierarchical thinking in network biology: the unbiased modularization of biochemical networks. Trends Biochem Sci 29: 641–647.
88. Pržulj N, Corneil DG, Jurisica I (2004) Modeling interactome: scale-free or geometric? Bioinformatics 20:3508–3515.
89. Quinn NR, Breuer MA (1979) A force directed component placement procedure for printed circuit boards. IEEE Trans Circuits Syst CAS 26:377–388.
90. Rahman SA, Schomburg D (2006) Observing local and global properties of metabolic pathways: "load points" and "choke points" in the metabolic networks. Bioinformatics 22: 1767–1774.
91. Ravasz E, Somera AL, Mongru DA, Oltvai ZN, Barabási AL (2002) Hierarchical organization of modularity in metabolic networks. Science 297:1551–1555.
92. Rives AW, Galitski T (2003) Modular organization of cellular networks. Proc Natl Acad Sci USA 100:1128–1133.
93. Roessner U, Luedemann A, Brust D, Fiehn O, Linke T, Willmitzer L, Fernie A (2001) Metabolic profiling allows comprehensive phenotyping of genetically or environmentally modified plant systems. Plant Cell 13:11–29.
94. Rolletschek H, Radchuk S, Klukas C, Schreiber F, Borisjuk L (2005) Regulation of lipid biosynthesis in soybean seeds: evidence for a key role of photosynthetic oxygen release. New Phytologist 167:777–786.
95. Rost U, Kummer U (2004) Visualisation of biochemical network simulations with SimWiz. In: Iyengar R, Wolkenhauer O, Kolch W, Kwang-hyun Cho, Klingmuller U (eds) Systems Biology, IEE Proc. Vol. 1, pp. 184–189.

96. Schomburg I, Chang A, Ebeling C, Gremse M, Heldt C, Huhn G, Schomburg D (2004) BRENDA, the enzyme database: updates and major new developments. Nucleic Acids Research, Database issue 32:D431–D433.
97. Schreiber F (2002) High quality visualization of biochemical pathways in BioPath. In Silico Biol 2:59–73.
98. Schreiber F, Schwöbbermeyer H (2005) MAVisto: a tool for the exploration of network motifs. Bioinformatics 21:3572–3574.
99. Schuster S, Pfeifer T, Moldenhauer F, Koch I, Dandekar T (2002) Exploring the pathway structure of metabolism: decomposition into subnetworks and application to *Mycoplasma pneumoniae*. Bioinformatics 18:351–361.
100. SEPARATOR http://pinguin.biologie.uni-jena.de/bioinformatik/networks/separator/separator.html.
101. Shannon P, Markiel A, Ozier O, Baliga NS, Wang JT, Ramage D, Amin N, Schwikowski B, Ideker T (2003) Cytoscape: a software environment for integrated models of biomolecular interaction networks. Genome Res 13:2498–2504.
102. Shen-Orr S, Milo R, Mangan S, Alon U (2002) Network motifs in the transcriptional regulation network of *Escherichia coli*. Nature Genetics 31:64–68.
103. Snel B, Bork P, Huynen MA (2002) The identification of functional modules from the genomic association of genes. Proc Natl Acad Sci USA 99:5890–5895.
104. Sokal RR, Michener CD (1958) A statistical method for evaluating systematic relationships. Univ Kansas Sci Bull 38:1409–1438.
105. Spirin V, Mirny LA (2003) Protein complexes and functional modules in molecular networks. Proc Natl Acad Sci USA 100:12123–12128.
106. Sturn A, Quackenbush J, Trajanoski Z (2002) Genesis: cluster analysis of microarray data. Bioinformatics 18:207–208.
107. Sugiyama K, Tagawa S, Toda M (1981) Methods for visual understanding of hierarchical systems. IEEE Trans Syst Man Cybern 11:109–125.
108. Thimm O, Bläsing O, Gibon Y, Nagel A, Meyer S, Krüger P, Selbig J, Müller LA, Rhee SY, Stitt M (2004) MAPMAN: a user-driven tool to display genomics data sets onto diagrams of metabolic pathways and other biological processes. Plant J 37:914–939.
109. Tokimatsu T, Sakurai N, Suzuki H, Ohta H, Nishitani K, Koyama T, Umezawa T, Misawa N, Saito K, Shibatanenell D (2005) KaPPA-View. A web-based analysis tool for integration of transcript and metabolite data on plant metabolic pathway maps. Plant Physiol 138:1289–1300.
110. Usadel B, Nagel A, Thimm O, Redestig H, Blaesing OE, Palacios-Rojas N, Selbig J, Hannemann J, Piques MC, Steinhauser D, Scheible WR, Gibon Y, Morcuende R, Weicht D, Meyer S, Stitt M (2005) Extension of the visualization tool MapMan to allow statistical analysis of arrays, display of corresponding genes, and comparison with known responses. Plant Physiol 138:1195–1204.
111. Wagner A, Fell DA (2001) The small world inside large metabolic networks. Proc Royal Society London B 268:1803–1810.
112. Wall ME, Dunlop MJ, Hlavacek WS (2005) Multiple functions of a feed-forward-loop gene circuit. J Mol Biol 349:501–514.
113. Wegner K, Kummer U (2005) A new dynamical layout algorithm for complex biochemical reaction networks. BMC Bioinformatics 6:212.
114. Wernicke S, Rasche F (2006) FANMOD: a tool for fast network motif detection. Bioinformatics 22:1152–1153.
115. Wiese R, Eiglsperger M, Kaufmann M (2001) Visualization and automatic layout of graphs. In: Mutzel P, Jünger M, Leipert S (eds) Proc. 9th International Symposium on Graph Drawing (GD 2001). Springer, Berlin/Heidelberg, pp. 453–462.
116. Wolf G (2000) Visualising gene expression in its metabolic context. Briefings in Bioinformatics 1:297–304.

117. Wuchty S, Oltvai ZN, Barabási AL (2003) Evolutionary conservation of motif constituents in the yeast protein interaction network. Nature Genet 35:176–179.
118. Wuchty S, Stadler PF (2003) Centers of complex networks. J Theor Biol 223: 45–53.
119. Zamora Lopez G, Steuer R (2008) Global network properties. In: BH Junker, Schreiber F (eds.) Analysis of Biological Networks. John Wiley & Sons, Hoboken, NJ, pp. 31–63.
120. Zhang LV, King OD, Wong SL, Goldberg DS, Tong AHY, Lesage G, Andrews B, Bussey H, Boone C, Rot FP (2005) Motifs, themes and thematic maps of an integrated *Saccharomyces cerevisiae* interaction network. J Biol 4:6.
121. Zhao J, Yu H, Luo JH, Cao ZW, Li YX (2006) Hierarchical modularity of nested bow-ties in metabolic networks. BMC Bioinformatics 7:386.
122. Zhu D, Qin ZS (2005) Structural comparison of metabolic networks in selected single cell organisms. BMC Bioinformatics 6:851.

Chapter 8
Network Stoichiometry

Nanette R. Boyle, Avantika A. Shastri, and John A. Morgan

8.1 Overview

This chapter describes approaches to modeling metabolic pathways that are based on biochemical reaction stoichiometry. These methods have some advantages over kinetic models because they do not require the determination of complicated kinetic expressions and associated kinetic parameters. Although based only upon reaction stoichiometry and mass balances, the techniques can be quite powerful in exploring the capabilities of a metabolic network. Stoichiometry-based models enable efficient calculation of theoretical yields on any nutrient [78]. The models may be used to rationally select genes for addition and/or deletion in the genome which have the most promise to significantly improve desired product yield. New targets for herbicides can be selected through a mathematical analysis of the sensitivity of inhibiting specific enzymes on growth fluxes [87]. Perhaps their greatest promise in conjunction with optimization strategies is the ability to predict metabolic fluxes to specific products or growth as a function of the environment.

In this chapter, we first present the details of how to create the stoichiometric matrix, which serves as the basis for all following sections. Next, we cover flux balance analysis, which is based on constraint-based optimization to make predictions about metabolic fluxes and pathway utilization. Following this, we describe techniques of elementary mode analysis and extreme pathways that are broadly aimed at analyzing the stoichiometric matrix. Finally, we conclude with a section on genome-scale models, which hold great promise in systems biology. In each section, we follow the outline of presenting the theory first, followed by applications, and conclude with specific examples from the literature.

J.A. Morgan (✉)
School of Chemical Engineering, 480 Stadium Mall Dr., Purdue University, West Lafayette, IN 47907, USA
e-mail: jamorgan@purdue.edu

8.2 Stoichiometric Modeling

8.2.1 Fundamentals of Stoichiometric Modeling

As the name suggests, stoichiometric modeling approaches examine biological systems primarily by using information contained in biochemical reaction stoichiometry, namely the mole ratios of reactants and products in each reaction. By convention, the stoichiometric coefficient (s_{ij}) for a reactant is negative and that for a product is positive. Therefore, for a hypothetical biochemical reaction 2A + B → C, the stoichiometric coefficients of A, B, and C are -2, -1, and 1, respectively. The first step in stoichiometric modeling is to define the system of interest (such as a pathway, organelle, cell, or whole organism). This is done by annotating the reactions within the system from genomic data and biochemical literature. Reactions that connect the system to its external environment (such as transport reactions) must also be modeled. Reactants (products) that can be uptaken (excreted) by the system are called external metabolites, while species involved in reactions occurring within the system are known as internal metabolites. Species A_e and F_e in Fig. 8.1(a) are external metabolites, whereas A, B, C, D, and E and F are internal metabolites, with reactions 1 and 14 (fluxes v_1 and v_{14}) connecting the external and internal metabolites. The normalized rate of every reaction in the system is called a flux (v), which has units of moles/cell/time or moles/biomass unit/time.

Once the reaction network is annotated, mass balances are written around each metabolite. A rigorous mass balance that includes the effect of growth (and therefore system volume) on the concentration of a metabolite i is written as

$$\frac{dX_i}{dt} = r_i - \mu X_i \tag{8.1}$$

where X_i is the concentration of metabolite "i" in the system, t is time, r_i is the net rate of production (or consumption for negative r_i) of the metabolite, and μ is the specific growth rate (units of time^{-1}). The term μX_i is known as the dilution factor, as it represents the change in concentration of the metabolite due to cell growth. This term can typically be ignored because it is much smaller than the rate of production of the metabolite. In many physiological situations, the dilution factor can justifiably be omitted [83]. In all the material discussed in this chapter, the dilution factor is assumed to be zero unless specifically stated. Ignoring the dilution factor, Eq. 8.1 can be rewritten as

$$\frac{dX_i}{dt} = r_i = \sum_j s_{ij} v_j \tag{8.2}$$

where s_{ij} is the stoichiometric coefficient of the ith metabolite in the jth reaction, as described above. As a demonstration, the mass balance on the metabolite B from Fig. 8.1(a) can be written as

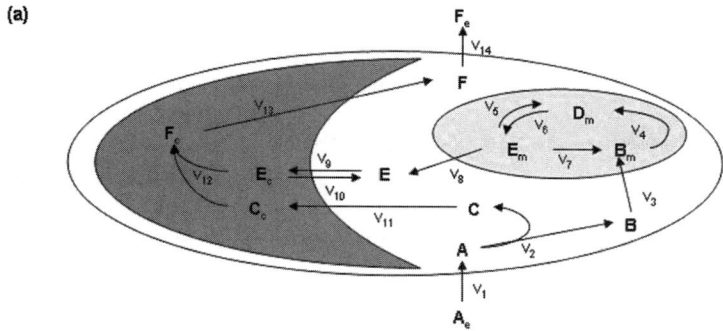

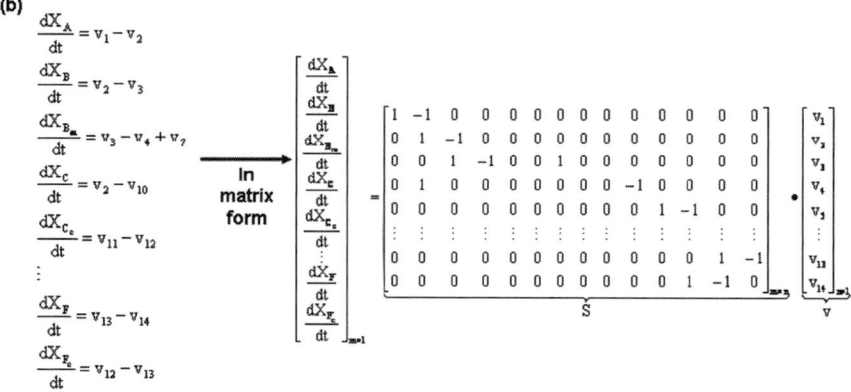

Fig. 8.1 Formulation of a stoichiometric model. (a) The first step is to collect reaction stoichiometry for the system to be modeled. (b) Next, a mass balance on each intracellular metabolite is written and then put into matrix-vector form. **v** is an nx1 vector of fluxes, **S** is an mxn matrix of coefficients, where m is the number of metabolites and n is the number of reactions

r_i = net rate of formation (or consumption) of B = rate of formation of B – rate of consumption of B, which is mathematically expressed as

$$\frac{dX_B}{dt} = v_2 - v_3 \tag{8.3}$$

As shown in Fig. 8.1(b), the set of mass balances on the entire set of intracellular metabolites can then be written compactly in matrix-vector form as

$$\frac{dX}{dt} = \mathbf{S} \cdot \mathbf{v} \tag{8.4}$$

where **S** is the $m \times n$ stoichiometric matrix such that an element s_{ij} is the stoichiometric coefficient of the ith metabolite in the jth reaction, **v** is $n \times 1$ column vector of metabolite fluxes, and **X** is an $m \times 1$ column vector of metabolite concentration in the system. The total number of metabolites in the system is m and the total number of fluxes is n. At a metabolic steady state, or under pseudo-steady-state conditions (see Chapter 11), the concentrations of intracellular metabolites do not change with time. As a result, Eq. 8.4, which is a system of ordinary differential equations, reduces to a system of linear algebraic equations written in matrix-vector form as

$$\mathbf{0} = \mathbf{S} \cdot \mathbf{v} \qquad (8.5)$$

The system of algebraic equations can be considerably easier than (8.4) to solve for the flux vector **v**. In general, the equations can be written to include balances on extracellular species, or species known to accumulate/deplete at a certain rate within the system. At steady state, the modified set of equations can be written as

$$\mathbf{b} = \mathbf{S} \cdot \mathbf{v} \qquad (8.6)$$

where **b** is an $m \times 1$ column vector, such that $b_i = 0$ if the ith metabolite is an internal metabolite, $b_i \geq 0$ for a metabolite that accumulates in the system or is uptaken, and $b_i \leq 0$ for a metabolite that is depleted in the system or is exported out. These non-zero values of b_i are equal to the rates of accumulation/depletion/transport of the respective metabolites and are often measured flux values. Metabolic flux analysis (MFA) is the approach used to solve stoichiometric models to determine unknown fluxes [83].

Three situations may be encountered once the stoichiometric matrix is set up:

1. Fully determined system – The matrix S has full rank (rank $r = m = n$). This means that the balances written on the m metabolites lead to m linearly independent reactions. Further, the number of independent equations equals the number of unknown fluxes ($m = n$). Such a system can be solved exactly to estimate the flux vector **v**:

$$\mathbf{v} = \mathbf{S}^{-1}\mathbf{b} \qquad (8.7)$$

   The vector **b** contains information from the measured fluxes, such as uptake or secretion rates, as stated above.

2. Underdetermined system – If $r < n$, as is frequently the case, the system of equations is underdetermined. This means there are not enough independent mass balances to determine the unknown fluxes. In such cases, the system can have infinitely many solutions that satisfy the mass balances. The degrees of freedom equals $n - r$, and the number of independent mass balances is r. Experimental flux measurements can be used to set the values of some fluxes, and reduce the number of unknown fluxes until $r = n$. The system can then be solved for exactly, like (8.1) above. Alternately other approaches like linear optimization (Section 3) or metabolic pathway analysis (Section 4) can be used. Typically in biological systems, even if the number of metabolites equals or is greater than the number of unknown fluxes, the system can contain linearly dependent reactions ($m \geq n$, but $r < n$). Some typical examples that arise in biological systems are

8 Network Stoichiometry

- Nitrogen assimilation pathways: Two pathways of nitrogen assimilation that frequently occur in many organisms are via glutamate dehydrogenase (Eq. 8.8) and the glutamine synthase-glutamate synthase pathway (Eqs 9a,b):

$$\alpha KG + NADPH + NH_3 \rightarrow GLU + NADP^+ \quad (8.8)$$

$$\alpha KG + NADPH + GLN \rightarrow 2GLU + NADP^+ \quad (8.9a)$$

$$GLU + \ + NH_3 + ATP \rightarrow GLN + ADP \quad (8.9b)$$

  In many situations, especially when modeling only a section of metabolism, ATP balances cannot be written since all reactions involving ATP are not modeled. In such cases, Eq. 8.9a + 9b leads to a net reaction equal to Eq. 8.8. Thus the three equations are linearly dependent.
- Cofactor pairs: Whenever ATP is consumed in a reaction, ADP is formed. The mass balance equation on ATP is therefore the negative of the mass balance on ADP, and the two equations are linearly dependent. . The same holds true for other cofactor pairs like NADP/NADPH and NAD/NADH.
- Transhydrogenases: These enzymes interconvert NADPH ↔ NADH, and thereby make the mass balances on NADPH and NADH linearly dependent.
- Other examples include the presence of isoenzymes in the model, which catalyze the same reaction (with or without the same cofactors). Linear dependency can also arise in more complicated ways by the interaction of a large number of reactions.

3. Over-determined system – If $r > n$, we obtain an over-determined system. These are typically encountered when there are plenty of experimental flux measurements available. The extra information can therefore be used for consistency checks between measured and experimental values, leading to more accurate estimation of fluxes. In such situations, the left-handed Moore–Penrose pseudo-inverse of S can be used to obtain a best fit for the unknown fluxes from the measured fluxes. These situations are not encountered frequently and will not be discussed in detail here. The reader is referred to chapter 8 in [83] for more details.

8.2.2 Applications of Stoichiometric Modeling

Although the stoichiometric description of a metabolic system is relatively simple compared to kinetic models, it can nevertheless be very useful in several situations. It can be used to find maximum theoretical yields and estimate the costs of various metabolic processes ([33, 39], also see examples in [83]). It can help in the identification of rigid or flexible branch points [82] and comparing flux maps under different environmental conditions or mutations [57] to gain insight into metabolic processes.

Examples of the application of stoichiometric modeling to photosynthetic systems include the work by Yang et al. [95] who used a stoichiometric model of central metabolism to investigate carbon and energy metabolism of *Chlorella pyrenoidosa* under autotrophic, mixotrophic, and cyclic light (autotrophic) and dark (heterotrophic) conditions (assuming steady state under both light and dark conditions). The model was divided into two compartments: a chloroplast and a combined cytosol and mitochondrion, with a total of 67 reactions and 61 metabolites describing glycolysis/gluconeogenesis, Calvin cycle/pentose phosphate cycle, the TCA cycle, nitrate assimilation, amino acid synthesis, and reactions related to energy transduction via cofactors such as NAD(P)H and ATP. Measured inputs to the model included glucose and nitrate uptake rates, biomass composition and growth rate, and incident light flux. Care was taken to remove singularities (linearly dependent reactions) from the metabolic network, so that the system of equations could be solved. Their analysis showed considerable activity of glycolysis, TCA cycle, and mitochondrial oxidative phosphorylation even during autotrophic growth, but very little cytosolic pentose phosphate activity (only sufficient to produce E4P and R5P as biosynthetic precursors). They evaluated and compared the energy efficiency of the three trophic conditions described above. Interestingly, they found that the cyclic mode had a higher biomass yield/unit energy supplied than the pure mixotrophic mode, which has implications for industrial mixotrophic growth schemes of photosynthetic microbes. Some other studies of photosynthetic metabolism using stoichiometric models include the study of cofactor limitation during growth and exopolysaccharide production in the cyanobacterium *Arthospira platensis* [16] and PHB production with and without phosphate limitation in *Synechococcus* sp. MA19 [53].

Specific shortcomings of the work described in this section include the need to (sometimes arbitrarily) restrict the network to remove singularities [95] and impose flux values or flux ratios that are not entirely justified [53]. The general shortcomings of simple stoichiometric analysis include obtaining flux distributions where certain physiologically irreversible reactions do not operate in the correct direction. For underdetermined systems, there might not be sufficient external flux measurements that can be made, in order to solve for the fluxes. Other methods of metabolic network analysis that overcome these limitations are described in the following sections.

8.3 Flux Balance Analysis

8.3.1 Introduction to Linear Optimization

Flux balance analysis (FBA) is a technique that explores the feasible solution space of the reaction network using linear optimization. Early examples of the application

8 Network Stoichiometry

of linear optimization in stoichiometric analysis include the work described in [27, 52, 94]. It was developed in great detail by Palsson and co-workers [69, 90, 91] in the early 1990s and has been applied extensively to study various organisms including *E. coli* [38] human mitochondria [65], *Haemophilus influenzae* [22], *Saccharomyces cerevisiae* [29], and *Synechocystis* [79].

The concept of FBA can be understood by the general concepts of linear optimization. Linear optimization allows us to obtain a feasible solution to a system of linear algebraic equations and inequalities, such that it maximizes (or minimizes) a selected linear function in the solution space. It also enables us to find particular solutions in underdetermined systems of algebraic equations (case b from section 2). Illustrating with a very simple 2-dimensional example – consider two variables x_1 and x_2, and the following problem statement (see Fig. 8.2):

$$\mathbf{maximize}\, x_1 + 2x_2 \qquad (8.10)$$

subject to

$$x_1 - x_2 - 2 \leq 0 \qquad (8.11)$$

$$x_1 + x_2 - 5 \leq 0 \qquad (8.12)$$

$$x_1 \geq 0, x_2 \geq 0 \qquad (8.13\text{a,b})$$

Equation 8.10 is called the objective function, and Eqs 8.11–8.13 are called the constraints. The optimum point occurs at $x_1 = 1.5$, $x_2 = 3.5$, which satisfies Eqs. 8.11–8.13 with the maximum value of the objective function (Eq. 8.10) equal to 8.5.

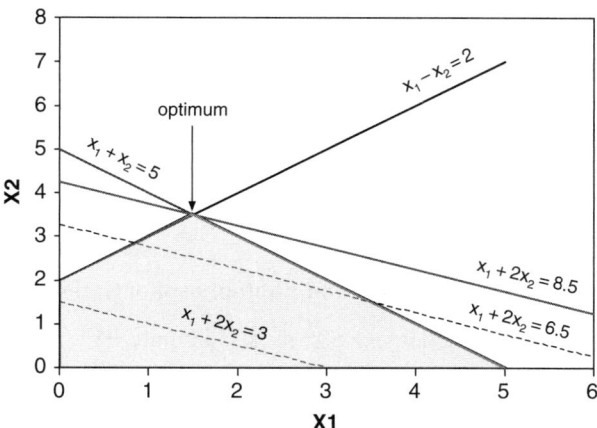

Fig. 8.2 Graphical representation of a linear optimization problem (see text). For a reaction network with n free fluxes, the same optimization principle would be performed in an n-dimensional space

Some important properties of a linear optimization problem are that the solution space is convex, and the optimum always occurs at a vertex or boundary of the feasible region. It must be noted that the optimum solution need not be unique. The reader may refer to an introductory text on linear programming for mathematical details [14].

In the context of stoichiometric models, there may be several flux distributions (values of vector **v**) that give the same (optimum) objective function value. Application of linear optimization allows us to obtain *one* such flux distribution. For stoichiometric models, the method of Lee et al. aims to enumerate all the alternative solutions [48]; however, the algorithm is computationally intensive for large reaction networks. Alternatively, the use of metabolic pathway analysis can help to find all possible alternatives by enumerating elementary modes or extreme pathways and selecting optimal solutions (see Section 8.4.2).

8.3.2 Formulation of Linear Optimization Problem for a Metabolic Network

In order to apply linear optimization to a metabolic network, each unknown variable (in our case, the fluxes v_i) must be non-negative (see Eqs 8.13a and 8.13b above). This is a requirement to apply the theory of linear optimization. Thus, reversible reactions must be redefined as two reactions, one forward and one reverse reaction, such that both the fluxes are non-negative. If there are a total of n reactions, the flux through each reaction (v_i) can be represented by an axis in an n-dimensional space. Since the fluxes are non-negative, the axes meet at zero (i.e., $\mathbf{v} = 0$). The typical constraints that define the solution space include

1. The mass balance equations on the internal metabolites (Eq. 8.5)
2. Inequality constraints on fluxes setting maximum or minimum values, if known/equality constraints for values of any known fluxes
3. Uptake rates of substrates or maximum uptake rates (if known)
4. Constraints modeling cellular energetics, biomass formation, and maintenance.

These constraints require careful modeling and are discussed in more detail below.

8.3.2.1 Oxidative Phosphorylation and Photophosphorylation

Along with metabolic reactions, FBA models require the input of reactions accounting for cellular energetics. For plants, this typically involves the addition of oxidative phosphorylation in the mitochondria and photophosphorylation in the chloroplast. Oxidative phosphorylation is the process that transfers electrons from NADH to O_2, the main electron acceptor to produce ATP from protons in the mitochondria that get pumped through an ATP synthase [54]. This entire process is often lumped into one reaction for simplicity of modeling [29, 79]. Although the exact

H⁺/ATP ratio for the mitochondria ATP synthase has not yet been agreed upon, it is assumed to be somewhere between 3 and 4 [2, 29, 85]. Therefore, oxidative phosphorylation can be lumped into one equation and modeled as follows:

$$XH^+ + Pi + ADP \rightarrow ATP \\ \text{where} \quad 3 \leq X \leq 4 \tag{8.14}$$

Photophosphorylation is the process by which light is converted into energy (ATP) and reducing power (NADPH) in photosynthetic organisms. This process occurs through two different electron transport chain (ETC) systems, the cyclic and the non-cyclic. Like oxidative phosphorylation, both the cyclic and non-cyclic are fairly complicated multiple-step processes that involve several cofactors and proteins (both membrane and non-membrane bound). To simplify modeling these processes, in a recent paper on *Synechocystis* [79], these were modeled as two reactions. The cyclic ETC was modeled as

$$1 \text{ absorbed photon} \rightarrow 2H^+ \tag{8.15}$$

And the non-cyclic ETC was modeled as

$$4 \text{ photons} + NADP^+ + H_2O \rightarrow NADPH + 6H^+ + 0.5\, O_2 \tag{8.16}$$

Despite the use of common electron carriers in both the cyclic and non-cyclic ETCs (ferredoxin, cytochrome b_6f complex, plastoquinone, cytochrome c_6, etc.), researchers have postulated that due to spatial separation in the thylakoid membranes, they do not interact with each other [1]; therefore, the reactions are modeled as separate, non-interacting processes. The simplified version of the photophosphorylation process shown above is just one way to model the process. If a more complete model of this process is the goal, each cofactor (ferredoxin, cytochrome b_6f complex, plastoquinone, cytochrome c_6, etc.) should be included in the model. Both the cyclic and non-cyclic ETCs are coupled to the chloroplast ATP synthase complex, coupling the translocation of protons with the production of ATP from ADP. Recent studies have shown that the H⁺/ATP ratio for the chloroplastic ATPase is 14/3 [2], which can then be modeled as [79]

$$4.67 H^+ + Pi + ADP \rightarrow ATP \tag{8.17}$$

8.3.2.2 Maintenance Energy

In order to fully capture the energetics of the organisms to be modeled, maintenance energy must be included in the model. There are generally two types of maintenance energy: growth-associated energy, which lumps partially unknown energy requirements for transport, biosynthesis, and polymerization [83], and non-growth-associated energy, which is used for cellular maintenance operations such as DNA repair, cell wall maintenance, and pH control. For microbial systems,

Table 8.1 Typical growth and non-growth-associated maintenance values used for FBA modeling

Organism	Maintenance energy		Source
	Growth (mmole ATP/g DW)	Non-growth (mmole ATP/ g DW hr)	
Streptomyces coelicolor	47.00	3.80	[8]
Escherichia coli	45.70	7.60	[66]
Staphylococcus aureus	40.00	5.00	[32]
Lactobacillus plantarum WCFS1	27.40	0.36	[85]
Saccharomyces cerevisiae	35.36	1.00	[25]
Lactococcus lactis	18.15	1.00	[55]

both types of maintenance energy are typically determined experimentally by measuring the growth rate and biomass yield. The data are then plotted (1/growth rate versus 1/yield); the slope of the line is growth-associated maintenance energy, and the intercept is non-growth-associated energy. Although the experiments needed for this calculation are relatively simple to perform for microorganisms grown in continuous bioreactors at different growth rates, this type of experiment is not applicable to plant systems. The first challenge in determining maintenance energy in plants is controlling plant growth rates; the second is that energy and carbon sources are not coupled in autotrophic growth, so ATP production is not as constrained as in heterotrophic growth regimes. Typical maintenance values measured experimentally are in grams substrate per gram dry weight per hour and are converted into moles ATP per gram dry weight per hour by determining how many ATPs are produced per gram substrate. Growth-associated maintenance is typically modeled as part of the biomass formation equation (see Section 3.2.4) in order to couple it to growth, while non-growth associated maintenance is modeled in a separate reaction which is essentially just an ATP drain (ATP + H_2O → ADP + P_i). Typical growth and non-growth-associated maintenance values for FBA modeling in different organisms can be found in Table 8.1; however, because many of these organisms are prokaryotes, they may not be representative of maintenance energy for a typical plant cell.

8.3.2.3 Biomass Formation Equation

Modeling the metabolism of an organism using FBA requires knowledge of the biochemical composition of the cell, which is then used to construct a biomass formation equation from the four main macromolecules that make up biomass: protein, lipid, DNA, and RNA. Ideally, data on both the percent dry weight of each of these components in the cell and the relative composition of monomers in the macromolecules should be included in the model. For protein, this requires measuring the mole ratio of each amino acid in the cell, which can then be used to create a protein formation equation that represents an average protein. The formation of a macromolecule must also include the polymerization energy needed to convert the monomer. Formation equations are also generated for lipids (chain length, mole ratio of each type of lipid), carbohydrates (which sugars and mole ratios), and nucleotides, RNA and DNA (mole ratio of each nucleotide is typically available from genome databases), in the same manner described above. Another important molecule in the cell that can readily be measured [3] is chlorophyll, which should also be included in the model. Finally, the growth-associated maintenance should be added to the biomass formation equation. Combining all of these components should give rise to a biomass formation equation in the following form:

$$\alpha_1 \text{ protein} + \alpha_2 \text{ carbohydrate} + \alpha_3 \text{ lipid} + \alpha_4 \text{ DNA} + \alpha_5 \text{ RNA} + \alpha_6 \text{ chlorophyll} + \beta \text{ ATP} \rightarrow \gamma \text{ biomass} + \beta \text{ ADP} + \text{products} \quad (8.18)$$

where the α_i are the calculated mole ratios of each macromolecule in the cell; β is the growth-associated maintenance energy coefficient; and γ is a given mass of biomass (grams, kilograms, etc.). It is important to note that biomass composition is known to change with changing environmental conditions [61] and must therefore be determined for each growth condition in order to have an accurate model.

8.3.2.4 Choice of Objective Function

The final step for the formulation of the mathematical model for FBA is to select an appropriate objective function. The most widely used objective functions for modeling wild-type microbial cells with FBA are maximize biomass [9, 20, 21, 50], maximize production of ATP [61, 65], or maximize the production of a particular product of interest [61, 89]. Although maximization of biomass is a widely used objective function, it does not always produce fluxes that agree with experimental data for mutants and other slow-growing organisms [77, 80]. Even in wild-type organisms, the assumption of optimality can be incorrect, for example, it was shown in *Bacillus subtilis* that several mutants had improved biomass productivity [28]. It is theorized that this is due to the resources the cell invests in being able to adapt quickly to changing environmental conditions instead of optimizing growth. Therefore, it is imperative that the type of organism and its growth patterns be considered fully before selecting an objective function because it determines the outcome of the model and how well it agrees with experimental data. The effect of objective function on the outcome of a simulation was examined by Schuetz et al. for the central metabolic network in *E. coli* [76]. Using 11 objective functions, 8 adjustable constraints and 6 environmental conditions, the authors found only 2 sets of objectives that agreed with experimental results without the need for additional artificial constraints. In reality, multiple objective functions may be required to best capture the organisms' goals, although this would likely require optimization methods more complex than linear programming (e.g., quadratic programming).

8.3.3 Applications of FBA

8.3.3.1 Synechocystis Metabolism

Flux balance analysis was used to study the central metabolic pathways in a model cyanobacterium, *Synechocystis* sp. PCC 6803, to predict flux distributions for optimal growth under autotrophic, heterotrophic (growth on glucose), and mixotrophic (growth on light and glucose) conditions [79]. The central metabolic network was reconstructed from genomic data, and it consisted of 70 reactions (after splitting of reversible reactions) and balances on 36 metabolites. The processes of photophosphorylation and ATP formation were modeled as described in Section 3.2.1. The extracellular reactants and products were glucose, CO_2, O_2, and light energy. Carbon and energy in form of ATP and NAD(P)H are two of the main components required for growth. However, in contrast to heterotrophic organisms where sugars such as glucose provide both the carbon skeletons as well as NADPH and the ATP by catabolism, photoautotrophic metabolism has distinct sources of carbon (CO_2) and energy (light). While the uptake rate of the carbon source as a constraint for growth is a natural choice for heterotrophic metabolism, the selection of a basis for photoautotrophic metabolism could be based on either carbon or light limitation. The authors approached the problem by using a two-step optimization process. They first maximized the biomass production rate, to obtain a certain biomass yield on carbon (Y_b). A second optimization was then performed by adding the maximal value of Y_b as constraint and minimizing the utilization of light energy. Thus, it was possible to obtain a flux distribution that maximized biomass production, while utilizing the minimum required light input. The flux distribution pattern (topology) was calculated for the autotrophic and heterotrophic cases from which several useful observations could be made. The predicted photosynthetic quotient, which is the ratio moles of oxygen produced compared to the moles of carbon dioxide assimilated, of 1.46 was found to fall in a physiological range for several photosynthetic organisms. Further, it was found that both the cyclic and non-cyclic electron transport chains were needed to meet the desired ATP/NADPH ratio for optimum autotrophic growth. As expected, for mixotrophic metabolism, it was found that as the supplied light energy increased, the biomass yield on glucose increased. Further, the utilization of the oxidative pentose phosphate reactions and the anapleurotic PEP carboxylase changed as a function of input light energy (Fig. 8.3). The effect of gene knockouts on optimal growth of *Synechocystis* by the FBA model revealed that the malic enzyme and the NADH-dependent glyceraldehyde phosphate dehydrogenase are essential for heterotrophic growth, while the NADPH-dependent glyceraldehyde phosphate dehydrogenase was essential for autotrophic growth. They also found glyoxylate shunt activity under heterotrophic conditions. All these predictions were consistent with experimental evidence in literature. Interestingly, although *Synechocystis* has an incomplete TCA cycle, the addition of the missing αKG dehydrogenase did not lead to a significant increase in biomass yield under any condition.

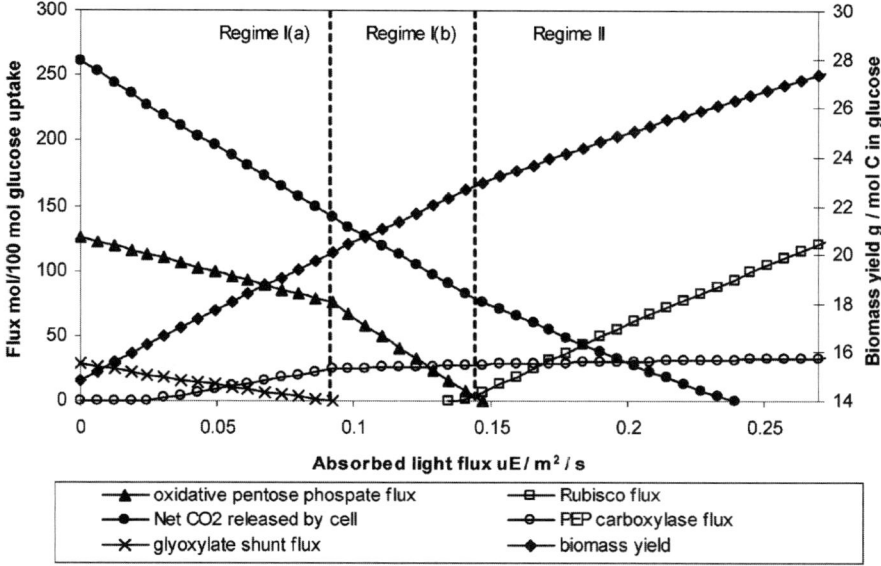

Fig. 8.3 Variation of mixotrophic growth yields and fluxes as a function of light flux utilization. The *left* y-axis is flux (mol/100 mol glucose uptake), and the *right* y-axis scale is biomass yield/mol C contained in glucose. Reproduced from [79] with permission of the American Chemical Society

8.3.3.2 Application of FBA to Metabolic Engineering

Models of metabolism are integral to rational metabolic engineering of organisms to produce specific chemicals of interest. One such tool that has been developed is OptStrain [58], which is a software tool designed to aide in metabolic engineering by designing the optimal strain to produce the metabolite of interest by determining the best microbial host, substrate choice, and genes that need to be added/deleted to achieve optimal production. While designing the host, the program seeks to minimize the number of non-native pathways as well. For the addition of non-native reactions, the software program searches a universal database of approximately 4000 enzymatic reactions which is based on reactions listed in the KEGG database [40] and automatically updated. OptStrain automatically adds the selected genes to the stoichiometric matrix of the chosen host and updates the constraints to reflect the choice of substrate. The in silico mutant is then subjected to further optimization using another metabolic engineering tool, OptKnock [10], which is integrated into OptStrain. OptKnock [10] is another software tool, which is designed to couple the production of a metabolite to growth and suggest gene knockouts to achieve the best product yield. OptKnock uses a bilevel optimization procedure to do this; the engineering objective (maximize product) is maximized simultaneously with the cellular objective (maximize biomass, maximize ATP, etc.). While performing this optimization, OptKnock systematically evaluates all the reactions in the network and removes metabolic reactions that uncouple cellular growth from chemical production. The software then provides a list of enzymes that can be removed to increase the production of the metabolite. In an example of the utility of OptStrain, it was used to design the optimal microbial host for vanillin production. The program selected *E. coli* grown on glucose with three additional enzymes (formaldehyde dehydrogenase, vanillate monooxygenase, and vanillin dehydrogenase). After the addition of the three non-native enzymes, OptKnock suggested several knockout strategies, but the highest yield was for the knockout of four native enzymes (acetate kinase, pyruvate kinase, PTS transport, and fructose-6-phosphate aldolase). The calculated yield for this mutant was 0.57 g/g glucose, which is 90% of the theoretical yield. Although this has not yet been experimentally validated, a previous study produced vanillin in *E. coli* by the addition of the three non-native enzymes listed above; however, the yield was reported to be much lower than theoretical, 0.15 g/g glucose [49]. Rational metabolic engineering is greatly aided by the development of tools such as OptKnock and OptStrain due to the ability to run models and check the feasibility of genetic manipulations before any experiment is performed. These tools are especially useful in the area of secondary metabolite production because interactions with primary metabolism and competition for intermediary metabolites may be less apparent. A recent review by Kim et al. [43] provides more examples of the use of FBA for metabolic engineering and highlights some of the constraints discussed below.

8.3.4 Regulatory Constraints

Although maximization of biomass works well for wild-type microbial strains, artificially created mutants typically do not exhibit optimal metabolic states because these mutants are not subjected to the same evolutionary pressure as wild-type strains. In order to accurately model suboptimal metabolism in these strains, two variations of FBA have been developed: (1) minimization of metabolic adjustment (MOMA) and (2) regulatory on/off minimization (ROOM). MOMA is a method to determine the flux distribution for a mutant that minimizes the difference between the mutant and the wild-type solution [77]. This is based on the assumption that a mutant will undergo a minimal redistribution of fluxes to remain as close to the wild-type flux distribution as possible. Mathematically, this difference is measured by the Euclidean distance between the two solutions in the solution space (Fig. 8.4) and is minimized by quadratic programming. MOMA is able to more accurately predict essential genes in metabolic networks than FBA. In a study performed on *E. coli* [77], MOMA predicted that the loss of triosphosphate isomerase, fructose-1,6-bisphosphate aldolase, or phosphofructokinase was lethal, which agrees with published literature, while FBA did not identify these genes as essential. The flux distributions for mutants using MOMA were also found to agree more closely with experimentally determined fluxes than FBA for the *E. coli* pyruvate kinase knockout, and the correlation coefficients between the calculated and experimental fluxes was found to be 0.59 for MOMA compared to –0.064 for FBA. This illustrates that the assumption of a metabolic optimum state being reached in mutants is not appropriate. A second method to address non-optimality in mutants is regulatory

8 Network Stoichiometry

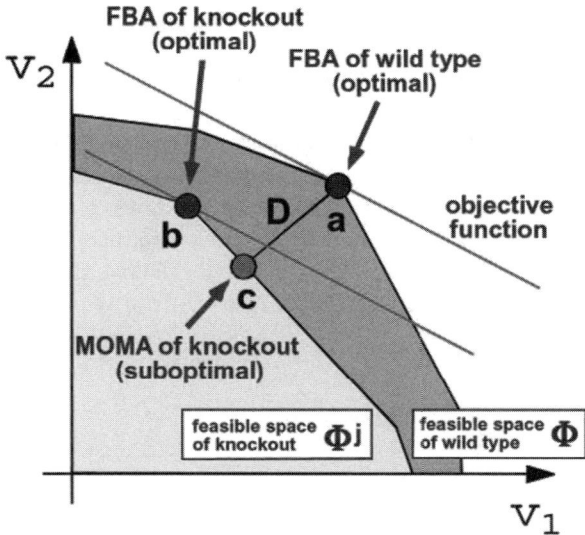

Fig. 8.4 Optimization principle underlying FBA and MOMA. The wild-type solution space is shown in green (Φ), and the mutant solution space is superimposed in yellow (Φ^j). The solution to the FBA problem is the point that maximizes the objective function (*red line*). The MOMA solution, calculated using quadratic programming, is the point in Φ^j that minimizes the distance to the FBA optimum. Figure is reprinted from Shlomi et al. [80], Copyright (2005) National Academy of Sciences, USA

on/off minimization (ROOM). In contrast to MOMA, ROOM seeks to minimize the number of significant flux changes with respect to the wild-type strain [80]. MOMA, by how it is defined, favors many small changes in fluxes instead of a few larger changes in fluxes. However, this may not accurately capture in vivo conditions because large modifications in fluxes can result in the redistribution of carbon through alternative pathways, which has been shown to occur in mutant strains [24]. The basic assumption for ROOM is that the cell minimizes genetic regulatory changes to minimize the adaptation costs because protein synthesis is so energetically costly to the cell. The mathematical formulation of ROOM is similar to FBA, except that constraints are applied to distinguish significant changes in fluxes and minimize them. The allowable changes in the fluxes for the initial study was ±3% of the wild type, anything more than this was defined as significant and subject to minimization. ROOM was tested against both FBA and MOMA to determine its ability to predict accurate fluxes. It was found in eight of nine knockouts studied, ROOM performed as well or better than its counterparts; in the case of the pyuravte kinase knockout, ROOM was able to predict fluxes with 13%, 1%, and 4% higher correlations than MOMA for continuous growth on low glucose (dilution rate = 0.08 h^{-1}), medium glucose (dilution rate = 0.40 h^{-1}), and ammonia-limited growth, respectively . Due to the inherent differences in these two methods, it has been suggested that MOMA more accurately predicts fluxes for transient post-perturbation states because these are usually characterized by large changes in expression patterns, while ROOM is more accurate for adapted strains [80].

Transcriptional regulation has also been coupled to FBA models using regulatory flux balance analysis (rFBA) [17, 70]. rFBA requires knowledge of transcriptional regulation in the cell under the conditions to be modeled. The regulation of the genes in the model are experimentally determined and then converted to Boolean logic statements. For example, it is known that the presence of glucose (GLC) represses the transcription of the *lac* operon, which controls the genes for the utilization of lactose (LAC) as a metabolite – causing the diauxic growth of *E. coli* when grown on media containing both substrates. This regulatory function can be written in Boolean logic as

$$lac = \text{IF(LAC) AND NOT(GLC)}$$

Or simply stated, the *lac* operon is only transcribed when lactose is present in the media and glucose is not. These statements are then coupled to the FBA model by saying that the flux through a given enzyme is zero for a given time interval if that enzyme is not transcribed. A time delay is also added, so if a gene is turned on under a given set of conditions, it takes a certain amount of time before that enzyme can participate in metabolic reactions. Dynamic conditions, such as depleting substrate concentrations in the media, can be modeled using rFBA using small time intervals and a pseudo-steady-state assumption. This can be done because the turnover rate of metabolic reactions is much faster than that for transcription; so, for small-enough time intervals, the cell can be assumed to be at metabolic steady state. To predict dynamic growth, each time interval is calculated separately with different starting conditions to account for the changing environment. For more on dynamic FBA models, see Section 8.3.6. Using this dynamic modeling approach, rFBA was used to model the diauxic growth of *E. coli* on glucose and lactose and compared to two other models: kinetic model and FBA. FBA doesn't account for transcriptional regulation, and therefore inaccurately predicts the simultaneous utilization of both substrates, resulting in a faster depletion of substrate and a much higher growth rate. In contrast, rFBA was able to accurately predict the utilization of glucose first, followed by lactose as well as the growth rate, comparable to the full kinetic model [70]. This is significant because rFBA requires far fewer parameters to predict this dynamic growth than kinetic models, while still being able to calculate the growth accurately. Due the ability to predict dynamic behavior in changing conditions, rFBA has the potential to be useful for modeling circadian rhythms in plants because it has the ability to capture the transition from night to day and day to night.

8.3.5 Thermodynamic Considerations for FBA

One of the limitations of the FBA methods discussed so far is that thermodynamic constraints are accounted for only via explicit constraints on directionality of physiologically irreversible reactions. It is therefore possible to arrive at flux distributions which contain loops such that there is no net reaction. This allows material to circulate infinitely in the loop and violates the laws of thermodynamics, although it satisfies the law of mass conservation. In this section, we describe two approaches that attempt to eliminate flux distributions that violate thermodynamics. The first of these approaches involves the calculation of the in vivo free energy change in every single reaction of the network and use the data to ensure thermodynamic feasibility of the network. The second approach eliminates infeasible loops without the explicit use of data on free energy (or chemical potential) changes, and it directly examines the *structural* properties of the network instead. This type of approach is often used in conjunction with extreme pathway or elementary modes analysis (see Section 8.4).

The first approach by Henry et al. [34, 35] is based on calculations of free energy change of each reaction within the cell in which they utilize a group contribution method [35] developed by Mavrovouniotis [53] to estimate the standard free energy change (ΔG^0) for every reaction in a genome-scale *E. coli* model. They integrated this thermodynamic data with regular FBA; for every reaction essential for optimal *E. coli* growth, metabolite concentrations that ensured a negative free energy change (ΔG) (thermodynamically feasible) were allowed. This method, named thermodynamics-based metabolic flux analysis (TMFA), imposes simultaneous mass balance and thermodynamic constraints. As opposed to the work described in [4, 64] where the thermodynamic constraints are non-linear, this work is formulated as a mixed integer linear program. In addition to the usual constraints of an FBA formulation described in Section 8.3, the following are added as thermodynamic constraints:

$$0 \leq v_i \leq z_i v_{\max} \quad (8.19)$$

$$\Delta G_i - K + K z_i < 0 \quad (8.20)$$

$$\Delta G_i = \Delta G_i^0 + RT \sum s_{ij} \ln c_j \quad (8.21)$$

where Eq. 8.19 ensures that z_i is equal to zero if v_i is zero, and z_i is equal to one if v_i is non-zero. $v_{\max}$ is a physiologically reasonable limit on the maximum flux value. K is a very large number so that Eq. 8.20 is always satisfied when z_i equals zero. ΔG_i is calculated from Eq. 8.21, using the estimated values of the standard free energy change (ΔG_i^0) for the ith reaction, c_j is concentration (or activity) of the jth metabolite, and s_{ij} is the stoichiometric coefficient as described before. The unknown parameters to optimize are v_i, c_j, and z_i, and the objective function maximizes biomass production. Using this framework, Mavrovouniotis evaluated reactions that must carry zero flux for optimum growth ("blocked reactions"). They found 606 blocked reactions, compared to only 576 blocked reactions when no thermodynamic constraints were used. These extra 30 reactions are therefore part of infeasible loops. The biggest impediment in using this approach is the estimation of ΔG_i^0 for all reactions. Even using group contribution methods, it is not possible to evaluate ΔG_i^0 for every reaction of interest. Also, there are errors and uncertainty associated with the estimations of metabolite activities [34]. Further, ΔG_i is dependent on factors such pH and ionic strength, which if not captured correctly can affect the results.

A different approach was taken by Beard et al. [4, 5] where thermodynamic constraints were added to the stoichiometric network, by calculating whether or not a chemical potential can exist across a reaction (loop), without explicitly using values for chemical potentials. The constraints are obtained using the theory of oriented matroids. In simplified terms, a matroid is a mathematical entity that consists of a finite set of elements (such as rows of a matrix) and subsets called circuits, which satisfy certain properties (such as linear independence) [31]. In this work, the elements of the oriented matroid are the sign patterns of a flux vector (i.e., whether each flux in a given vector is positive, negative, or zero) that belongs to the row space of the stoichiometric matrix. The thermodynamic constraint on a flux vector J (that satisfies mass balance) is that its sign pattern must be an element of the matroid described above. The detailed mathematical description can be found in [4]. A limitation of this method is that the algorithm used scales exponentially with increase in network complexity.

Price et al. [64] analyzed a genome-scale stoichiometric model (476 intracellular reactions and 411 intracellular metabolites) of *Helicobacter pylori* using non-linear thermodynamic constraints to eliminate infeasible loops. They too did not make explicit use of chemical potential values. Instead, the thermodynamic constraints were expressed in the form of a "Loop Law" which states that the net flux through a loop must be zero, since there is no thermodynamic driving force [5, 62] and is expressed mathematically as

$$\sum_{\text{loop}} \Delta u_i = 0 \qquad (8.22)$$

$$v_i \Delta u_i \leq 0 \qquad (8.23)$$

where Δu_i is the change in chemical potential in the ith reaction, and Eq. 8.22 ensures that the net change in chemical potential in looped pathway is zero. Eq. 8.23 ensures that the net free energy change of each reaction is zero. If a flux v_i is positive, the chemical potential change must be negative, and vice versa. Equations 8.22 and 8.23 together ensure that net flux in a loop equals zero. Since constraint 23 is non-linear, the feasible space becomes non-convex and much harder to solve than with all linear constraints (e.g., FBA). In order to explore this feasible space, they first used extreme pathway analysis to list the Type III extreme pathways (see Section 8.4). These are the stoichiometrically feasible pathways, which violate thermodynamics, because they are loops. Four such non-trivial loops were found for the network. For some reactions, the loop law stated above could be expressed as two linear constraints. This was done as follows: for each reaction r_i of a loop, FBA was used to maximize (v_{max}) and minimize (v_{min}) each flux through the other reactions of the loop, while setting the flux through r_i equal to zero, which helped to identify reactions that could not carry any flux unless the loop operated. For these reactions, a linear constraint could be written, setting flux through it to zero. Thus, they were able to eliminate two loops by this process. For the other two loops, the values of v_{max} and v_{min} values obtained by FBA were used to further constrain the non-convex space. Monte Carlo methods were then used to uniformly sample a convex feasible flux space that enclosed the non-convex space. Roughly one in a thousand points (flux distributions) had to be discarded for violation the loop law. This was done by the method described in [4]. The remaining set of sampled points is the set of feasible flux distributions for the *H. pylori* network (candidate states). Imposition of loop law had a marked effect on the flux through reactions involved in the loops but only a small number of non-loop reactions were affected. They showed that genome-scale metabolic networks can have loops which violate thermodynamics, and it is possible to remove these systematically by imposing the loop law. Their method of using FBA to reduce certain loop law constraints to linear constraints and place bounds on v_{max} and v_{min} of certain reactions was a significant contributor in reduction of the problem complexity. The model used in this work was based on the most recent network reconstruction, including several irreversible reactions. A major hindrance to perform the same analysis on a network in the absence of any a priori irreversibility constraints is the efficient calculation of all possible loops (Type III extreme pathways) [64].

Kümmel et al. [47] describe a systematic computational method to assign reaction directions in a newly reconstructed metabolic network. Demonstrating the technique on the *E. coli* iJR904 model, they first assigned reaction directions based on available free energy change data for typical intracellular concentrations. Then, cycles were identified by an analysis of the null space of the stoichiometric matrix. Futile cycles converting high-energy cofactors into their low-energy forms were identified (e.g., net reaction ATP → ADP), and the direction of the cycle was constrained such that the high-energy form of the cofactor could not be formed from the operation of the cycle (e.g., net reaction of ADP → ATP is disallowed). These cycles are similar to Type II extreme pathways described in Section 8.4. Although the methodology was not exhaustive in identifying all cycles like the methods of Beard et al. [5], it enabled the assignment of directions to 130 reactions (out of 920 total reactions) using a very short computation time. This corresponds to about 70% of all irreversible reactions that are required to disable thermodynamically infeasible energy production.

8.3.6 Dynamic FBA

All the FBA methods described so far are methods for pseudo-steady-state analysis. However, a lot of physiologically as well as industrially important phenomena occur under non-steady-state conditions. Several efforts have been made to utilize the stoichiometric framework for dynamic situations. In all these works, once the substrate is taken up by the cell, it is assumed that the intracellular dynamics responds immediately to adjust to a new pseudo-steady-state condition. The simplest conceptual approach was adopted by Varma and Palsson to describe batch and fed batch growth of *E. coli* on glucose [92]. They divided the experiment into small time steps and used the FBA model to predict extracellular concentrations at every time step. They could successfully predict the secretion and re-utilization of acetate in batch aerobic growth and the secretion profiles of acetate, ethanol, and formate for anaerobic batch growth. The rFBA approach described in Section 8.3.4 is also an example of dynamic flux balance analysis which combines FBA to known transcriptional regulatory phenomena [17, 70]. An alternative optimization framework was shown to match experimental data better for diauxic growth of *E. coli* on glucose and acetate [50] compared to [92]. Yeast fermentations in batch reactors [68] and fed-batch reactors [36, 37] have also been studied by dynamic FBA. In these studies, the investigators combine unstructured kinetic models and FBA by modeling the dynamics of substrate uptake by a kinetic model. Additionally, effects of product inhibition can also be modeled. The FBA model predicts internal fluxes at every time instant, as well as the fluxes to excreted products, which can affect uptake dynamics via the kinetic model. Unsteady-state mass balances (differential equations) on reactor volume, biomass, and external metabolite concentrations along with the FBA model are used as constraints to obtain operating parameters that maximize product yields. An objective function (such as maximization of biomass) is also chosen for the inner FBA model. Different approaches are available for solving the bi-level optimization problem [36, 50, 68] – a non-linear problem that is considerably harder to solve than simple FBA formulations.

8.4 Metabolic Pathway Analysis

The previous section described various approaches to find a flux distribution based on optimizing user-defined criteria. In this section, we describe metabolic pathway analysis which is another widely used approach to explore the feasible flux space. In this methodology, the properties of a metabolic network are explored without the need of any optimality criteria. Also, compared to FBA which provides only one flux distribution, all permissible flux distributions and pathways can be analyzed by these methods. Extreme pathway analysis (EP) [71] and elementary mode analysis (EM) [72, 74] are two similar methods to study the properties of metabolic networks, each having their own advantages and disadvantages [45]. They are both developed from the theory of convex analysis [67] and extend the idea of "extreme currents" in stoichiometric networks developed by Clarke [15]. Both methods involve the calculation of a basic set of (feasible) pathways which enable the description of all feasible flux distributions by a suitable linear combination of the basic pathways.

To calculate EMs, one starts with building a stoichiometric matrix, as described in Section 8.2. Although it is necessary to know which reactions are irreversible, the reversible reactions are not split into a forward and reverse reaction. Further, for this discussion, an exchange flux is defined as a reaction that transports a metabolite into

8 Network Stoichiometry

or out of the system (fluxes v_1 and v_{14} in Fig. 8.1a). An elementary mode is defined as a vector **e** of fluxes $(v_1, v_2, ... v_n)$ such that the three following conditions are met:

S·e = 0 (pseudo-steady-state condition with no accumulation of internal metabolites)

$v_i \geq 0$ for all irreversible reactions – this ensures thermodynamic feasibility of the pathway

Non-decomposability – there exists no other vector $\mathbf{e_1}$ such that $\mathbf{e_1}$ is a subset of e and also satisfies conditions 1 and 2. Thus, if any (non-zero) flux v_i is removed from the elementary mode **e**, the mode will no longer carry a balanced flux through it (violation of condition 1). This condition is also known as genetic independence, since this implies that different sets of enzymes (and hence genes) required for the operation of each elementary mode.

Elementary modes enumerate all possible pathways from a substrate A to a product B using a minimum set of enzymes. The intermediates in the pathways are formed and consumed in a balanced manner such that there is no accumulation of the intermediates. For a given system, EMs are uniquely determined (up to a scaling factor). All feasible flux distributions can be expressed as a linear combination of these EMs ($\mathbf{v} = \lambda_1 \mathbf{e_1} + \lambda_2 \mathbf{e_2} + ... + \lambda_{n1} \mathbf{e}_n$ such that $\lambda_i \geq 0$). The λ_i are not necessarily unique, implying that different combinations of the elementary modes can lead to the same flux distribution. The λ_i must be non-negative for any EM that contains even one irreversible reaction in order to ensure that no flux proceeds in a forbidden direction. The detailed algorithm for computing the EMs is described in [73], which is implemented in the freely available software METATOOL [93] and FluxAnalyzer [46].

Extreme pathways are a subset of the elementary modes. They satisfy conditions 1–3 stated above, in addition to possessing the following properties:

1. Each reversible internal reaction is split into a forward and a reverse reaction. Thus all reaction fluxes can only be non-negative. This is similar to the splitting performed for FBA analysis in Section 8.3.1.
2. Mathematically, EPs form a convex basis of the reaction network. This means that they are a set of "systemically independent" pathways, such that all other (mathematically) permissible metabolic pathways are obtained by a non-negative linear combination of the EPs.

In a graphical representation, with each flux forming an axis, the mass balance equations (similar to constraints described for FBA in Section 8.2) can be plotted to obtain the permissible flux space or flux cone. The EPs are equations (pathways) that form the edges of this flux cone. We shall not enter into detailed comparison of the two methods, which can be found in literature [44, 45, 56]. It is sufficient to note that the techniques are conceptually very similar, and allow a very similar analysis with the networks. Free software for calculation of extreme pathways is also available [7]. Extreme pathways can be classified into three types [71] as shown

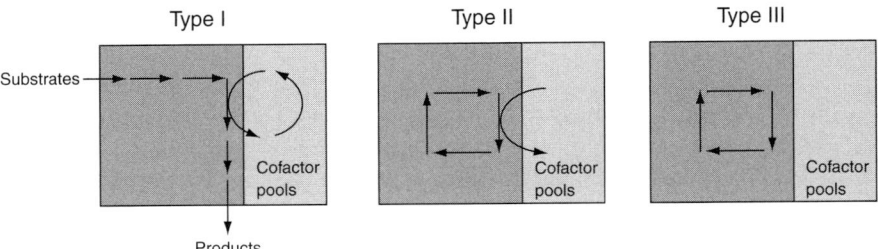

Fig. 8.5 Types of extreme pathways. Type III extreme pathways are those that have exchange fluxes across the system boundaries. Type II extreme pathways have only cofactors that cross system boundaries. These cofactors can thus be thought of as being accumulated or depleted in the cell. Type III extreme pathways do not contain any exchange fluxes, and thus correspond to internal cycles. Reproduced from [62] with permission from the Biophysical Society

in Fig. 8.5. Type I denotes the class of pathways that have non-zero values for net exchange fluxes. Thus, the overall stoichiometry of the pathway converts a substrate into a product. Type II pathways are cycles that only lead to net conversion of cofactors. These therefore include futile cycles that dissipate ATP. Type III pathways are cycles that lead to no net conversion of any metabolite.

The calculation of elementary modes is combinatorial in nature and therefore increases nonlinearly with increasing network size. For example, in a biochemical network of *E. coli* consisting of 89 metabolites and 110 reactions, about 5×10^5 elementary modes were found [81]. An increase in the number of elementary modes leads to longer computational times as well as a non-trivial process of sorting and interpreting the elementary modes. However, extreme pathways have been used successfully to study several large metabolic networks, including genome-scale networks [63].

8.4.1 Applications of Metabolic Pathway Analysis

One of the powers of metabolic pathway analysis is that it can be used to enumerate and subsequently rank the modes (or pathways) according to the yield of desired product. Thus, all alternative pathways (or modes) for the same yield can be identified. The techniques have been applied to a wide variety of systems, and a few are listed here. Carlson and Srienc [12] enumerated the elementary modes for an *E. coli* central metabolism model and identified four modes which gave maximum biomass yield. From analysis of these modes, they predicted five gene knockouts that would eliminate flux through suboptimal modes. Knockout mutants of all five genes were created, and the prediction was experimentally verified as the mutant strain of *E. coli* growing on glucose had a higher biomass yield [88]. In another example, extreme pathways and elementary modes have been used in conjunction with FBA to identify alternative optimal pathways [51] and study the genome-scale metabolic network of *Lactobacillus plantarum* [85]. Elementary modes have also been used to study of nucleotide salvage pathways in red blood cells [75].

8.4.2 Application of Elementary Modes Analysis in Plant Systems

This section describes two applications of elementary modes analysis in plant systems. Both studies used elementary modes in combination with other modeling or experimental approaches, in order to verify results or test hypotheses.

Poolman et al. used elementary modes analysis to study photosynthetic metabolism in the chloroplast [60]. The analysis of elementary modes helped to answer questions raised during the authors' previous work on the same system using kinetic modeling [62, 63]. Their model of the chloroplast metabolism had a total of 23 reactions covering the pentose phosphate pathway, the Calvin cycle, and starch synthesis (Fig. 8.6). The external metabolites (which were allowed to accumulate) were starch, CO_2, NADPH, ATP, and cytosolic metabolites (the exported PGA, GAP, and DHAP). Three exchange reactions transported the triose phosphates out of the chloroplast. The model simulates the dynamics of starch synthesis, degradation, and triose phosphate export out of the stroma in the presence of the light reactions (i.e., when Rubisco is active). It is known that the reaction catalyzed

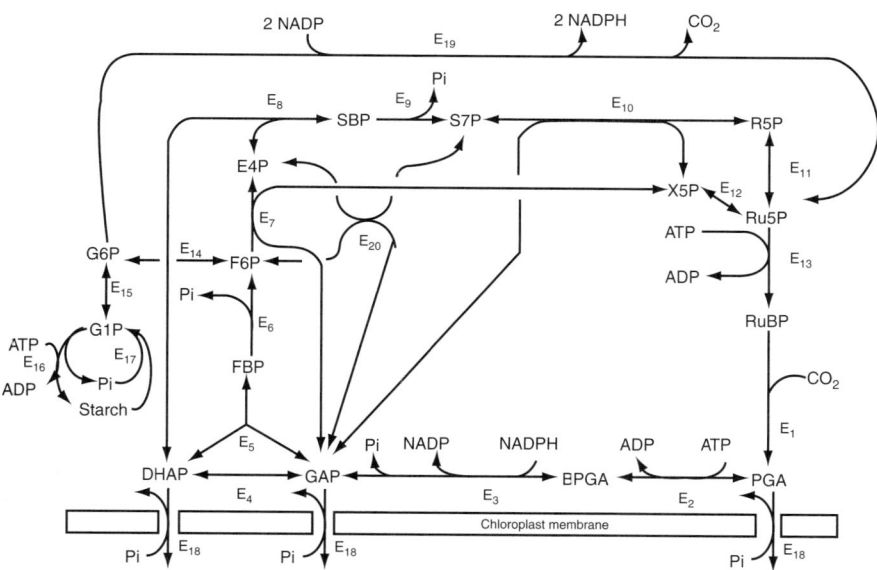

Fig. 8.6 Reactions of the Calvin cycle and oxidative pentose phosphate pathway as considered by Poolman et al. Bidirectional arrows indicate reversible reactions and unidirectional *arrows*, irreversible reactions. The light reactions, assumed to catalyze ADP+Pi to ATP, and processes consuming E4P, Ru5P, or G6P are omitted for clarity. Metabolite abbreviations: PGA, 3-phosphoglycerate; BPGA, glycerate 1,3-bisphosphate; GAP, glyceraldehyde 3-phosphate; DHAP, dihydroxyacetone phosphate; FBP, fructose 1,6-bisphosphate; F6P, fructose 6-phosphate; E4P, erythrose 4-phosphate; SBP, sedoheptulose-1,7-bisphosphate; S7P, sedoheptulose 7-phosphate; R5P, ribose 5-phosphate; X5P, xylulose 5-phosphate; Ru5P, ribulose 5-phosphate; RuBP, ribulose 1,5-bisphosphate; G6P, glucose-6-phosphate; G1P, glucose-1-phosphate. Reproduced from [60] with permission from Blackwell Publishing

by ADPG-pyrophosphorylase (lumped with starch synthesis as E16 in Fig. 8.6) is activated by PGA and deactivated by stromal phosphate concentration. The kinetic model showed that the flux to starch synthesis (E16) showed the same qualitative response to stromal phosphate concentration irrespective of whether the kinetics of ADPG-pyrophosphorylase was sensitive to PGA and stromal phosphate concentration. Further, as the stromal phosphate concentration rose, starch degradation also set in (flux E15 is negative). Thus, under these circumstances, the net export of triose phosphate exceeds the net carbon fixed by Rubisco (with the extra supplied from starch degradation). The authors then wished to evaluate if this was a valid mode of starch degradation during the night also (i.e., absence of light reactions and Rubisco). Setting the Rubisco flux to zero and running the kinetic model resulted in all fluxes going to zero. Since this result from the kinetic model could well be due to a poor choice of kinetic equations or parameters, they performed elementary modes analysis on the system. In the absence of the Rubisco, they found that no mode could solely degrade starch into triose phosphate (using the reactions of the Calvin cycle). Thus, they confirmed that the result was a *structural* feature of the network and not due to a poor kinetic model. It must be noted that triose phosphate export from starch breakdown in the dark with concomitant CO_2 release was possible when the unique reactions of the pentose phosphate pathway (E19 and E20) were active.

The elementary modes analysis also led to several other interesting observations. For example, sedoheptulose 1, 7-bisphosphatase (SBPase, E9) is regulated by thioredoxin such that it is inactivated in the absence of light. All modes of sugar phosphate export (triose phosphates, E4P, and R5P) in the dark were found to be concomitant with NADPH and CO_2 production (a consequence of using the oxidative pentose phosphate reactions, E19). Addition of SBPase activity in the dark led to the creation of a new mode which could completely oxidize starch to CO_2, the stoichiometry of the net reaction being STARCH + 12 NADP → 6 CO_2 + 12 NADPH. This new mode is hypothesized to have implications in the de-coupling of sugar phosphate export and NADPH production in the dark.

A criticism of this analysis is that the model of chloroplast metabolism did not include the glycolytic enzymes which are known to be present in chloroplasts [42, 59, 84]. In the presence of these enzymes (phosphofructokinase, glyceraldehyde 3-phosphate dehydrogenase, and phosphoglycerate kinase), it is possible for starch to be converted to triose phosphate in the absence of Rubisco. Since they modeled the Calvin cycle/pentose phosphate in isolation from the glycolytic/gluconeogenic enzymes, the scope of the results must be treated with regards to the limitation of the model. Addition of the glycolytic enzymes could lead to new modes which use a combination of reactions that are traditionally assigned to glycolysis or the pentose phosphate pathway. The power of stoichiometric models allows us to model highly interactive metabolism in which many cycles and pathways share several intermediates and must not be treated in isolation, for a comprehensive understanding of metabolic processes.

In a second example of the use of elementary mode analysis, Schwender et al. validated results from ^{13}C and ^{14}C labeling experiments in developing embryos of

Brassica napus which convert sucrose into fatty acids [76]. The experimentally observed yield of fatty acid (oil) on sucrose was much lower than expected if all the acetyl CoA were derived solely from glycolysis, indicating the presence of a more efficient carbon utilization pathway. They found that the developing embryo could simultaneously utilize Rubisco to fix CO_2 and break down sucrose via glycolysis, to form acetyl CoA, the metabolic precursor of fatty acids. Further, based on the ^{13}C labeling pattern in PGA, it was found that the complete Calvin cycle (regeneration of PGA) was not active. Elementary modes analysis of the system confirmed the existence of elementary modes which converted hexose phosphates into pentose phosphates via non-oxidative reactions of the pentose phosphate pathway, followed by CO_2 fixation into PGA by Rubisco, and conversion of PGA into acetyl CoA, with the release of CO_2 (Fig. 8.7(C)). It is this released CO_2 that can be fixed by Rubisco, thereby improving the oil yield. Other elementary modes involving Rubisco and the oxidative pentose phosphate reactions (which release CO_2) could not explain the observed CO_2 to oil ratio.

Modes involving the complete Calvin cycle could not explain the isotopic labeling patterns. Only the set of elementary modes shown in Fig. 8.7(C) could explain all the data.

8.5 Genome-Scale Networks

8.5.1 Model Reconstruction

Reconstruction of a genome-scale metabolic network requires the collection of data from a variety of sources and is laborious due to the iterative nature of creating a high-quality, accurate network. Despite the development of software to facilitate reconstruction [18, 41], automatically created models still require manual curation [8] to fill in gaps. The first step in any reconstruction effort (Fig. 8.8) is to search the genomic database of the plant species to be modeled; of course, this requires the genome to be fully sequenced and annotated. The initial effort of network reconstruction requires a complete list of metabolic reactions known to be present in the cell to be collected. This first draft of the network should include (for each reaction) enzyme, stoichiometry, cofactors needed and, if possible, direction and localization. In eukaryotic organisms, it is also important to include transport and carrier enzymes to account for the movement of metabolites across intracellular membranes. Accounting for compartmentation in the cell is a key aspect of modeling higher eukaryotes, such as plants, because the distribution of flux between compartments is one of the more interesting results from such models. Despite the availability of many sequenced genomes, none have been fully annotated. Even the genome of the most widely studied organism, *E. coli*, is constantly being updated and annotated as new information becomes available. Due to the lack of complete annotation, reconstruction of a genome-scale model requires additional information to be incorporated with information garnered from the genomic database, in order

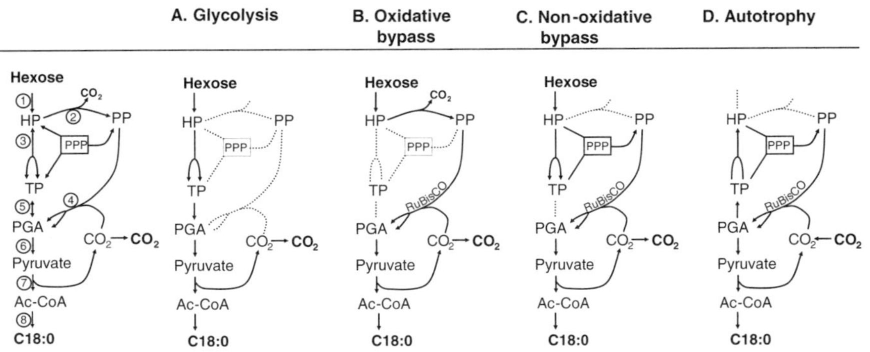

	A. Glycolysis	B. Oxidative bypass	C. Non-oxidative bypass	D. Autotrophy
Carbon in C18:0 / Carbon uptake (glucose)	66.7 %	66.7 %	80 %	–
ATP balance	+1	–8	–8	–71
NADPH balance	+2	+2	–7	–52
Number of modes of this type	1	9	7	3

Fig. 8.7 Elementary flux-modes analysis of seed metabolism from glucose to fatty acids using a network of 26 reactions. Carbon use efficiency (carbon stored in oil/carbon uptake as glucose), cofactor balances, and four characteristic fluxes are shown relative to the formation of 1 mol C18:0. Of the 28 elementary modes, 20 produce stearic acid (C18:0). These modes fall into 4 categories (**A** to **D**) and for each category one representative mode is shown. Each category comprises modes that are very similar; with small differences in CO_2 and ATP balance due to ambivalent enzyme functions present in the model (e.g., transaldolase can replace sedoheptulose bisphosphate aldolase/sedoheptulose bisphosphatase). NADH is balanced in all cases. Modes with negative ATP or NADPH balance require cofactor supply from photosynthetic light reactions. Numbers refer to enzymes: hexokinase (1); glucose-6-phosphate dehydrogenase, 6-phosphoglucono-lactonase, 6-phosphogluconate dehydrogenase (2); phosphofructokinase, fructose-1,6-bisphosphate aldolase (3); phosphoribulokinase, ribulose-1,5-bisphosphate carboxylase/oxygenase (4); NADH-glyceraldehyde-3-phosphate dehydrogenase, NADPH-glyceraldehyde-3-phosphate dehydrogenase, phosphoglycerate kinase (5); phosphoglycerate mutase, enolase, pyruvate kinase (6); pyruvate dehydrogenase complex (7); synthesis of stearic acid by fatty acid synthethase complex (8); transketolase, transaldolase, ribose-5-phosphate isomerase, ribulose-5-phosphate epimerase (PPP). Metabolites: Ac-CoA, acetyl coenzyme-A; HP, hexose phosphates (glucose 6-phosphate, fructose 6-phosphate); PP, pentose phosphates (ribulose 5-phosphate, ribose 5-phosphate, xylulose 5-phosphate); TP, triose phosphates (dihydroxy acetone phosphate, glyceraldehyde 3-phosphate). Figure obtained from [76] with permission from the Nature Publishing Group

to have a complete model of metabolism. Additional information and assumptions will have to be made in order to fill in gaps in known pathways and add missing pathways. Gaps in the metabolic pathway, such as missing enzymes, pathways, and transporters can be filled in with information found in biochemistry books, archival journal articles, and pathway databases [13, 40]. Another common problem in genome-scale metabolic networks is the presence of dead-end metabolites. Dead-end metabolites arise from two sources: (1) a byproduct of a given reaction

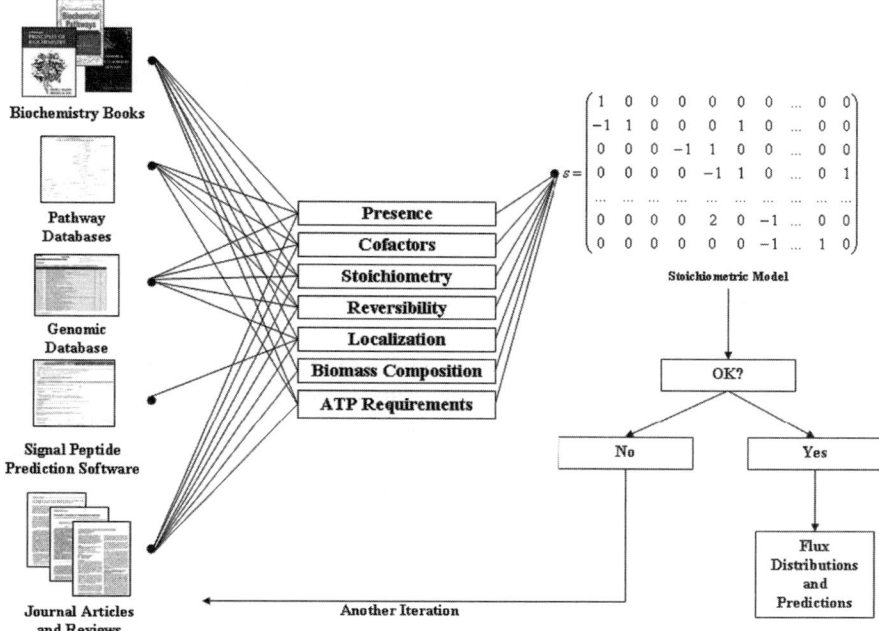

Fig. 8.8 Genome-scale metabolic network reconstruction. Genome-scale metabolic models typically require the use of several sources to capture the entire network including genomic database, pathway databases, journal articles, biochemistry textbooks, and signal peptide prediction software. The picture of the pathway was taken from MetaCyc (http://www.metacyc.com); the signal peptide prediction picture was taken from TargetP (http://www.cbs.dtu.dk/services/TargetP/), and the figure itself is an adaptation of Fig. 1 of [29]

is not used anywhere else in the network and therefore accumulates in the cell or (2) a particular metabolite is needed as a reactant in a reaction, but is not synthesized in any reaction in the network. To avoid having accumulation in the cell, either a synthesis or degradation reaction must be added to the network. Another important aspect of genome-scale modeling is the definition of a quantitative biomass equation. Development of the biomass equation requires experimental determination of the macromolecular content and composition of the cell (DNA, RNA, lipid, carbohydrate, and protein). The next step in network reconstruction for eukaryotic organisms is to determine the localization of each enzyme to the compartments to be modeled. Localization in the cell is determined by signal peptides (SP), which are located on the N-terminal end of the protein; the composition of the SP determines where the protein is targeted to. Free online software, such as TargetP [23], can be used to predict the presence of SPs and to determine the localization of the enzyme to a particular compartment (secretory pathway, chloroplast, mitochondria, or other).

Once a first draft of the network is collected, the iterative process of network refinement and completion begins (Fig. 8.8). The assembled list of reactions can then be used to construct the stoichiometric matrix. The model is then tested by using a modeling technique, such as FBA, to check (1) if the model runs and (2) if the mass balance closes. FBA has been used in several genome-scale models [6, 8, 22, 29, 34] because it requires no additional data or information and can quickly predict the flux distribution in the cell once the network has been reconstructed. Typically, the test of the first draft of the genome-scale model does not match experimental results (such as biomass yield per mole substrate); this can be corrected by rigorous investigation of flux distributions to determine where gaps or dead-end metabolites occur and/or to identify imbalances in the network (such as protons or maintenance). This process of iterative refinement of the network continues until the results produced by the model matches experimental data (yield, growth rate, substrate uptake, and product excretion rates). Although model reconstruction is iterative, even with a complete network, the model results may not correlate well with experiments. This, in part, may be due to invalid assumptions, over-simplification of the network, or an inappropriate choice for an objective function. It is imperative that throughout the process of model reconstruction, each of these is reevaluated to ensure agreement with the organism being modeled.

8.5.2 Application of Genome-Scale Models

Despite the large number of genome sequences available, only a handful of genome-scale metabolic models have been completed to date (Table 8.2). Some of these organisms are industrially important, such as *E. coli, L. lactis, S. cerevisiae, S. coelicolor*, and *L. plantarum*, and these models can be used to optimize the productivity of natural products. Others are human pathogens, such as *H. influenzae*,

Table 8. 2 List of organisms with genome-scale metabolic models. *If multiple versions exist, only the most recent version is listed.* The naming convention for genome-scale models is as follows: "*i*" represents the in silico strain, the initials represent he principal author for the reconstruction, and the number is the number of genes included in the reconstruction

Organism	Version	Reference
Haemophilus influenzae	N/A	[22]
Lactococcus lactis	iAO385	[55]
Helicobacter pylori	iIT341	[86]
Saccharomyces cerevisiae	iND750	[19]
Escherichia coli	iAF1260	[26]
Staphylococcus aureus N315	iSB619	[6]
Lactobacillus plantarum WCFS1	iBT721	[85]
Streptomyces coelicolor A3(2)	iIB711	[8]
Bacillus subtillus	iYO844	[54]

H. pylori, and *S. aureus*, and the completion of their genome-scale models opens up a new way to select antibiotic targets by selecting enzymes found to be essential from the model. In much the same way, completion of genome-scale models in plant species can have similar applications. Genome-scale metabolic models represent the most complete model of cellular metabolism available. A genome-scale model of crop plants, like corn, soybean, or wheat, could be used to identify metabolic engineering targets to optimize growth or product (e.g., fruit, seed) yield. New pathways could also be introduced and optimized in plants for the production of chemicals. Similar to the example above of human pathogens, genome-scale models of undesirable plants could be used to identify essential genes which can be the target for new species-specific herbicides (same would be true for broad spectrum herbicides). Although purely stoichiometric genome-scale models have great potential to aid in metabolic engineering efforts, the addition of regulatory information will greatly enhance the model. Finally, genome-scale models serve as a starting point for the integration of systems biology data into a mathematical model.

8.6 Future Directions for Stoichiometric Modeling in Plants

As shown in this chapter, the stoichiometric modeling of metabolic networks is expanding to larger networks and more complex organisms. However, as the systems become more complex, the predictive value of the models by themselves is limited. The modeling approaches' strongest application is in combination with experimental methods such as metabolic flux analysis and kinetic modeling (see Chapters 9 and 11). For example, one could use the stoichiometric framework to map systems biology data (e.g., transcriptome, proteome, metabolome, and fluxome) with genome-scale models [11].

With the completion of genome sequencing of *Arabidopsis* and several major agronomically important crops, it begs the question of whether a genome model of a higher plant is possible and what would be gained? For stoichiometric analysis, this model would serve as the basis for unifying the analysis of large sets of metabolomic, transcriptomic, and proteomic data. The effort to complete the genome-scale models would be an enormous undertaking, but the benefits realized in understanding plant metabolic physiology as large interacting networks of transcripts, enzymes, and metabolites clearly justify such an effort. A logical starting point is to first model individual cell types of higher plants. For example, one could envision a model of a single photosynthetic cell as the basis for modeling an "average" leaf. An appropriate analogue to the leaf cell could be to model a single-celled algae, such as *Chlamydomonas reinhardtii*, which has a completed genome. Similar models could be constructed for root cells. The next step would be to average over these individual cell models to represent a tissue. Finally, the tissues could be connected with transport reactions, in a similar fashion to how individual organs are modeled in physiological compartmental pharmacokinetic studies [30].

References

1. Albertsson PA (2001) A quantitative model of the domain structure of the photosynthetic membrane. Trends Plant Sci 6:349–358.
2. Allen JF (2002) Photosynthesis of ATP – electrons, proton pumps, roters and poise. Cell 110:273–276.
3. Arnon DI (1948) Copper enzymes in isolated chloroplasts polyphenoloxidase in *Beta vulgaris*. Plant Physiol 24:1–15.
4. Beard DA, Babson E, Curtis E, Qian H (2004) Thermodynamic constraints for biochemical networks. J Theor Biol 228:327–333.
5. Beard DA, Liang SD, Qian H (2002) Energy balance for analysis of complex metabolic networks. Biophys J 83:79–86.
6. Becker S, Palsson B (2005) Genome-scale reconstruction of the metabolic network in *Staphylococcus aureus* N315: an initial draft to the two-dimensional annotation. BMC Microbiology 5:8.
7. Bell SL, Palsson BO (2005) expa: a program for calculating extreme pathways in biochemical reaction networks. Bioinformatics 21:1739–1740.
8. Borodina I, Krabben P, Nielsen J (2005) Genome-scale analysis of *Streptomyces coelicolor* A3(2) metabolism. Genome Res 15:820–829.
9. Burgard A, Maranas C (2001) Probing the performance limits of the *Escherichia coli* metabolic network subject to gene additions or deletions. Biotechnnol Bioeng 74: 364–375.
10. Burgard AP, Pharkya P, Maranas CD (2003) Optknock: A bilevel programming framework for identifying gene knockout strategies for microbial strain optimization. Biotechnol Bioeng 84:647–657.
11. Çakir T, Kirdar B, Ülgen KÖ (2004) Metabolic pathway analysis of yeast strengthens the bridge between transcriptomics and metabolic networks. Biotechnol Bioeng 86:251–260.
12. Carlson R, Srienc F (2004) Fundamental *Escherichia coli* biochemical pathways for biomass and energy production: Creation of overall flux states. Biotechnol Bioeng 86:149–162.
13. Caspi R, Foerster H, Fulcher C, Kaipa P, Krummenacker M, Latendresse M, Paley S, Rhee S, Shearer A, Tissier C, Walk T, Zhang P, Karp P (2006) MetaCyc: A multiorganism database of metabolic pathways and enzymes. Nucleic Acids Res 34:D511–D516.
14. Chvatal V (1983) Linear Programming. W. H. Freeman and Company, New York.
15. Clarke B (1988) Stoichiometric network analysis. Cell Biophy 12:237–253.
16. Cogne G, Gros JB, Dussap CG (2003) Identification of a metabolic network structure representative of *Arthrospira* (Spirulina) *platensis* metabolism. Biotechnol Bioeng 84:667–676.
17. Covert MW, Schilling CH, Palsson B (2001) Regulation of gene expression in flux balance models of metabolism. J Theor Biol 213:73–88.
18. DeJongh M, Formsma K, Boillot P, Gould J, Rycenga M, Best A (2007) Toward the automated generation of genome-scale metabolic networks in the SEED. BMC Bioinformatics 8:139.
19. Duarte N, Herrgard MJ, Palsson BO (2007) Reconstruction and validation of *Saccharomyces cerevisiae* iND750, a fully compartmentalized genome-scale metabolic model. Genome Res 14:1298–1309.
20. Edwards J, Palsson B (2000) The *Escherichia coli* MG1655 *in silico* metabolic genotoype: its definition, characteristics, and capabilities. Proc Natl Acad Sci USA 97:5528–5533.
21. Edwards J, Ramakrishna R, Palsson B (2002) Characterizing the metabolic phenotype: a phenotype phase plane analysis. Biotechnnol Bioeng 77:27–36.
22. Edwards JS, Palsson BO (1999) Systems properties of the *Haemophilus influenzae* Rd metabolic genotype. J Biol Chem 274:17410–17416.
23. Emanuelsson O, Nielsen H, Brunak S, Heijne Gv (2000) Predicting subcellular localization of proteins based on their N-terminal amino acid sequence. J Mol Biol 300:1005–1016.
24. Emmerling M, Dauner M, Ponti A, Fiaux J, Hochuli M, Szyperski T, Wuthrich K, Bailey JE, Sauer U (2002) Metabolic flux responses to pyruvate kinase knockout in *Escherichia coli*. J Bacteriol 184: 152–164.

25. Famili I, Forster J, Nielson J, Palsson BO (2003) *Saccharomyces cerevisiae* phenotypes can be predicted by using constraint-based analysis of a genome-scale reconstructed metabolic network. Proc Natl Acad Sci USA 100:13134–13139.
26. Feist AM, Henry CS, Reed JL, Krummenacker M, Joyce AR, Karp PD, Broadbelt LJ, Hatzimanikatis V, Palsson BO (2007) A genome-scale metabolic reconstruction for *Escherichia coli* K-12 MG1655 that accounts for 1260 ORFs and thermodynamic information. Mol Syst Biol 3: 121.
27. Fell DA, Small JR (1986) Fat synthesis in adipose tissue – an examination of stoichiometric constraints. Biochem J 238:781–786.
28. Fischer E, Sauer U (2005) Large-scale *in vivo* flux analysis shows rigidity and suboptimal performance of *Bacillus subtilis* metabolism. Nature Genet 37:636–640.
29. Forster J, Famili I, Fu P, Palsson BO, Nielsen J (2003) Genome-scale reconstruction of the *Saccharomyces cerevisiae* metabolic network. Genome Res 13:244–253.
30. Gibaldi M, Perrier D (1983) Pharmacokinetics. Marcel Dekker, New York.
31. Harary F (1994) Graph Theory. Perseus Books, Reading Westview Press, Boulder, CO.
32. Heinemann M, Kümmel A, Ruinatscha R, Panke S (2005) In silico genome-scale reconstruction and validation of the *Staphylococcus aureus* metabolic network. Biotechnol Bioeng 92:850–864.
33. Henriksen CM, Christensen LH, Nielsen J, Villadsen J (1996) Growth energetics and metabolic fluxes in continuous cultures of *Penicillium chrysogenum*. J Biotechnol 45: 149–164.
34. Henry CS, Broadbelt LJ, Hatzimanikatis V (2007) Thermodynamics-based metabolic flux analysis. Biophys J 92:1792–1805.
35. Henry CS, Jankowski MD, Broadbelt LJ, Hatzimanikatis V (2006) Genome-scale thermodynamic analysis of *Escherichia coli* metabolism. Biophys J 90:1453–1461.
36. Hjersted JL, Henson MA (2006) Optimization of fed-batch Saccharomyces cerevisiae fermentation using dynamic flux balance models. Biotechnol Prog 22:1239–1248.
37. Hjersted JL, Henson MA, Mahadevan R (2007) Genome-scale analysis of *Saccharomyces cerevisiae* metabolism and ethanol production in fed-batch culture. Biotechnol Bioeng 97:1190–1204.
38. Ibarra RU, Edwards JS, Palsson BO (2002) *Escherichia coli* K-12 undergoes adaptive evolution to achieve *in silico* predicted optimal growth. Nature 420:186–189.
39. Jorgensen H, Nielsen J, Villadsen J, Mollgaard H (1995) Metabolic flux distributions in *Penicillium chrysogenum* during fed-batch cultivations. Biotechnol Bioeng 46:117–131.
40. Kanehisa M, Goto S, Hattori M, Aoki-Kinoshita KF, Itoh M, Kawashima S, Katayama T, Araki M, Hirakawa M (2006) From genomics to chemical genomics: new developments in KEGG. Nucleic Acids Res 34:D354–357.
41. Karp PD, Paley S, Romero P (2002) The pathway tools software. Bioinformatics 18: S225–232.
42. Kelly GJ, Latzko E (1977) Chloroplast phosphofructokinase .1. Proof of phosphofructokinase activity in chloroplasts. Plant Physiol 60:290–294.
43. Kim HU, Kim TY, Lee SY (2008) Metabolic flux analysis and metabolic engineering of microorganisms. Mol BioSyst 4:113–120.
44. Klamt S, Stelling J. 2003. Stoichiometric analysis of metabolic networks. in: Proceedings of the Tutorial at the 4th International Conference on Systems Biology, Heidelberg, Germany.
45. Klamt S, Stelling J (2003) Two approaches for metabolic pathway analysis? Trends Biotechnol 21:64–69.
46. Klamt S, Stelling J, Ginkel M, Gilles ED (2003) FluxAnalyzer: exploring structure, pathways, flux distributions in metabolic networks on interactive flux maps. Bioinformatics 19: 261–269.
47. Kummel A, Panke S, Heinemann M (2006) Systematic assignment of thermodynamic constraints in metabolic network models. BMC Bioinformatics 7.

48. Lee S, Palakornkule C, Domach MM, Grossmann IE (2000) Recursive MILP model for finding all the alternate optima in LP models for metabolic networks. Comput Chem Eng 24: 711–716.
49. Li K, Frost JW (1998) Synthesis of vanillin from glucose. J Am Chem Soc 120:10545–10546.
50. Mahadevan R, Edwards JS, Doyle FJ, III (2002) Dynamic flux balance analysis of diauxic growth in *Escherichia coli*. Biophys J 83:1331–1340.
51. Mahadevan R, Schilling CH (2003) The effects of alternate optimal solutions in constraint-based genome-scale metabolic models. Metabol Eng 5:264–276.
52. Majewski RA, Domach MM (1990) Simple constrained optimization view of acetate overflow in *Escherichia coli*. Biotechnol Bioeng 35:732–738.
53. Mavrovouniotis ML (1990) Group contributions for estimating standard Gibbs energies of formation of biochemical compounds in aqueous solution. Biotechnol Bioeng 36: 1070–1082.
54. Nelson DL, Cox MM (2005) Lehninger Principles of Biochemistry. W.H. Freeman and Company, New York.
55. Oliveira A, Nielsen J, Forster J (2005) Modeling *Lactococcus lactis* using a genome-scale flux model. BMC Microbiology 5:39.
56. Papin JA, Stelling J, Price ND, Klamt S, Schuster S, Palsson BO (2004) Comparison of network-based pathway analysis methods. Trends Biotechnol 22:400–405.
57. Park SM, Sinskey AJ, Stephanopoulos G (1997) Metabolic and physiological studies of *Corynebacterium glutamicum* mutants. Biotechnol Bioeng 55:864–879.
58. Pharkya P, Burgard AP, Maranas CD (2003) Exploring the overproduction of amino acids using the bilevel optimization framework OptKnock. Biotechnol Bioeng 84:887–899.
59. Plaxton WC (1996) The organization and regulation of plant glycolysis. Annu Rev Plant Physiol Plant Mol Biol 47:185–214.
60. Poolman MG, Fell DA, Raines CA (2003) Elementary modes analysis of photosynthate metabolism in the chloroplast stroma. Eur J Biochem 270:430–439.
61. Pramanik J, Keasling JD (1997) Stoichiometric model of *Escherichia coli* metabolism: Incorporation of growth-rate dependent biomass composition and mechanistic energy requirements. Biotechnol Bioeng 56:398–421.
62. Price ND, Famili I, Beard DA, Palsson BO (2002) Extreme Pathways and Kirchhoff's Second Law. Biophys J 83:2879–2882.
63. Price ND, Papin JA, Palsson BO (2002) Determination of redundancy and systems properties of the metabolic network of *Helicobacter pylori* using genome-scale extreme pathway analysis. Genome Res 12:760–769.
64. Price ND, Thiele I, Palsson BO (2006) Candidate states of *Helicobacter pylori*'s genome-scale metabolic network upon application of "loop law" thermodynamic constraints. Biophys J 90:3919–3928.
65. Ramakrishna R, Edwards J, McCulloch A, Palsson B (2001) Flux balance analysis of mitochondrial energy metabolism: consequences of systemic stoichiometry constraints. Am J Physiolo Regul Integr Comp Physiol 280:R695–R704.
66. Reed J, Vo T, Schilling C, Palsson B (2003) An expanded genome-scale model of *Escherichia coli* K-12 (iJR904 GSM/GPR). Genome Biol 4:R54.
67. Rockafellar RT (1970) Convex Analysis. Princeton University Press, Princeton, NJ.
68. Sainz J, Pizarro F, Perez-Correa JR, Agosin E (2003) Modeling of yeast metabolism and process dynamics in batch fermentation. Biotechnol Bioeng 81:818–828.
69. Savinell JM, Palsson BO (1992) Optimal selection of metabolic fluxes for *in vivo* measurement .1. Development of mathematical methods. J Theor Biol 155:201–214.
70. Schilling CH, Covert MW, Famili I, Church GM, Edwards JS, Palsson BO (2002) Genome-scale metabolic model of *Helicobacter pylori* 26695. J Bacteriol 184:4582–4593.
71. Schilling CH, Letscher D, Palsson BO (2000) Theory for the systemic definition of metabolic pathways and their use in interpreting metabolic function from a pathway-oriented perspective. J Theor Biol 203:229–248.

72. Schuster S, Dandekar T, Fell DA (1999) Detection of elementary flux modes in biochemical networks: a promising tool for pathway analysis and metabolic engineering. Trends Biotechnol 17:53–60.
73. Schuster S, Fell DA, Dandekar T (2000) A general definition of metabolic pathways useful for systematic organization and analysis of complex metabolic networks. Nat Biotech 18:326–332.
74. Schuster S, Hilgetag C (1994) On elementary flux modes in biochemical reaction systems at steady state. J Biol Syst 2:165–182.
75. Schuster S, Kenanov D (2005) Adenine and adenosine salvage pathways in erythrocytes and the role of S-adenosylhomocysteine hydrolase. A theoretical study using elementary flux modes. FEBS Journal 272:5278–5290.
76. Schwender J, Goffman F, Ohlrogge JB, Shachar-Hill Y (2004) Rubisco without the Calvin cycle improves the carbon efficiency of developing green seeds. Nature 432:779–782.
77. Segre D, Vitkup D, Church GM (2002) Analysis of optimality in natural and perturbed metabolic networks. Proc Natl Acad Sci USA 99:15112–15117.
78. Shastri A, Morgan J (2004) Calculation of theoretical yields in metabolic networks. Biochem Mol Biol Educ 32:314–318.
79. Shastri AA, Morgan JA (2005) Flux balance analysis of photoautotrophic metabolism. Biotechnol Prog 21:1617–1626.
80. Shlomi T, Berkman O, Ruppin E (2005) Regulatory on/off minimization of metabolic flux changes after genetic perturbations. Proc Natl Acad Sci USA 102:7695–7700.
81. Stelling J, Klamt S, Bettenbrock K, Schuster S, Gilles ED (2002) Metabolic network structure determines key aspects of functionality and regulation. Nature 420:190–193.
82. Stephanopoulos G, Vallino JJ (1991) Network rigidity and metabolic engineering in metabolite overproduction. Science 252:1675–1681.
83. Stephanopoulos GN, Aristidou AA, Nielsen J (1998) Metabolic Engineering Principles and Methodologies. Academic Press, San Diego.
84. Stitt M, Heldt HW (1981) Physiological rates of starch breakdown in isolated intact spinach chloroplasts. Plant Physiol 68:755–761.
85. Teusink B, Wiersma A, Molenaar D, Francke C, Vos WMd, Siezen RJ, Smid EJ (2006) Analysis of growth of *Lactobacillus plantarum* WCFS1 on a complex medium using genome-scale metabolic model. J Biol Chem 281:40041–40048.
86. Thiele I, Vo TD, Price ND, Palsson BO (2005) Expanded metabolic reconstruction of *Helicobacter pylori* (*i*IT341 GSM/GPR): an *in silico* genome-scale characterization of single- and double-deletion mutants J Bacteriol 187:5818–5830.
87. Trawick JD, Schilling CH (2006) Use of constraint-based modeling for the prediction and validation of antimicrobial targets. Biochem Pharmacol 71:1026–1035.
88. Trinh CT, Carlson R, Wlaschin A, Srienc F (2006) Design, construction and performance of the most efficient biomass producing *E. coli* bacterium. Metabol Eng 8:628–638.
89. Varma A, Boesch BW, Palsson BO (1993) Biochemical production capabilities of *Escherichia coli*. Biotechnol Bioeng 42:59–73.
90. Varma A, Palsson BO (1993) Metabolic capabilities of *Escherichia coli*. 1. Synthesis of biosynthetic precursors and cofactors. J Theor Biol 165:477–502.
91. Varma A, Palsson BO (1993) Metabolic capabilities of *Escherichia coli*. 2. Optimal-growth patterns. J Theor Biol 165:503–522.
92. Varma A, Palsson BO (1994) Stoichiometric flux balance models quantitatively predict growth and metabolic by-product secretion in wild-type *Escherichia coli* W3110. Appl Environ Microbiol 60:3724–3731.
93. von Kamp A, Schuster S (2006) Metatool 5.0: fast and flexible elementary modes analysis. Bioinformatics 22:1930–1931.
94. Watson MR (1986) A discrete model of bacterial metabolism. Comput Appl Biosci 2:23–27.
95. Yang C, Hua Q, Shimizu K (2000) Energetics and carbon metabolism during growth of microalgal cells under photoautotrophic, mixotrophic and cyclic light-autotrophic/dark-heterotrophic conditions. Biochem Eng J 6:87–102.

Chapter 9
Isotopic Steady-State Flux Analysis

Jörg Schwender

9.1 Introduction

Metabolic flux analysis (MFA) provides an integrated view of the function of biochemical pathways within a cell and is an important methodology in systems biology and metabolic engineering [102]. Originally developed to study microbial metabolism, it has also been applied to plants [67, 74, 83, 85, 88, 106]. A key concept in MFA is the biochemical reaction stoichiometry, which is used to mathematically describe a cellular reaction network at metabolic steady state. For example, the theoretical capability of metabolic network can be explored by exhaustive enumeration of all possible distinct routes in a metabolic network [96, 101; see also Chapter 8]. With additional consideration of some physiological data and boundaries for fluxes (e.g., maximal uptake rates), flux balance analysis [35, Chapter 8] allows prediction of metabolic states (flux distributions) which are optimal with respect to cellular objectives such as growth speed or product yield. While this methodology has proven valuable in devising strategies for metabolic engineering of microorganisms [58], it often cannot exactly predict the reaction rates of particular reactions, since a variety of different intracellular flux states may describe the cells physiological behavior equally well. Without intracellular flux measurements by isotopic tracers, this methodology has the limitations of a "black-box" analysis. Another way to study the intracellular flux distribution is to apply dynamic simulation of metabolic networks, based on kinetic rate expressions of enzyme reactions. Since this approach requires both the stoichiometries and an enormous amount of data about in vivo reaction mechanisms, kinetic parameters, and regulatory properties [129], it is often not practicable. Therefore, as a valuable addition to purely stoichiometric analysis or dynamic modeling, ^{13}C-metabolic flux analysis (^{13}C-MFA) allows measurements of the intracellular flux distribution under a particular physiological condition or in consequence of transgenic alteration. In addition, while pure stoichiometric models rely on the comprehensive balance of cofactors, ^{13}C-tracer techniques typically

---

J. Schwender (✉)
Brookhaven National Laboratory, Biology Department, Upton, NY 11973, USA
e-mail: schwend@bnl.gov

allow intracellular fluxes to be determined without cofactor balances. This is useful since often processes like ATP consumption by futile cycles, ATP yield of the respiratory system, the specificity of some enzymes toward NADH or NADPH, or the presence of transhydrogenases cannot be exactly defined [21].

This chapter is intended to give an overview of the insights that can be gained by use of ^{13}C-MFA in plants and to give details on the method that help to design, perform, and analyze the experiments.

9.1.1 Some History on the Use of Isotopic Tracers

Isotopic tracer techniques have contributed greatly to the elucidation of pathways in plant central metabolism (see, e.g., [1, 11, 12, 14, 17, 22, 24, 118]). While the most prominent parts of central metabolism in plants were described in the 1950s and 1960s, some intricate details were not recognized until several decades later, and this may be due in part to the high complexity of plant central metabolism, with metabolic pathways organized in different subcellular compartments. For example, acetyl-CoA, a membrane impermeable metabolite, is involved in metabolic processes in multiple subcellular compartments [39]. Mainly due to evidence from tracer experiments and the presence in plastids of acetyl-CoA synthethase, an enzyme that transforms acetate into acetyl-CoA, it has been widely believed that free acetate is a direct precursor of plastidic acetyl-CoA, in particular for fatty acid synthesis in leaves [84, 89]. However, by use of labeled CO_2 as a tracer, Bao et al. [16] could show that in leaves, the flow of photosynthetically fixed carbon into fatty acids is not via the free acetate pool, but via plastidic pyruvate dehydrogenase, which transforms pyruvate into acetyl-CoA. The central metabolite acetyl-CoA was also believed to be the general precursor of isoprenoids in plants via the acetate mevalonate pathway. Only by the end of the 1990s, by use of stable isotopic tracers, was it shown that plastids have their own pathway to generate biosynthetic isoprene units via a pathway starting from pyruvate and glyceraldehyde 3-phosphate [66]. In both examples, the use of isotopic tracers made major contributions to the detailed elucidation of plant metabolic pathways.

In addition to the elucidation of pathways, the in vivo function of segments of central metabolism was quantified with the help of isotopic tracers. By feeding ^{13}C-labeled glucose to wheat endosperm tissue and to intact plants, Keeling et al. [56] studied redistribution of label between the C-1 and C-6 positions of hexose units in starch and sucrose isolated from the endosperm. Randomization of label between C-1 and C-6 of hexoses is caused by a cyclic process of breakdown of hexoses to trioses, interconversion of the trioses and re-synthesis of hexoses. The authors found that the same degree of label redistribution was consistently found in hexose units derived from both starch and sucrose. This can only be explained by attributing triose/hexose cycling mainly to the cytosolic compartment. The data also speak for the uptake of hexoses by the amyloplasts and against uptake of trioses. In similar, using different plant cell cultures, Hatzfeld and Stitt [47] measured randomization of ^{14}C-label between the C-1 and C-6 positions in hexose units in starch and sucrose. In similar it was concluded here that the primary route to plastidic starch synthesis involved the import of hexose phosphates and not triose phosphates into the amyloplasts. In this study, labeling signatures were also detected that could be ascribed to the operation of the oxidative pentose phosphate pathway (OPPP). Similar studies of central metabolism in heterotrophic tissues used ^{13}C-labeled glucose and ^{13}C-NMR analysis of positional enrichment in free sugars, sucrose and starch, metabolism [130, 62]. Fernie et al. [40] studied the effect of elevated levels of fructose 2,6-bisphosphate (F2,6P$_2$) in heterotrophically grown tobacco cells on pyrophosphate:fructose 6-phosphate 1-phosphotransferase, an enzyme in glycolysis. Feeding [1-^{13}C]glucose, the label redistribution into different carbon positions of free soluble sugars indicated that in the presence of F2,6P$_2$, there is a higher in vivo flux through the enzyme in direction of the formation of hexose phosphates.

A larger segment of plant central metabolism was first studied by Salon et al. [92]. In this study, germinating lettuce embryos were fed five different ^{14}C-labeled tracers, and in each case, the specific ^{14}C-activity was determined in glutamate and aspartate. Key fluxes associated with the TCA cycle and the glyoxylate cycle were resolved. Radioactive tracers such as acetate, palmitate, or hexanoate were fed in low concentrations so that the assumption could be made that the tracers had no significant mass contribution to metabolism, i.e., that they did not disturb the metabolic steady state. In contrast to the use of trace amounts of ^{14}C, the ^{13}C-labeled tracers used in steady-state flux studies are typically the main sources of carbon in the culture medium. Growing, excised maize root tips were intensively studied by steady-state flux analysis using ^{13}C-glucose as a substrate [3, 4, 6, 30, 34], and in this way, Dieuaide-Noubhani et al. [30] first resolved a large compartmentalized model with 20 fluxes. With the measurement of 13 positional enrichments in 5 metabolites and the measurement of 5 biosynthetic fluxes, a system of 26 stoichiometric equations was solved.

9.1.2 Insights Gained from Steady-State Flux Studies in Plants

A number of ^{13}C-MFA studies on plant central metabolism have been published and will be reviewed in short here. Typically, tissues isolated from plants, developing embryos, or plant cell cultures are kept under constant culture conditions and grow heterotrophically on a ^{13}C-labeled substrate (glucose or sucrose). Using models of central metabolism, typically around 30 fluxes can be estimated based on the labeling experiments. For example, tomato suspension cells were grown in batch culture and intracellular fluxes were compared for different growth phases [87]. *Arabidopsis thaliana* cell cultures were studied under different oxygen availability to measure changes in flux and metabolite levels [141]. Besides cell cultures, maize root tips detached from germinating seeds were extensively studied [3, 4, 6, 30] and may be taken as a model for fast-growing non-photosynthetic tissues. Furthermore, developing seeds (embryos) of rapeseed [55, 103, 104, 105, 107], soybean [54, 111], sunflower [5], or developing maize kernels [37, 110] have been isolated and cultured and studied mainly with the aim of understanding the processes of carbon partitioning and storage synthesis in seeds. The study of *Catharanthus roseus* hairy root culture [113] has implications for the production of secondary metabolites by cell cultures. As an example for a prokaryotic photosynthetic organism, cultures of *Synechocystis* have been compared for heterotrophical and photo-mixotrophical conditions [143].

In difference to flux balance analysis and related approaches (Chapter 8), the use of tracers in ^{13}C-MFA allows to quantify the in vivo function of parallel or cyclic pathways in central metabolism. Referring to the validation of flux models discussed in later sections of this chapter, it is often advisable that major biological conclusions arising from flux studies should be validated by independent experimental approaches, if possible. For example, estimations of flux through RubisCO in developing rapeseed embryos by ^{13}C-tracer techniques were corroborated by mass balances of carbon uptake and net CO_2 release [105].

9.1.2.1 Alternative Pathways

One example of quantification of in vivo function of alternative pathways is the bypass of pyruvate kinase (PK) [79] via the reaction sequence phosphoenol pyruvate (PEP) carboxylase, malate dehydrogenase, and malic enzyme. Flux studies in several seed models revealed significant fluxes through the bypass. Plastidic as well as cytosolic PK can be bypassed by respective plastidic or cytosolic isoforms of malic enzyme. Accordingly, ^{13}C-MFA showed that mitochondrial malic enzyme provides 14%, 20%, and 40% of mitochondrial pyruvate in developing soybean, sunflower, and *Brassica. napus* embryos, respectively [5, 107, 111]. The plastidic bypass could not be detected in *B. napus* embryos [107] but may account for up to 20% of the plastidic pyruvate in soybean embryos [54, 111]. In developing sunflower embryos, the bypass of plastidic PK accounts for less than 10% of total carbon flux into fatty acids [5]. This low contribution is in contrast to the dominant role of the bypass suggested by in vitro tracer experiments. By incubating plastids isolated from developing sunflower seeds with ^{14}C-malate and other isotope-labeled tracers, malic acid was identified as the preferred precursor of fatty acid synthesis [80]. This difference may be a consequence of the plastids being studied away from the cellular environment and demonstrates the value of in vivo flux analysis as being a method to study flux in a non-invasive way on an unperturbed cell culture.

The GABA shunt [124] is a reaction sequence that can be understood as a bypass to the mitochondrial conversion of 2-ketoglutarate (KG) to succinate, which is part of the tricarboxylic acid (TCA) cycle. First, Glu can be derived from KG by transamination and then be converted to GABA by cytosolic glutamate decarboxylase. GABA in turn is transformed in the mitochondria to succinate by GABA transaminase and succinic semialdehyde dehydrogenase [38]. The potential of the GABA shunt to bypass the TCA cycle in plants has been demonstrated with mutants deficient in the mitochondrial conversion of KG to succinate [119]. Different additional observations suggest that the GABA shunt could carry substantial carbon flux under many different conditions, i.e., the GABA shunt could be understood as part of the TCA cycle in plants [38]. In fact, substantial flux from glutamate via γ-aminobutyrate (GABA) into the TCA cycle has been reported in developing soybean embryos [54, 111]. With glutamine being used as sole nitrogen source in the embryo cultures, the GABA shunt provides an alternative route for the entry of glutamate into the tricarboxylic acid cycle in addition to entry as 2-oxoglutarate after deamination or transamination.

During storage deposition, oilseeds receive sugars from the mother plant, and in *B. napus* a big part of the carbon input is converted into storage lipids (triacylglycerols). As a classical paradigm for fatty acid synthesis in seeds, the conversion of sugars to pyruvate was assumed to take place via glycolysis and the oxidative pentose phosphate pathway (OPPP), while in developing *B. napus* embryos the capacity for photosynthetic CO_2 fixation had been recognized as well [45, 59, 91]. With *B. napus* embryos cultured under low light levels, in vivo operation of an alternative to glycolysis could be demonstrated [105]. Glycolysis is bypassed by the enzymes of the non-oxidative PPP, phosphoribulokinase, and ribulose 1,5 bisphosphate carboxylase/oxygenase (RubisCO). Labeling experiments using $^{13}CO_2$ as a tracer and measurement of the amount of CO_2 released by the embryos per mole fatty acid produced showed that >40% of the 3-phosphoglycerate (3PGA) is produced by the RubisCO bypass. Due to reduction in net CO_2 release during conversion of sugars to fatty acids, the carbon economy is substantially improved as compared to non-green seeds [105]. Elementary flux modes analysis of a reaction network comprising hexose catabolism and fatty acid synthesis confirmed that only the operation of the RubisCO bypass together with contribution of photosynthetic NADPH can explain both the results of both the ^{13}C-labeling and carbon-balancing approaches.

9.1.2.2 Futile Cycles

The above-mentioned in vivo functionality of different metabolic bypass routes as revealed by ^{13}C-MFA demonstrates the plasticity of plant metabolism. In addition to the observation of various bypass reactions, ^{13}C-MFA has also quantified futile substrate cycles. By cyclic interconversion of substrates under consumption of ATP, a substantial part of cellular ATP can be lost to these cycles. In particular, cyclic synthesis and degradation of sucrose and hexose phosphates has been observed, although its biological significance is not yet clearly understood [3, 4, 30, 44, 104]. Sucrose cycling was supposed to consume 69% and up to 80% of the ATP produced by mitochondrial respiration in maize root tips and tomato cell cultures, respectively [30, 87]. Later, by different experimental techniques, Alonso et al. [3] attributed most of the formerly described ATP cycling in maize root tips to cyclic interconversion of glucose and glucose 6-phosphate. Subsequently, by re-evaluation of the maize root flux models, Kruger et al. [64] showed that these very high values would be overestimates if separate subcellular pools of glucose that are differently labeled are extracted and modeled as one pool. The controversy on the quantification of sucrose cycling demonstrates one of the possible pitfalls in ^{13}C-MFA. However, even with possible overestimation in some cases, it appears that futile cycles are a common feature in plants.

9.1.2.3 Estimate for Cofactor Supplies to Biosynthetic Reactions

Furthermore, ^{13}C-MFA leads to estimates of the contribution of cofactor (NADPH, ATP) generating pathways to the cofactor requirements of biosynthetic processes [5, 104, 107]. Although cofactor balances are typically not included in ^{13}C-MFA (see Section 9.4), the fluxes through particular pathways allow cofactor demands and supplies to be assessed without describing and analyzing the cofactor mass balances comprehensively for the whole cell. For example, the OPPP is likely a major source of NADPH for fatty acid synthesis in lipid-storing seeds [84]. If one could observe in different lipid-producing systems that the OPPP delivers just the right amount of reductant needed for fatty acid synthesis, one could argue that the flux capacity of the OPPP likely limits lipid synthesis. However, the flux through the OPPP in *B. napus* developing embryos accounts for at most half of the demand of NADPH for fatty acid synthesis [104]. In addition to the OPPP, the photosynthetic electron transport could account for the difference in reductant requirement [105]. In case of soybean embryos studied under different conditions [54, 111], similar flux-based calculations[1] show that the plastidic OPPP flux provides between 200% and 400% of the demands for fatty acid synthesis, while for non-green sunflower embryos [5] this number is 240%. In consequence, for soybean and sunflower, the OPPP over-satisfies the NADPH demands of fatty acid synthesis and it can therefore be concluded that it provides reductant for other biosynthetic processes as well. Comparing cofactor balances for storage oil synthesis in different seeds gives no support for the idea that oil synthesis is limited by the OPPP, i.e., that increase in OPPP flux would force more carbon to enter the lipid biosynthesis pathway.

9.1.2.4 Robustness

The insight into the organization of central metabolism given by ^{13}C-MFA should allow for rational design of metabolism to achieve changes in the amounts of end products. From metabolic engineering of microbial cells, it is known that cells regularly resist genetic manipulations intended to redirect flux. Branch-point rigidity, resulting mainly from feedback control of enzymes, stabilizes flux ratios at metabolic branch points [116]. Also, changes in levels of single enzymes usually have limited effect on flux due to the frequently encountered distribution of flux

---

[1] Flux values from the cited publications were used: NADPH production is calculated as $2\times$ the plastidic OPPP flux. NADPH demand is calculated as $(8/9)\times$ the flux of acetyl-CoA into fatty acids, assuming the condensation of 9 acetate units into stearic acid requires 8 NADPH + 8 NADH. It is also assumed, as in *B. napus* embryos, that one reduction step in fatty acid chain elongation has high affinity for NADPH, while the other preferentially uses NADH. [104].

control over most of the enzymes of a pathway. This kind of resistance of flux to fluctuations in enzyme and metabolite levels has also been described as stochastic robustness of metabolic networks [15, 93, 115], i.e., the capacity to maintain a flux status against small perturbations. Recent plant flux studies address the question of robustness in central metabolism for cell cultures [87, 141] and cultivated maize kernels [110]. It was shown that net fluxes through glycolysis, the tricarboxylic acid (TCA) cycle, and the OPPP do not change substantially relative to each other in response to physiological perturbations [87, 141] or different genetic backgrounds affecting starch content in kernels [110]. While for *A. thaliana* cell cultures, increased oxygen availability leads to increased fluxes with little change of fluxes relative to each other, the levels of several amino acids, sugars, and organic acids changed differently [141].

9.1.2.5 Flux Analysis for CO_2 Autotrophy

While ^{13}C-MFA has been useful to gain insight into cellular metabolism in great detail, it is currently not advanced enough to analyze plant metabolism on the whole organism scale. Plants are organized into different organs, tissues, and cell types, and their metabolism has marked diurnal and circadian oscillations. Therefore, a whole-plant steady-state labeling approach does not appear to be very promising. Typically, multi-carbon tracers like glucose are used in ^{13}C-MFA, and informative labeling signatures in metabolites result from positional differences in ^{13}C-enrichment (positional labeling) or tracing of contiguous ^{13}C nuclei through the network (bond labeling) [106, 83]. However, the principal carbon source of an intact green plant is atmospheric CO_2. At steady state, feeding $^{13}CO_2$ (in mixture with $^{12}CO_2$) will result in perfectly randomized carbon isotope distributions in metabolites. One way to achieve informative labeling pattern from labeling with $^{13}CO_2$ was explored by Schaefer et al. [95]. They exposed soybean leaves to $^{13}CO_2$, resulting in bond-labeled sucrose being transported into the developing seeds. Analysis of labeled bonds in sugars and fatty acids by ^{13}C-NMR allowed estimates on the contribution of the involvement of the OPPP in central metabolism. A similar concept is based on formation of highly labeled internal carbohydrate pools [86]. During a short $^{13}CO_2$ pulse to whole plants, highly ^{13}C enriched starch reserves are formed and stored in leaves. Later, during growth for several days in ambient air, hexose units derived from the highly ^{13}C-enriched transitory starch are mixed in central metabolism with newly produced unlabeled photosynthetic intermediates [86]. The approach resulted in bond labeling in metabolic end products that can be interpreted with respect to metabolic history in central metabolism.

$^{13}CO_2$ can also be used in transient labeling, where the labeling in metabolites is time resolved. Transient (isotopic in-stationary) labeling was pioneered by Calvin and Benson to analyze the biochemical steps of primary fixation of CO_2 in microalgae [23]. An analytical and computational framework for transient labeling with $^{13}CO_2$ was recently presented by Shastri et al. [108]. Prokaryotic photosynthetic microalgae are cultured in a chemostat, and after, the ^{13}C pulse cells are to be sampled over a period of about 10 min. Sampled cells are extracted after fast quenching of metabolic activity. By GC/MS analysis of metabolite extracts the time-resolved labeling in free amino acids can be obtained. Theoretically, the approach resolves central carbon metabolism in similar detail to ^{13}C-MFA studies at steady state in other microbes. Another concept based on $^{13}CO_2$ pulse labeling with whole plants, kinetic metabolic phenotyping, has been developed by Huege et al. [53]. The method follows the labeling kinetics in different plant parts based on estimation of ^{13}C half life in free metabolites. How far this approach can be developed in the direction of tissue specific resolution of flux is an open question.

9.2 Overview of ^{13}C-Metabolic Flux Analysis

In microbes and plants the steady-state labeling approach of ^{13}C-MFA [106, 133] is usually applied to a network of central "core" metabolism, describing the large fluxes with many redundant connections, substrate cycles, and variability in flux directionality [93]. In general, only ^{13}C-labeling signatures in metabolites of converging reactions yield information about fluxes [125] (Fig. 9.1). Reactions that split or join carbon backbones may produce differences in labeling signatures that mix at converging reactions (Fig. 9.1A). In a similar was, if the cell culture under study uses co-substrates, then isotopic dilution effects can be used to derive intracellular flux [25, 103, 107] (Fig. 9.1B).

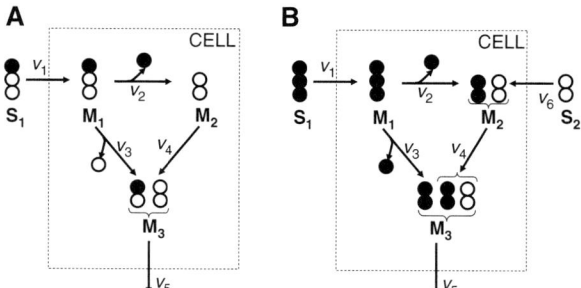

Fig. 9.1 Hypothetical metabolic network and determination of intracellular fluxes by the use of carbon-isotopic tracers under steady state. (**A**) Carbon-1 of the substrate S_1 is labeled and traced through the network. Via reactions 2 and 4 metabolite M_3 is formed by loss of carbon-1, while by formation of M_3 via reaction 3, carbon 3 of the substrate is lost (carbon loss could be a decarboxylation). Therefore, the labeled carbon in S_1 reaches M_3 only via reaction 3, and the label of molecules in metabolite M_3 will be dependent on the flux ratio v_3/v_4. (**B**) In this example, all carbon positions of substrate S_1 are labeled and the carbon transitions within the network do not cause any differential labeling pattern in M_3. However, due to additional uptake of an unlabeled co-substrate S_2, isotopic dilution will allow the determination of v_3/v_4

9.2.1 Steady-State vs. Dynamic Labeling Approaches

While modification of carbon backbones in central metabolism delivers informative labeling pattern, reactions of peripheral biosynthetic pathways and secondary metabolism typically introduce chemical modifications to the formerly established carbon structure. The ^{13}C-labeling signature in a whole class of compounds may be the same and therefore not informative for determining the rate of interconversion between them. Therefore, for secondary metabolism, typically transient labeling techniques are used (e.g., [20, 49, 70, 72, 77]; see Chapter 10). Transient labeling involves measuring isotopic enrichment multiple times after application of an isotopic tracer, and it also requires the quantification of metabolic pool size. Transient labeling flux analysis also helps to refine the in vivo structure of such networks [20]. For example, promiscuous substrate specificity of enzymes in secondary metabolism observed in vitro may suggest more metabolic connections than are actually active in vivo – hence, there is some debate in the literature on the existence of "metabolic grids" in secondary metabolism [31].

9.2.2 Isotopomer Concept

A central concept in ^{13}C-MFA is the description of labeling states of metabolites by using isotopomer fractions (isotope isomers). ^{13}C-isotopomers describe all possible ^{12}C/^{13}C isotope isomers of a molecule, usually with a binary notation where ^{13}C can be represented by "1" and ^{12}C by "0". For example, Ser#110 describes a serine molecule that is ^{13}C in carbon positions one and two, and ^{12}C in position three. For a metabolite with n carbon atoms, 2^n isotopomers can be distinguished. The relative abundance of isotopomers is given in percent (isotopomer fractions) so that the sum of all isotopomer species is always 100%. The probably first use of the isotopmer concept in modeling of central metabolism can be found in Malloy et al. [69].

9.2.3 Steady-State Isotopic Labeling Experiment

General considerations for ^{13}C-MFA experiments with microbes have been outlined before, e.g., by Wiechert [134]. Using the isotopic steady-state approach, a typical ^{13}C-MFA experiment can be described by the following steps (Fig. 9.2):

1. Cells are grown under constant physiological conditions in presence of a ^{13}C-labeled carbon source. Physiological parameters, including substrate uptake rates, synthesis rates of biomass compounds, and growth rate, are measured.
2. After the cells are harvested, metabolites and biomass compounds are extracted. Free metabolites are fractionated and/or biomass polymers are hydrolyzed into their building blocks (Chapter 5).

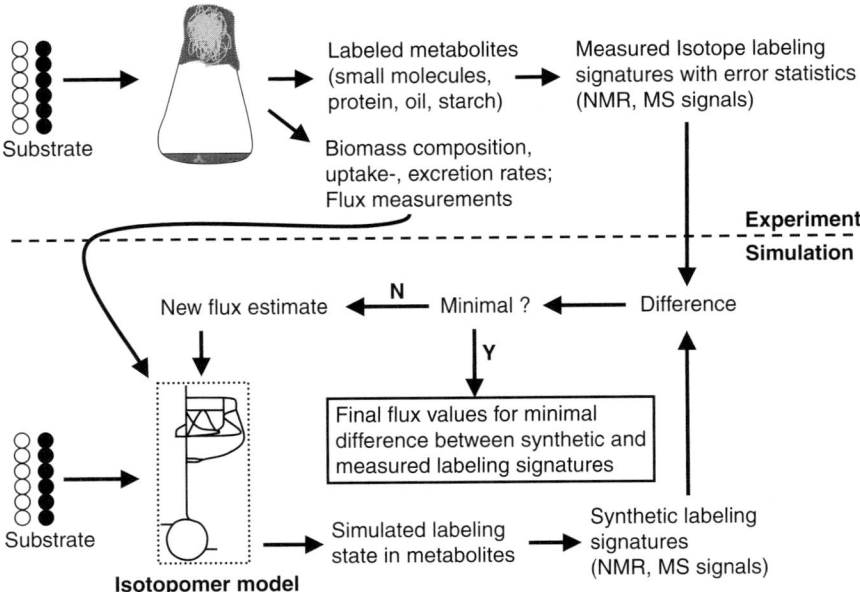

Fig. 9.2 Procedure of a steady-state ^{13}C-labeling experiment and flux parameter fitting. Cells are grown heterotrophically on a carbon source containing unlabeled (*open circles*) and ^{13}C-labeled molecule species (*closed circles*). For more details see text

3. Labeling signatures are measured in the different metabolites and monomers by NMR or MS (Label measurements, Chapter 5).
4. The labeling signatures are translated into intracellular fluxes (see Section 9.5). At the center of this process is a network model that simulates steady-state label signatures. Flux values are determined for which the model can best explain the experimentally observed labeling signatures.
5. For flux values determined in this way, statistical errors are computed based on the standard errors in the label and physiological measurements.

This procedure leads to a static view of the intracellular fluxes that relates to the metabolic state for the physiological condition in the experiment. The validity of this picture depends on several basic conditions that should be met [126, 134, 136]:

1. The cells approximate metabolic and isotopic steady state. As a precondition, the labeling experiment must be conducted under well-defined constant physiological conditions (see Section 9.3).
2. If the study aims at describing metabolism of a certain plant organ or tissue, the physiological condition of the cell culture should be defined appropriately to avoid perturbations of metabolism caused by such factors as osmotic stress or the use of labeled precursors that are not major carbon sources *in planta*.

3. All reactions in the metabolic network that carry significant flux are represented in the flux model, and the fate of each carbon atom is formulated in complete detail (see Section 9.4).
4. Also, it has to be assumed that each metabolite represented in the model is in reality one homogeneous intracellular pool of molecules. This may be problematic considering the complex subcellular compartmentation of plant cells (see Section 9.5). Also, substrate channeling has been found in plants [46] and would lead to errors if not properly accounted for in the model [60, 126].
5. There are only insignificant isotope discrimination effects in metabolism. The modeling approach assumes that enzyme reactions do have no preference or bias against any of the possible $^{12}C/^{13}C$ molecule isomers. In a study using *A. thaliana* cell cultures, possible effects of ^{13}C-labeled substrates on metabolism were studied; it was concluded that fluxes through central metabolism are not perturbed by the presence of ^{13}C isotope [63].

9.3 ^{13}C-Labeling Experiments and Steady-State Assumption

In living cells, different processes take place on very different time scales. While changes in protein levels can occur over hours, allosteric control of enzymes should be faster than 1 s, and changes in enzyme-catalyzed reaction rates due to change in substrate levels should occur within about 10^{-4} s [117]. If a system is observed on a certain time scale, the dynamics of much slower processes can be neglected. This means that such slow processes can be assumed to be in "quasi-steady state" [48].

For the experimental observation of steady-state levels of cellular compounds – as, e.g., in metabolomics – cells can be harvested and immediately extracted. Then the levels of metabolites can be assumed to represent steady-state cellular concentrations at the moment of harvest. However, for experimental observation of flux in steady-state ^{13}C-flux analysis, cellular conversion rates have to be in quasi-steady state for hours or longer. This is because, after feeding the labeled precursor, the labeling experiment must continue long enough for metabolic pools to turn over multiple times until the ^{13}C-labeling signatures in metabolites become time invariant (isotopic steady state). For metabolism to be kept continuously in steady state, strict control of the environment (extracellular nutrient concentrations, pH, etc.) is a precondition. For microorganisms, this is typically accomplished in chemostat culture where cells grow continuously under constant conditions and cell density. However, the validity of the steady-state assumption should always be critically assessed. For example, substantial cell-cycle-dependent variations of flux have been reported in yeast [26]. Therefore, flux maps derived from continuously grown cells will represent an average over potentially very different metabolic states.

In plant cell or tissue cultures, isotopic steady state may be reached within hours for central metabolites [83]. By in vivo NMR studies, it has been found that for developing linseed embryos kept in a liquid growth medium, isotopic steady state of key central metabolites is reached within less than 3 h, while large pools like sucrose and free hexoses need up to 18 h for complete isotopic equilibrium [123]. For tomato cells growing on ^{13}C-glucose in batch cultures, Rontein et al. [87] described metabolic fluxes in the exponential phase, the arrest of cell division phase and the pre-stationary phase. They observed distinct metabolic (and isotopic) steady states during culture. They assessed steady state and the possible presence of "memory effects" of isotope label in metabolic pools. Turnover times[2] for all measured free metabolites were determined to be less than 1 h. During 5 h, five turnovers will take place meaning that 97% of the molecules of a pool will be renewed. In conclusion, no "memory" of the labeling state 5 h ago will be present. Since cells were sampled in intervals of 24 h or longer, the samples taken at different time points are likely to represent distinct steady states. The turnover time suggests that the distribution of isotopic label in the network may be a faster process than the changes in the fluxes, suggesting that for the three samplings three distinct steady states were observed characterizing the different phases in batch culture. The authors also assessed limitations to the validity of the steady-state assumption in their experiment. In particular, turnover of polymers can add inaccuracy. The authors showed that about 20% of the free glutamate comes from protein, i.e., Glu that had been labeled long before it was stored in protein and was later freed by protein degradation.

While Rontein et al. measured label in free metabolites during batch culture, many studies are based on label information from protein and other biomass compounds. In batch culture, where cells make a transition through different physiological states, the labeling information of different metabolic states will continuously accumulate in biomass. This kind of non-stationary state has recently been addressed by a modified methodological MFA-framework for bacterial fed-batch fermentation [10]. During culture, cells are sampled repeatedly and metabolic non-steady state is resolved into time profiles of metabolic fluxes [10].

In studies on developing seeds [5, 37, 54, 55, 103, 104, 105, 107, 110, 111], metabolic steady state is typically assumed for the duration of the whole culture of up to 2 weeks. For developing embryos of *Brassica napus* in liquid culture [107], the labeling pattern of free amino acids after 3 d in culture was compared with the labeling pattern in protein-bound amino acids after 2 weeks of culture, and no significant difference could be found (supplemental text of [107]). This suggests that the labeling state of amino acids accumulated in storage protein continuously accumulated with approximately the same labeling signature.

9.4 Definition of the Biochemical Network

A critical step in the MFA is the development of a biochemical reaction network. Figure 9.3 summarizes the main features of reaction networks of central metabolism, representative for several studies on developing seeds [5, 54, 104, 111].

---

[2]Ratio of pool size to total flux into the pool, i.e., the time needed for the molar amount of molecules present in the pool to enter (and exit) the pool.

9 Isotopic Steady-State Flux Analysis

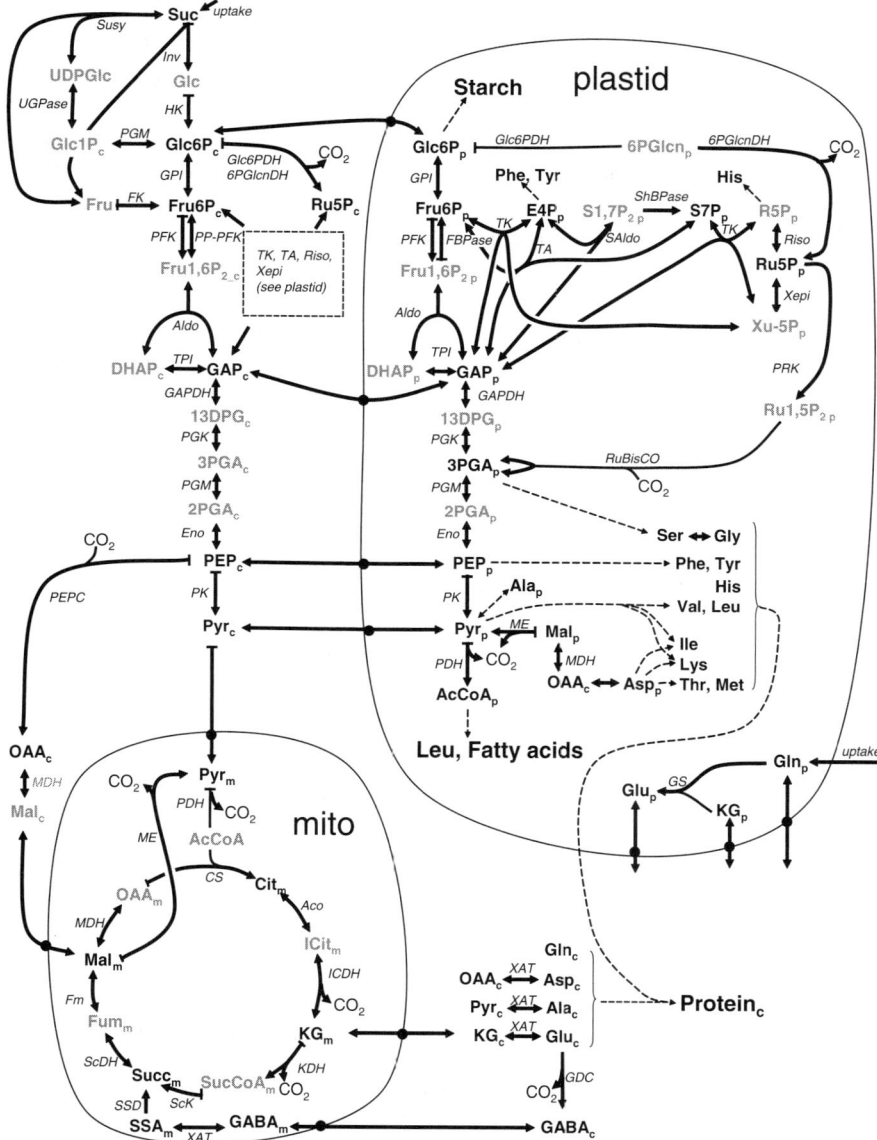

Fig. 9.3 Generic metabolic network, representative for reactions of central metabolism often considered in plant ^{13}C-MFA. Compartmentation is indicated by "c", "m," and "p" for cytosol, mitochondria, and plastid, respectively. Biosynthetic pathways (*dashed arrows*) into proteinogenic amino acids and other products are lumped into single carbon transitions. Metabolites shown in grey typically need not be explicitly represented in the model due to lumping of reactions (see text)

Abbr. mito: mitochondrion. **Enzymes**: Aco: aconitase; Aldo: fructose 1,6-bisphosphate aldolase; AAT: alanine transaminase; CS: citrate synthase; Eno: enolase; FBPase: fructose 1,6-bisphosphatase; FK: fructokinase; Fm: fumarase; GAPDH: glyceraldehyde 3-phosphate dehydrogenase; GDC: glutamate decarboxylase; Glc6PDH: glucose 6-phosphate dehydrogenase; GPI:

Such networks typically contain the central "core" metabolism, describing the large fluxes with many cycles, redundant connections, and variability in flux directionality. Typically, only carbon transitions are modeled, since cofactor balances do not need to be included, which is advantageous if exact knowledge on the cofactor usage is limited [98]. Reactions that split and join carbon backbones of metabolites often result in ^{13}C-labeling signatures that are informative with respect to cyclic, parallel, and reversible fluxes [133]. In particular, only labeling information in metabolites of converging fluxes yields information about fluxes while nodes with only one influx store labeling information identical to the preceding metabolite pool [125]. This means that usually all carbon transitions in linear pathways can be lumped into one reaction. Biosynthetic reactions, e.g., of amino acids or fatty acids are typically unified into unidirectional single reactions. For reversible reactions, lumping of metabolite pools can be justified if fast interconversion between metabolites is found [125]. For example, in a *B. napus* embryo flux model, the different hexose phosphates, pentose phosphates, triose phosphates, and C4 acids (malate, oxaloacetate, fumarate, succinate) were each modeled as single pools, in each case based on observations of labeling pattern that suggested high interconversion [104, 107]. Altogether, lumping of metabolite pools can speed up computational performance. The remaining essential carbon transitions in the network must be formulated in detail. While most of the reactions shown in Fig. 9.3 are well documented in biochemical text books, some enzymes can catalyze a variety of similar reactions, which is sometimes not realized in ^{13}C-MFA simulations. The reactions of the non-oxidative pentose phosphate pathway are usually formulated by three reversible reactions, two transketolase and one transaldolase reaction (Fig. 9.3). However, it

◀──

Fig. 9.3 glucose 6-phosphate isomerase; GS: glutamate synthase; HK: hexokinase; ICDH: isocitrate dehydrogenase; Inv: invertase; KDH: ketoglutarate dehydrogenase; MDH: malate dehydrogenase; ME malic enzyme; PDH: pyruvate dehydrogenase; PEPC: phosphoenol pyruvate carboxylase; PFK: phospho fructokinase; PGM: phosphoglycerate mutase; PGK: phosphoglycerate kinase; PK: pyruvate kinase; PRK: phosphoribulokinase; 6PGlcnDH: 6-phosphogluconate dehydrogenase; Riso: ribose 5-phosphate isomerase; RuBisCO: ribulose 1,5-bisphosphate carboxylase/oxygenase; SAldo: sedoheptulose 1,7-bisphosphate aldolase; ScDH: succinate dehydrogenase; ScK: succinate thiokinase; ShBPase: sedoheptulose bisphosphatase; SSD: succinate semialdehyde dehydrogenase; SuSy: sucrose synthase; TK: transketolase; UGPase: UDP-glucose pyrophosphorylase; Xepi: xylulose 5-phosphate epimerase; TPI: triose phosphate isomerase; XAT: diverse aminotransferase activities. **Metabolites**: AcCoA: Acetyl-CoA; Cit: citrate; DHAP: dihydroxy acetone phosphate; E4P: erythrose 4-phosphate; Fru1,6P2: fructose 1,6-bisphosphate; Fru: fructose; Fru6P: fructose 6-phosphate; Fum: fumarate; GABA: γ-aminobutyric acid; GAP: glyceraldehyde 3-phosphate; Glc: glucose; Glc1P: glucose 1-phosphate; Glc6P: glucose 6-phosphate; ICit: isocitrate; KG: ketoglutarate; Mal: malate; OAA: oxaloacetate; PEP: phosphoenol pyruvate; 13DPG: 1,3-bisphosphoglycerate; 2PGA: 2-phosphoglycerate; 3PGA: 3-phosphoglycerate; 6PGlcn: 6-phosphogluconate; Pyr: pyruvate; R5P: ribose 5-phosphate; Ru1,5P2: ribulose 1,5-bisphosphate; Ru5P: ribulose 5-phosphate; Sh1,7P2: sedoheptulose 1,7-bisphosphate; Sh7P: sedoheptulose 7-phosphate; SSA: succinate semialdehyde; Suc: Sucrose; Succ: succinate; SucCoA: Succinyl-CoA; UDPGlc: UDG-glucose; Xu5P: xylulose 5-phosphate

has been shown that due to the enzyme mechanisms and substrate affinities of the two enzymes, actually six additional reactions can take place, which have neutral reaction stoichiometry but impact redistribution of ^{13}C-label [126]. Including the additional reactions into the network model can make a significant difference in the overall flux distributions finally obtained for the network [60, 126].

The network shown in Fig. 9.3 is certainly a simplification of what would be considered to be the major fluxes in a plant cell. For example, amino acid degradation pathways are mostly not considered in flux studies on microbes or plants, an assumption that may be adequate if growing cells are studied, but one that is generally not rigorously verified.

9.4.1 Reaction Reversibility

While biochemical reactions always proceed in two directions, the reaction rate in one direction is often very small. In this case, a reaction can be considered irreversible (unidirectional) for modeling purposes. If in a ^{13}C-network, a reaction is considered to be reversible (bi-directional), then two reactions have to be modeled, one as forward and one as reverse flux, and the net conversion could be in either direction [133]. The definition of reaction directionality is important for the validity of the flux values resulting from flux modeling. An incorrect choice of reaction directionality in a single reaction may not allow the model network to assume a flux distribution that is fully consistent with the experimental data – i.e., a flux distribution that reflects the real in vivo fluxes. Also, the presence of reversible interconversions typically has a large impact on the labeling signatures that are measured in metabolites. The flux model must therefore properly account for reversible reactions in order to avoid misinterpretation of the data. This problem has been described and discussed in particular for the pentose phosphate pathway [60, 106, 126].

Although the reaction reversibility is important for the validity of the modeling results, it is often of most interest for the investigator to know the net flux rates, i.e., the difference between forward and reverse flux. Therefore, forward and reverse reaction rates are typically expressed as net and exchange fluxes (Fig. 9.4).

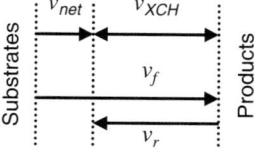

Fig. 9.4 Quantification of flux reversibility of a biochemical reaction in ^{13}C-MFA. As shown, the forward and reverse flux (v_f, v_r) can be expressed by net and exchange fluxes (v_{net}, v_{XCH}). Redrawn from Schmidt et al. [99] with permission from Elsevier

9.4.2 Reaction Directionality and Thermodynamics

Detailed description and discussion of the thermodynamics of biochemical reactions can be found, e.g., in Heinrich and Schuster (1996) [48]. In the following, it is shown how the Gibbs free energy and metabolite concentrations relate to reaction directionality and reversibility, while later more practical approaches for their definition are discussed.

The in vivo feasibility and reversibility of a generic reaction "A + B ↔ C + D" is dependent on its thermodynamic potential, or free energy (ΔG). For the reaction to proceed in the forward direction, ΔG has to be negative. A reaction is said to be reversible if ΔG is small (see below, [48]). ΔG is given by the sum of the standard Gibbs free energy ($\Delta G^{0'}$) and a term that accounts for the in vivo concentrations of the reactants ([A], [B], [C], and [D]):

$$\Delta G = \Delta G^{0'} + RT \ln \left(\frac{[C][D]}{[A][B]}\right); \quad \text{with } \Delta G^{0'} = -RT \ln \left(\frac{[C]_{eq}[D]_{eq}}{[A]_{eq}[B]_{eq}}\right) \quad (9.1)$$

with the temperature T in Kelvin and $R = 8.314$ J K^{-1} mol^{-1}. While the term for $\Delta G^{0'}$ is defined by the concentrations of the reactants at equilibrium ([A]$_{eq}$, etc.)[3] and therefore is characteristic for a reaction, the in vivo concentrations of the reactants may differ between organisms or change with the physiological condition. This means that for a particular reaction, the value of ΔG and with it the directionality and reversibility cannot be generally defined.

Importantly for flux analysis, ΔG can be related to the ratio of forward and reverse fluxes of a reaction. This was analyzed in detail by Wiechert [135]. In short, considering that an enzyme-catalyzed reaction typically can be broken down into several distinct binding and transformation steps, the ratio of forward (v_f) and reverse fluxes (v_r) cannot be exactly given but within the limits of the following inequality [135]:

$$\frac{v_f}{v_r} \leq e^{\frac{-\Delta G}{RT}} \quad (9.2)$$

According to Eq. (9.2), for $\Delta G = -5.7$ kJ/mol, v_f/v_r would be 10 or less. This means that for -5.7 kJ/mol $\leq \Delta G < 0$, the forward and reverse fluxes are expected to be of the same order of magnitude.

---

[3]Considering for Eq. (9.1) that the in vivo concentrations of the reactants are close to the equilibrium concentrations, then ΔG would be small. Therefore, saying that ΔG is small or that a reaction is close to equilibrium refers to reversibility of the reaction.

9.4.3 Definition of Reaction Reversibility in Metabolic Networks

The generic reaction network in Fig. 9.3 shows typical definitions of reaction reversibilities for flux studies in microbes and plants, although such definitions sometimes differ between studies, e.g., for isocitrate dehydrogenase [107]. In the following, an overview is given on how reaction directionality can be defined in spite of lack of accurate knowledge of the necessary data. As stated above, in order to define reversibility of a reaction in a rigorous way, one needs to consider the standard Gibbs free energy and the steady-state concentrations of all reactants and products. Since metabolite concentrations are involved, a reaction could be reversible in one organism or cell type while irreversible in another. Recent approaches consider metabolome data for the definition of thermodynamic constraints in flux balance analysis [52, 65]. Therefore, it should be possible to derive reaction reversibility in such a way. However, coverage of all metabolites of a network by quantitative metabolome data would be necessary, and in particular in plants and other eukaryotes, data on subcellular concentration of metabolites would have to be generated.

Without such exact data, some rough rules or guidelines can be formulated based on ΔG^0. An enzyme-catalyzed reaction is likely to be in vivo unidirectional if it is highly exergonic under standard conditions (large negative value for $\Delta G^{0'}$). The same applies if only a part of an enzyme-catalyzed reaction – such as transfer of γ-phosphate from ATP to a substrate – is known to be highly exergonic. Also, hydrolysis reactions may be mostly irreversible because the reactant H_2O is present at a very high concentration. Following the rules given by Ma and Zeng [68] and in large agreement with Fig. 9.3, the following types of reactions are likely to be irreversible in vivo:

1. Reactions that include transfer of γ-phosphate from ATP to a substrate (e.g., hexokinase)
2. Reactions that are coupled to hydrolysis of phosphate esters or thioesters of coenzyme-A (e.g., fructose bisphosphatase, citrate synthase)
3. Other hydrolysis reactions (e.g., invertase, inorganic pyrophosphatase)
4. Reactions that include production of CO_2 (e.g., 6-phosphogluconate dehydrogenase).

In contrast, with increasing similarity between substrates, group-transfer reactions (e.g., by transketolase, transaldolase, or transaminase reactions) or isomerizations (e.g., glucose 6-phosphate isomerase, triose phosphate isomerase) tend to be reversible in vivo.

9.4.4 Measurement of Label Signatures

In ^{13}C-MFA, the labeling signatures in metabolites store information that is translated into intracellular fluxes. Different methods are available to extract and analyze different metabolites (see Chapter 5). Measurement of label in protein-derived amino acids by NMR or GC/MS has proven to be the method of choice in multiple plant flux studies [5, 103, 104, 105, 107, 111, 113], since the amino acids store the labeling information of their various respective central metabolite precursors (Fig. 9.3) [122]. However, some of the amino acids cannot be recovered by acidic hydrolysis of protein. Cysteine and tryptophan are lost, while glutamine and asparagine are deaminated to glutamate and aspartate, respectively. Typically labeling signatures in 16 amino acids can be analyzed in protein hydrolysate by GC/MS or 2D-NMR (see Chapter 5) and provide information about the label in the biosynthetic precursors. Ideally all isotopomers of the amino acids would be quantified, but in case of both NMR and MS, less than 10 % of the total isotopomer information is accessible (Table 9.1). However, if the labeling information measured by MS and NMR is combined [143, 144], the amount of accessible isotopomer information increases (Table 9.1). The amount of isotopomer information extracted from one metabolite can also be increased by using tandem MS, since this technique allows

Table 9.1 Number of labeling states resolved by GC/MS or 2D-NMR of protein hydrolysates. The total number of ^{13}C-isotopomers that define the labeling state of amino acids is given with the percentage of those that can be quantified. In short, for each amino acid, the MS and NMR signals were mapped to isotopomers as illustrated in Fig. 9.6. The algebraic rank of such mapping matrices equals the number of isotopomers or groups of isotopomers that can be determined from the label measurements. GC/MS refers to 150 signals in 32 fragments of t-butyl-dimethylsilyl (TBDMS) derivatives of 15 amino acids. NMR refers to data retrieved by 2-D HSQC NMR of protein hydrolysates [111]

Amino acid	Precursor	Total # of amino acid Isotopomers	MS: % of isotopomers	2D-NMR: % of isotopomers	combined: % of isotopomers
Ala	Pyr$_c$	8	75%	63%	100%
Arg	KG	64	0%	8%	8%
Asp/Asn	OAA$_c$	16	69%	44%	94%
Glu/Gln	KG$_c$	32	31%	22%	53%
Gly	3PGA	4	100%	75%	100%
His	R5P$_p$	64	17%	16%	33%
Ile	OAA$_p$, Pyr$_p$	64	8%	17%	25%
Leu	Pyr$_p$, AcCoA$_p$	64	9%	17%	27%
Lys	OAA, Pyr	64	17%	17%	34%
Met	OAA$_p$	32	16%	6%	22%
Phe	E4P$_p$, PEP$_p$	512	2%	2%	4%
Pro	KG	32	28%	16%	44%
Ser	3PGA$_p$	8	100%	75%	100%
Thr	OAA$_p$	16	63%	38%	94%
Tyr	E4P$_p$, PEP$_p$	512	4%	2%	6%
Val	Pyr$_p$	32	28%	25%	53%
Total:		1524 (100%)	9%	8%	15%

to deal with problems arising from overlaps of fragments in the MS spectra [57, 81, 90] (see also Chapter 5).

9.4.5 Subcellular Compartments

Besides protein-derived amino acids, the label in abundant free sugars and amino acids is often measured in plants [3, 4, 30, 87]. Recently, sensitive techniques have been developed allowing access to carbon labeling in low abundance intracellular metabolites in microbes [57, 61, 75, 128]. In plants, central metabolites like organic acids and sugar phosphates are typically present and metabolized in different subcellular compartments. In a ^{13}C-MFA experiment such subcellular pools may be labeled different, but a cell extract will contain a mixture of such differently labeled subcellular pools, which is a problem for ^{13}C-MFA in plants [64] and other eukaryotes (see Section 9.2.1). Therefore, the applicability of analysis of low abundance intracellular metabolites to plants has yet to be determined. Instead, to attain compartment specific labeling information, one may take advantage of the subcellular localization of certain biosynthetic pathways. For example, there is evidence that in higher plants the biosyntheses of His and of the branched-chain amino acids are exclusively plastidic [76, 109]. This means that His, Val, and Leu as well as Ile store labeling information for the plastidic pools of pentose phosphate, pyruvate, and oxaloacetate, respectively. Also, there is evidence that the key enzymes in aromatic amino acid synthesis in plants are exclusively plastid localized [50], and therefore, the label of plastidic PEP and erythrose 4-phosphate is found in the carbon chains of Phe and Tyr. In the case of serine, more care is needed to interpret labeling since different biosynthetic origins are possible. Serine can be formed from 3-phosphoglyceric acid by the plastidic phosphorylated serine biosynthetic pathway [51]. Also, in photorespiration, Ser formation from Gly takes place by a mitochondrial enzyme complex of serine hydroxymethyl transferase (SHMT) and glycine decarboxylase [33]. In the absence of photorespiration in developing cultured *B. napus* embryos [103], serine can be assumed to be formed primarily from plastidic PGA. However, SHMT may be present in different subcellular compartments, [71] and the labeling signature in Ser in *B. napus* embryos indicates reversible interconversion of Ser and Gly (Schwender, unpublished results).

Some amino acids can be formed in multiple compartments but the protein-derived amino acids may still store compartment-specific label. For example, as a result of the transaminase activities in different compartments, the amino acids Ala, Asp, and Glu can be interconverted with the central metabolites pyruvate, oxaloacetate, and α-ketoglutarate, respectively [100, 140]. In *B. napus* developing embryos, highly reversible transamination is evident by ^{15}N-labeling experiments [107] as well as enzyme profiles that show high abundance of Ala- and Asp-aminotransferase [55]. Therefore, it is very likely that in different compartments, Ala, Asp, and Glu are isotopically equilibrated with their respective organic acids. If it is further considered that *B. napus* storage protein is formed by cytosolic translation, then the labeling signatures in protein-derived Asp, Glu, and Ala will represent cytosolic pyruvate, OAA, and KG, respectively.

Another metabolite that stores compartment-specific label is starch. In higher plants, starch is synthesized inside the plastids from ADP-glucose. However, while ADP-glucose pyrophosphorylase, the enzyme producing ADP-glucose, is localized to the plastid compartment in *B. napus* embryos [27], the enzyme was localized to the cytosol in graminaceous endosperm [18]. Therefore, labeling in starch glucosyl units will represent different subcellular pools of hexose in the two systems. Furthermore, labeling in sugar monomers of protein glucans, obtained by hydrolysis of storage proteins or released from cell wall polymers, has been used to determine labeling in cytosolic pools of hexose phosphates [2, 111, 114]. In addition, fatty acids store compartment-specific labeling information. In plants, fatty acid synthesis is predominantly localized in plastids [29, 78]. Plastidic fatty acid synthesis produces fatty acids of 16 and 18 carbon chain length, whereas the elongation of C18:1 to longer chain length takes place by a cytosolic fatty acid elongation system [32, 132]. Thus, labeling in the carboxyl-terminal acetate units of C18 and C22 fatty acids represent plastidic and cytosolic acetyl-CoA pools, respectively [103, 107].

9.4.6 Choice of Substrate Label and Optimal Experimental Design

Glucose is the most frequently used ^{13}C-labeled substrate in ^{13}C-MFA. It can be purchased in many different ^{13}C-labeled forms. Among those, [U-^{13}C$_6$]glucose (labeled at all six carbons) and [1-^{13}C]glucose are the most inexpensive and have been used in most studies in microbes and plants. Other much more expensive forms like [2-^{13}C]glucose, [6-^{13}C]glucose, or [1,2-^{13}C$_2$]glucose have been used less frequently in flux studies (e.g., [104]).

The accuracy by which the value of a certain flux can be determined depends in big parts on the particular substrate label. Microbial and plant studies have given some general ideas about which fluxes can be measured with good accuracy for a given substrate label. By using [U-^{13}C$_6$]glucose in a mixture with unlabeled glucose as a substrate, good resolution of key fluxes in lower glycolysis and the TCA cycle has been found [28] (see Figure 9.3, reactions PEPC, PK, ME, CS). The same fluxes are determined with much less accuracy if [1-^{13}C]glucose is used, while this substrate is better suited to resolve the split between glycolysis and the OPPP (see Figure 9.3, net flux of reactions Glc6PDH, GPI) [28]. Also, [2-^{13}C]glucose or [1,2-^{13}C$_2$]glucose have been used to determine OPPP flux [104]. In developing seeds, sugar catabolism via glycolysis or the OPPP can be bypassed by flux via RubisCO [105]. It has been found that for different choices of labeled sugars, RubisCO flux is poorly resolved. Better experimental designs have to be developed according to optimal design strategies [67].

A satisfactory choice of substrate label may be possible using the principles of optimal design [73]. Prior to performing a labeling experiment, i.e., prior to obtaining real labeling data, a given flux model can be used to study the impact of substrate label on flux identifiability. By using the model simulation with arbitrary flux values, label measurements can be predicted for different substrate labels. Statistical measures of the fluxes are obtained in the form of a flux covariance matrix, which is then condensed into one parameter (relating to the volume of the multidimensional confidence interval). This optimality parameter is compared for the different substrates [73]. The difference in statistical performance tends to be highly dependent on the choice of substrate label but rather independent of the choice of fluxes for the simulation [73]. With this approach, mixtures of [U-^{13}C$_6$]glucose, [1-^{13}C]glucose, and unlabeled glucose have been found to perform better than [U-^{13}C$_6$]glucose or [1-^{13}C]glucose alone [13]. In addition to judging the overall statistical quality of the fluxes, substrate label can be optimized for the determination of fluxes of particular interest [13, 67, 73].

The resolution of flux between subcellular compartments is of particular importance for plant flux studies (see Fig. 9.3). Here, the use of different substrate labels has been helpful as well. For the analysis of cultured developing embryos of *Brassica napus*, the substrates glucose, alanine, and glutamine were used at the same time [107]. In separate experiments, each of the substrates was replaced by the uniformly ^{13}C-labeled form. All three experiments were analyzed in a combined simulation. The experiment with [U-^{13}C]Ala contributed most of the flux information for plastidic and cytosolic lower glycolysis. [U-^{13}C]Gln was key to resolve fluxes associated with the TCA cycle and the origin of cytosolic acetyl-CoA [107].

9.5 Estimation of Fluxes from Labeling Data

There are several different approaches for interpreting ^{13}C-labeling signatures, as summarized in Table 9.2 and discussed below (see Sections 9.5.1 and 9.5.2). Of consideration for the choice of method is, for example, whether a comprehensive flux map is of interest or only a specific metabolic branch point is to be studied. Also it should be considered whether one experimental condition is to be studied in detail, or whether many genotypes or conditions have to be compared. Furthermore, the availability of software packages for non-experts in computational/mathematical aspects may be important (see Section 9.5.5). Currently, ^{13}C-MFA in plants is often based on global isotopomer balancing following the protocol outlined in Section 9.2.1.

9.5.1 Direct Interpretation of Labeling Signatures

9.5.1.1 Model-Independent Comparison of Labeling Signatures (Fluxome Profiling)

By growing an organism on a ^{13}C-tracer, the labeling signature in metabolites depend on the cellular flux distribution, and therefore, changes in intracellular flux can be indicated simply by comparing labeling data from different experiments. Labeling profiles, e.g., from amino acids analyzed by GC/MS, can be compared by multivariate statistics [7, 147, 148]. This approach does not produce flux estimates but allows comparison of labeling patterns and identification of particular

Table 9.2 Different approaches related to ^{13}C-MFA with their main characteristics and some particular requirements. See also Section 9.5.1

Approach	Results/main characteristics	Need network stoichiometry?	Need physiological data?[1]	Cofactor balances	Reversible fluxes
Model-independent fluxome profiling [147]	Compares labeling signatures as "flux fingerprints". No flux information (flux values) obtained	N	N	N	N
Metabolic flux ratio analysis (METAFoR) [41, 82, 113, 121, 122]	Local flux ratios at metabolic nodes	N	N	N	N
^{13}C-constrained metabolic net flux analysis [36, 42]	Integrated flux map. Stoichiometric balances are solved by use of flux ratios from METAFoR.	Y	Y	Y	N
MFA based on carbon balances [30, 87]	Integrated flux map. System of carbon balance equations. Restricted to use of carbon-positional label.	Y	Y	N	Y
MFA based on global isotopomer balances [28, 83, 106, 133]	Integrated flux map. Uses comprehensive isotopomer model. Various substrate labels usable (positional, bond label). Optimal design of labeling experiments.	Y	Y	N	Y
Large-scale ^{13}C-MFA [19, 120]	Uses fluxes derived from ^{13}C-labeling data as additional constraints in flux balance analysis (search for flux values based on optimality). Genome scale models.	Y	Y	Y	N

[1] e.g., measured uptake rates.

metabolites that are labeled differently depending on the physiological condition or genotype [147]. The approach is not restricted to ^{13}C-labeled substrates, and substrates such as ^{2}H-labeled glucose can be used to highlight particular metabolic processes [147]. The fluxome profiling approach has also been applied to plants, comparing metabolic phenotypes for cultured maize kernels of 18 different genotypes grown on uniformly ^{13}C-labeled glucose or sucrose [110]. It requires less experimental and computational effort than the methods described below and is therefore suitable for fast screening of many different conditions for metabolic phenotypes.

9.5.1.2 Carbon and Isotopomer Balancing of Biochemical Reactions

As mentioned in Section 9.2, labeling signatures resulting from reactions that split or join the carbon backbones of metabolites are important for the quantification of flux by ^{13}C-tracers. Figure 9.5 shows the formation of the serine backbone by two different pathways. Serine may be formed by hydroxymethyl transferase (SHMT) in a reaction joining two molecules or from 3-phosphoglycerate (PGA) without changes in the carbon chain. If positional ^{13}C-enrichment is considered, then the ^{13}C-enrichment in C-1 of serine, for example, results from the enrichments in C-1 of glycine and in C-1 of PGA, weighted according to the ratio of the fluxes v_1 and v_2 (Fig. 9.5a, c). If isotopomer fractions are considered, a biochemical joining reaction can be described by probabilistic relations as shown in Fig. 9.5b, c. For example, the fractional abundance of the isotopomer Ser#101 formed by SHMT equals the product of the fractional abundances of Gly#10 and C1#1.

On the basis of the equations presented in Fig. 9.5c for serine formation, metabolic flux ratio (METAFoR) analysis (Table 9.2) has been developed to calculate flux ratios at isolated metabolic branch points [94]. Flux ratios in central metabolism can be defined based on the isotopomer definition (Fig. 9.5b) or based on formalisms that describe NMR labeling signatures or MS labeling signatures [41, 94]. For example, after growing *E. coli* in the presence of uniformly ^{13}C-labeled glucose, flux ratios at 10 central metabolites were derived [94]. In general, to calculate flux ratios ^{13}C-labeling information is required for the node metabolite and for the metabolites that are converted into the node. Most commonly, the flux ratio equations have to be manually derived while a framework for their automatic generation was published recently [82]. In contrast to global isotopomer balancing, in METAFoR, flux ratios can be determined without a completely defined network topology and without measurements of physiological data such as substrate uptake rates, secretion rates, or biomass composition (Table 9.2).

9.5.2 Network Scale Label Balancing

9.5.2.1 Network Simulation by Positional ^{13}C-Enrichment

Based on the principle outlined in Section 9.5.1 for one reaction, positional labeling can be balanced for networks of central metabolism (Table 9.2). Dieuaide-Noubhani

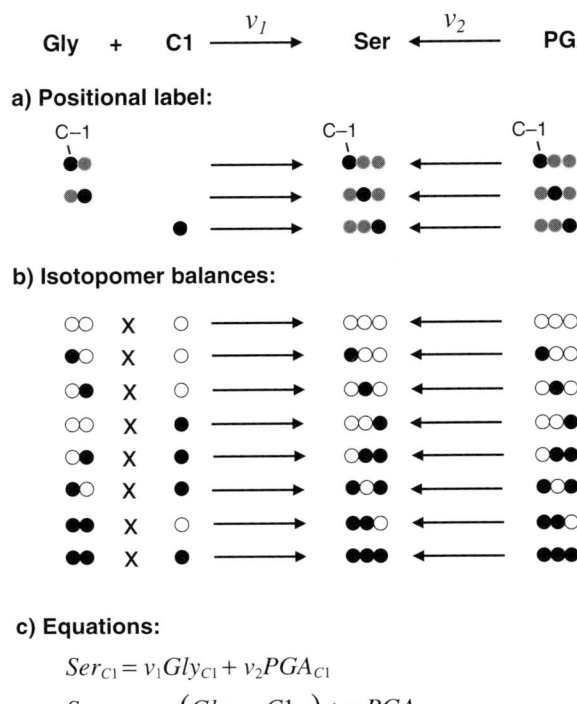

Fig. 9.5 Depiction of the carbon position and isotopomer balances for the synthesis of serine by serine hydroxymethyl transferase (rate v_1) and from 3-phosphoglycerate (PGA) with rate v_2. (**a**) For positional enrichments, one relation exists for each carbon (*grey shapes*, unspecified carbon isotope). (**b**) For isotopomer balancing, eight relations are to be considered. The bimolecular joining reaction is predicted by multiplicative terms, i.e., such joining reactions introduce non-linear structures, which necessitates an iterative computation process for the steady-state labeling in isotopomer networks [138]. (**c**) Example equations for the calculation of positional enrichments or isotopomer abundance dependent of flux values. C1, 5,10-methylenetetrahydrofolate; *open shapes*, ^{12}C; *black shapes*, ^{13}C

et al. [30] resolved a compartmentalized model of fast-growing maize root tips. By the use of positional ^{13}C-labeled glucose as a tracer, the values for 20 fluxes were obtained by solving a system of linear equations using the measured positional enrichments in five metabolites and the values of five measured biosynthetic fluxes and uptake rates.

9.5.2.2 Isotopomer Network Simulation

While the labeling balances for positional labeling can be solved in a straightforward way (calculate fluxes from labeling data and physiological measurements), isotopomer network simulation always uses a more complex process. In an iterative fitting process, the best fit between predicted and experimental labeling signatures is

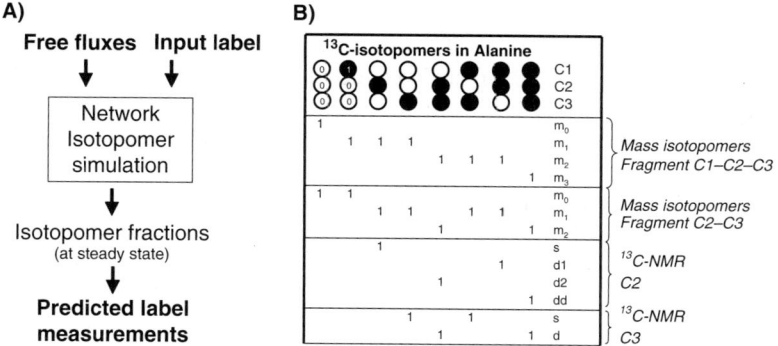

Fig. 9.6 Isotopomer network simulation and prediction of labeling measurements. (**A**) Dependent on the values for the free fluxes and label in the precursors, the steady-state abundances of all isotopomers in the network are computed. Model predicted isotopomer abundances are then used to predict labeling measurements. (**B**) Mapping of isotopomer species of the 3-carbon molecule alanine to signal intensities in MS or NMR. Open circles, ^{12}C; closed circles, ^{13}C; s, singlet; d, d1, d2 doublet; dd doublet of doublets

determined (see Section 9.5.3). Thus, for a typical biochemical network, hundreds of isotopomer equations have to be generated. Handling of the many equations is best done with dedicated software packages such as *13CFLUX* [139]. Since the isotopomer formalism comprehensively describes the labeling state in a network, one isotopomer model can be used to consider any kind of ^{13}C-labeled substrate (positional labeled, bond labeled, substrate mixtures), and all experimentally producible MS or NMR label signatures can be predicted (Fig. 9.6).

9.5.2.3 Cumomer Simulation: Computational Speedup

In order to compute steady-state isotopomer abundances, isotopomer balance equations have to be solved by iterative numerical algorithms [97, 138]. This procedure is computationally demanding and suffers from computational instability [138]. With a data transformation of the isotopomer fractions into *cumomers* (*cum*ulated isotop*omer*), Wiechert et al. (1999) introduced an elegant way to solve the respective equation systems of cumomer balances analytically and much quicker with typically about 1 s computation time [138]. Thus, cumomers are merely a computer-internal data representation that can be interconverted with isotopomer fractions without loss of information. For example, the software packages *13CFLUX* [139] and NMR2Flux [111] use the cumomer approach.

9.5.2.4 Alternative Descriptions of Labeling States Used in Network Flux Analysis

Besides the isotopomer formalism and the cumomer simulation, alternative concepts for representation and simulation of labeling states have been developed. All of them represent ways to reduce the number of labeling states that have to be simulated. The *bondomer* concept (*bond isomer*) [112, 127] is based on the usage of uniformly ^{13}C-labeled tracers (in mixture with unlabeled molecules). In NMR, directly neighboring ^{13}C-nuclei are usually clearly visible in the fine structure of the spectrum as ^{13}C/^{13}C coupling (Chapter 5). Bondomer simulation traces C–C bonds from the labeled substrates through the network. A C–C bond can have two states, either "unbroken" or "newly formed" in a biochemical reaction. Unbroken bonds retain neighboring ^{13}C nuclei while biosynthetically new bonds are most frequently formed between ^{12}C and ^{13}C atoms.[4] Based on the simulated bondomer labeling states, NMR labeling data can be predicted. One advantage of the bondomer concept is the reduced computational complexity relative to the isotopomer concept. While for a molecule with n carbon atoms the network simulation has to balance 2^n isotopomer species, bondomer simulation has only to balance half the number of molecule species (2^{n-1}) [112, 127]. As a disadvantage, the bondomer concept is limited to the use of uniform ^{13}C-labeled tracers.

Another description that is closely related to isotopomers and bondomers is the concept of "X-groups" [110]. Here uniformly labeled substrates (e.g., [U-^{13}C]glucose) are highly diluted with the unlabeled substrate (1:30) [110]. This not only reduces signal intensity in the NMR spectrum but also simplifies the isotopomer computation, since for the formation of a new carbon bond the probability of two ^{13}C atoms meeting can be neglected, and related isotopomer species will not be present and therefore need not to be considered in the mathematical formalism. For example, of the 64 isotopomers that describe the labeling in the six carbon compound glucose, 42 isotopomers can be excluded from network simulation [110].

In a typical isotopomer network simulation, the steady-state abundance of hundreds of isotopomer fractions is computed, comprehensively predicting the labeling state of the network. Afterward, this information can be used to predict NMR or MS measurements (see Fig. 9.6). Typically, the number of predicted label measurements is much smaller than the hundreds of isotopomers simulated in the network, and a reduction of the complexity of the calculations would reduce computation time. As mentioned in Section 9.4, the complexity of network simulation can usually be reduced by lumping metabolite pools. In addition, approaches have been developed that further reduce the number of simulated labeling states and therefore computation time [9, 43, 131]. In an approach by Antoniewicz et al. [9], network simulation is performed with a relatively small number of elementary metabolic units (EMU), which are identified by tracing molecule species back through the network, starting with those detected by label measurements and ending up in the labeled substrates [9]. In another approach, graph-based topological analysis and decomposition of isotope labeling networks (ILN) allow a reduction in the size of the network (and thus fewer cumomer equations) in a similar way [131]. Both approaches can be applied to any metabolic network, and they have been reported

---

[4]By probabilistic relations, a small fraction of the newly formed carbon bonds will also result in new ^{13}C–^{13}C pairs, which is accounted for in the bondomer approach.

to reduce computational times by three orders of magnitude, which opens the door for simulation of computationally expensive non-stationary labeling experiments [146]. In addition, EMU models allow simulation of the label redistribution of multiple isotopes of multiple elements, e.g., ^{2}H, ^{13}C, ^{15}N, and ^{18}O. Here, it should be noted that the number of isotopomer species increases exponentially with the number of labeled positions in a molecule. While glucose (6 carbons) has 2^6 (64) carbon isotopomers, it has 2^{24} isotopomers if C, H, and O isotopes are considered. It has been shown that a network simulation with ^{2}H-, ^{13}C-, and ^{18}O-label requiring 10^6 isotopomers can be simulated with less than 500 EMU species [9].

9.5.3 Finding Flux Values that Best Explain the Labeling Data

Due to the mathematical structure of isotopomer simulations, the direct computation of fluxes from labeling signatures is not possible. Conversely, labeling signatures can be computed from fluxes, i.e., central to the flux analysis process (Fig. 9.2) is an algorithm that simulates labeling signatures (NMR or MS data) based on estimated flux values. The flux values that explain the labeling data have to be found by a search algorithm. The determination of flux values requires some basic understanding of the search algorithms and can be computationally demanding. Also, the error statistics for the flux values cannot be obtained in a straightforward way (see Section 9.5.4). The principle of the "non-linear inverse problem" in ^{13}C-MFA is depicted in Fig. 9.7, simplified for the relation between one flux and one measurement. By varying the flux values in the metabolic model, a network flux state is sought that best explains all labeling measurements. For this purpose, different classical optimization algorithms are commonly applied, typically gradient-based search algorithms (e.g., [137, 142]), stochastic global optimization like simulated annealing (e.g., [111]), or combinations of these methods ([145, 150]). In general, these algorithms iteratively improve the quality of an initial guess for the fluxes until a stopping criterion is reached. In each iterative step, the quality of the flux values is judged by computing the following value (see also Fig. 9.7.):

$$X^2 = \sum_j \frac{\left(m_j^{\text{exp}} - m_j^{\text{pred}}\right)^2}{\sigma_j^2} \qquad (9.3)^5$$

where m_j^{pred} and m_j^{exp} are the predicted and measured value for the jth labeling measurement, respectively, with standard deviation σ_j. The formula sums the squares of differences between the predicted and measured labeling data – weighted by the standard errors in the measurements. The summands are also called residuals. Label measurements with large standard deviation have less weight in the sum and so the optimization will tolerate larger deviations for measurements with larger statistical

---

[5] In some cases, the formula also includes scaling factors for the measurement data.

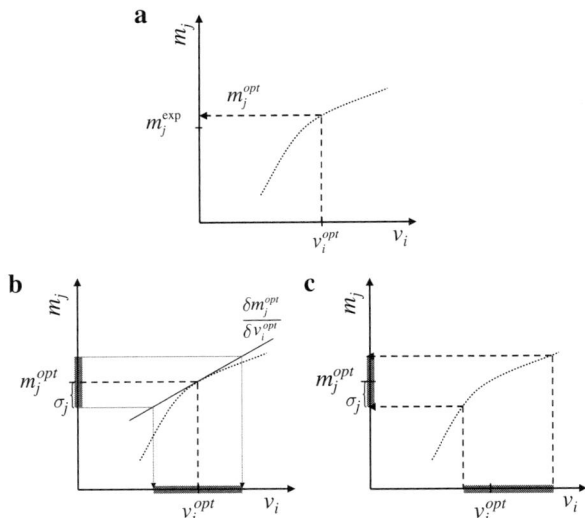

Fig. 9.7 Relations between multiple fluxes and label measurements, predicted by isotopomer simulation and illustrated by showing only two dimensions (v_i, m_j). For different fixed values for the flux v_i, the optimization process finds different values for the label measurement (*dashed curves*). (**a**) A best fit for all measurements (v_i^{opt}, m_j^{opt}), according to Eq. (9.4), will predict a value close to m_j^{exp}. (**b**) With the best fit, an approximation of the standard error for v_i^{opt} is possible by using the experimental standard deviation σ_j and the numerically derived local derivative for the optimum ($\delta m_j^{opt}/\delta v_i^{opt}$) (principle of linearized statistics, see Section 9.5.4). (**c**) Optimizing with stepwise increase of v_i allows to find a more exact confidence interval for v_i^{opt}

uncertainty. The optimization algorithm minimizes the χ^2 (chi-square) value by changing the values for the free fluxes in the system (within boundaries set in the model) and predicting the labeling signatures (Fig. 9.6). Gradient-based optimizations move only "down-hill" and can get stuck in local minima, thus missing to find the global optimum (global minimal χ^2). Therefore, especially for gradient-based algorithms, multiple optimization runs with random start points should be performed to increase confidence in the optimization outcome.

The reliability of the final flux values can be assessed by considering the following questions:

1. *With random start points, does the optimization consistently find the same optimal flux values?* Depending on the experimental data and the modeling configuration (e.g., the choice of precursor labeling), multiple optimization runs may result in different solutions with the same χ^2 value (degenerate optima). In particular, there could be high variability in some fluxes, while others are consistently estimated with the same value.
2. *Are important fluxes given by different pieces of data (redundant data)?* In an extreme case, the value for a particular flux could be dependent on the value of only one measurement. This kind of relationship is described in the measurement

9 Isotopic Steady-State Flux Analysis

sensitivity matrix, which can be obtained for example in the data outputs of the 13CFLUX software (see Section 9.5.5). Figure 9.7b depicts the sensitivity ($\delta m_j^{opt}/\delta v_i^{opt}$) between one measurement and one flux as the slope at a particular point of the system function. If the value of the slope is very low, there is low sensitivity of the measurement value to a change of the flux value, i.e., the flux value is not well defined by the measurement. The sensitivity matrix contains these values for all pairs of free fluxes and measurements.[6] As derived from the sensitivity matrix and the measurement statistics, the contribution matrix shows how many redundant measurements contribute to each flux [8].

3. *Are the fluxes statistically well determined?* In ^{13}C-MFA studies, the estimation of confidence intervals is very important in order to know how much precision there is in the flux values (see Section 9.5.4). If the confidence interval for a flux is of similar magnitude to the flux value itself, then there is low statistical confidence in the value.

4. *Can the difference between model prediction (best fit) and measurement data be attributed to random noise in the data only?* If it is assumed in Eq. (9.4) that the σ-values represent normal distribution of statistical noise in the measurement data, the final minimal χ^2 (chi-square) value can be used for a statistical test for goodness of fit. For this purpose, the degree of freedom in the model has to be known. If there are k signals in the labeling data that could assume values freely and independent from each other (independent measurements) and if in the model l fluxes can be varied independently (free fluxes), then the model has $n = k-l$ degrees of freedom. Now the theoretically allowed sum of errors $\chi^2_{n,1-\alpha}$ can be taken from a chi-square distribution table (n is the degrees of freedom; $1-\alpha$ is the confidence level of the test) [28, 137]. If the iteratively determined sum of errors (χ^2) exceeds that value, then the deviations between predicted and labeling data could be caused by inconsistencies in the assumed metabolic network or gross measurement errors. Inspection of the individual residuals (Sumands in Eq. 9.4) may reveal that some label measurements have outstandingly high deviation from the model. These "gross-error" measurements can be removed and the iterative fitting repeated in order to test if the modeling result can be obtained without the removed measurement and if the χ^2 sum will then be acceptable.

9.5.4 Statistical Analysis of Flux Results

For interpretation of the flux values obtained from ^{13}C-MFA, the numbers should always be given with statistical uncertainty. This is often done by computation of a standard deviation (also known as symmetrical 68% confidence interval). However, particularly for exchange fluxes, the real statistical uncertainty is mostly better

---

[6] Another property of the sensitivity matrix: its numerical rank indicates the number of independent fluxes that can be derived from the labeling data.

approximated by calculation of unsymmetrical confidence intervals [137]. The statistical uncertainty in the flux is derived from the statistical errors in the labeling measurements and the physiological measurements. This can be done in different ways (see below). In many flux studies, the importance of the statistical analysis for the interpretation of the results has been emphasized. The mathematical approaches introduced below are in part recent developments.

9.5.4.1 Measurement Statistical Errors

In the first place, the statistical reliability of the measurements has to be established, taking care to distinguish instrumental and experimental errors. Repetition of an entire labeling experiment should usually be less accurate than repeated measurement of a sample on the instrument. Errors in label measurements have been determined or estimated in different ways. In NMR experiments, where a relatively large sample size is needed and experimental repetition is cumbersome, the statistical errors may be estimated by using the larger of (a) the signal-to-noise ratio of the NMR spectrum or (b) the difference between intensities from duplicate spectra [111]. In GC/MS analysis, true experimental repetition is more easily done because of the much smaller sample size needed for GC/MS analysis, i.e., a labeling experiment can be repeated several times.

9.5.4.2 Linearized Statistics

The errors in the measurements can be translated into errors in the flux measurements by using the concept of linearized statistics. The contribution of the standard error in a measurement to the error in the flux can be derived from the local derivative (sensitivity) shown in Fig. 9.7b. [28, 136]. This approach suffers from significant inaccuracies in particular for the estimation of large exchange fluxes. This problem has been in part reduced by the use of [0, 1]-rescaled exchange fluxes in the statistical model, as described in detail by Wiechert et al. [137].

9.5.4.3 Monte Carlo Statistics

As a different approach, Monte Carlo stochastic simulation has been used to examine how the elucidated fluxes change when uncertainty in the form of normally distributed noise is added to the data [43, 111, 113, 142, 150]. This means that the original experimental data are re-sampled many times based on the standard deviations, and the flux parameter fitting procedure is repeated each time. This approach should be more accurate than the use of linearized statistics, but it has the disadvantage of being computationally more demanding. Also, its accuracy is limited by the number of times the data are re-sampled.

9 Isotopic Steady-State Flux Analysis

9.5.4.4 True Confidence Intervals

A more recent approach [8] derives the flux statistics as confidence intervals based on the χ^2 value used in flux parameter fitting (Eq. 9.4; see Fig. 9.7c). For a selected flux, the originally optimized value is increased step by step and each time the other free fluxes are re-optimized by minimization of χ^2 (the usual optimization procedure to find the optimum flux values). With deviation of the selected flux from its original optimum, the re-optimized χ^2 value will increase continuously. If the increase of the χ^2 value reaches a limit (given by a $\chi^2_{1,1-\alpha}$ distribution), the selected flux has reached the threshold of the upper confidence interval. Similarly, the lower bound of the confidence interval is derived by decreasing the selected flux. Antoniewicz and co-workers [8] showed that this method can produce confidence intervals that are identical to exactly known intervals.

9.5.5 Software Tools for ^{13}C-MFA

13CFLUX is a software package developed by the research group of W. Wiechert (University of Siegen, Germany) [139]. The software runs under Linux environments (PC). For academic use, the software is freely available from the authors upon request. The user has to define a model as a collection of data sheets following a specific format. The model files allow the configuration of any metabolic network along with free definition of network stoichiometry and carbon transitions, input label, extracellular flux measurements, and different types of NMR or MS data. Data extraction and processing, such as correction of MS data for naturally occurring isotopes, has to be done by the user. From the model text files, flux parameter fitting or statistical analysis can be performed. The generated data have to be collected from textual outputs of the software. The software does not require programming skills but is not particularly user-friendly (limited documentation, simple user interface, limited error messages).

Currently, a new version of 13CFLUX is under development. A feature of the new software (April 2008; W. Wiechert, personal communication) will be very fast simulation of large networks based on the reduction of network size as described in [131]. This will allow increased accuracy in the calculation of confidence intervals and the analysis of labeling experiments in high throughput. The software package will also support instationary flux analysis [75]. Model definition will be in extensible markup language (XML) format, allowing the model files to be read by other systems biology software packages.

FiatFlux [149] is an open source software package running under the MatLab© environment. FiatFlux is more user-friendly than 13CFLUX but does not have the capability for global isotopomer balancing of metabolic networks. It consists of two independently running modules: flux ratio analysis (RATIO) and ^{13}C-constrained flux analysis (NETTO). The RATIO software package is preconfigured for the analysis of protein-bound amino acids by GC/MS of t-butyl dimethylsilyl (TBS) derivatives. It can extract labeling information from raw MS data and automatically derives multiple flux ratios, including relevant statistics. The glucose input label can be defined as molar abundance of [1-^{13}C]glucose or [U-^{13}C$_6$]glucose.

NMR2Flux [111] is a software dedicated to flux analysis with NMR data. The user creates data sheets as input files that define the network stoichiometry and carbon transitions, input label, extracellular flux measurements, and NMR data. The program can then find optimal flux solutions by simulated annealing. To change model configurations, the software has to be modified by re-programming.

9.6 Concluding Remarks

In this chapter, the use of steady-state MFA with use of stable isotope tracers has been outlined with a focus on its application to plants. Isotopic tracers have been used from the pioneering days of research in plant metabolism to the more recent ^{13}C-MFA studies that give quantitative insights into the in vivo function of metabolic networks. Following a general overview on the major steps involved in a ^{13}C-MFA experiment, the steady-state condition, as well as some issues of network construction, experimental design, and flux estimation, was reviewed in some detail. It should be kept in mind that the validity of the results of flux analysis experiments is limited by the incompleteness of information available to build the network models.

In recent years, the number of published ^{13}C-MFA studies in plants increased substantially. Among the challenges for the further development in plant ^{13}C-MFA methods are the currently limited ability to resolve fluxes in-between subcellular compartments, the consideration of substrate channeling in network models, and the reconstruction of flux models by more exact data. Another challenge is the integration of flux data with data from metabolite profiling, enzyme profiling, and transcript or proteome analysis.

References

1. Agrawal PK, Canvin DT (1970) Contribution of pentose phosphate pathway in developing castor bean endosperm. Can J Bot 49:267–272.
2. Allen DK, Shachar-Hill Y, Ohlrogge JB (2007) Compartment-specific labeling information in ^{13}C metabolic flux analysis of plants. Phytochemistry 68:2197–2210.
3. Alonso AP, Vigeolas H, Raymond P, Rolin D, Dieuaide-Noubhani M (2005) A new substrate cycle in plants. Evidence for a high glucosephosphate glucosephosphate-to-glucose turnover from in vivo steady-state and pulse labeling experiments with [^{13}C]glucose and [^{14}C]glucose. Plant Physiol 138:2220–2232.
4. Alonso AP, Raymond P, Rolin D, Dieuaide-Noubhani M (2007) Substrate cycles in the central metabolism of maize root tips under hypoxia. Phytochemistry 68:2222–2231.
5. Alonso AP, Goffman FD, Ohlrogge JB, Shachar-Hill Y (2007) Carbon conversion efficiency and central metabolic fluxes in developing sunflower (*Helianthus annuus* L.) embryos. Plant J 52:296–308.
6. Alonso AP, Raymond P, Hernould M, Rondeau-Mouro C, de Graaf A, Chourey P, Lahaye M, Shachar-Hill Y, Rolin D, Dieuaide-Noubhani M (2007) A metabolic flux analysis to study

the role of sucrose synthase in the regulation of the carbon partitioning in central metabolism in maize root tips. Metab Eng 9:419–432.
7. Antoniewicz MR, Stephanopoulos G, Kelleher JK (2006) Evaluation of regression models in metabolic physiology: predicting fluxes from isotopic data without knowledge of the pathway. Metabolomics 2:41–52.
8. Antoniewicz MR, Kelleher JK, Stephanopoulos G (2006) Determination of confidence intervals of metabolic fluxes estimated from stable isotope measurements. Metab Eng 8: 324–337.
9. Antoniewicz MR, Kelleher JK, Stephanopoulos G. (2007) Elementary metabolite units (EMU): a novel framework for modeling isotopic distributions. Metab Eng 9:68–86.
10. Antoniewicz MR, Kraynie DF, Laffend LA, González-Lergier J, Kelleher JK, Stephanopoulos G (2007) Metabolic flux analysis in a nonstationary system: fed-batch fermentation of a high yielding strain of *E. coli* producing 1,3-propanediol. Metab Eng 9:277–292.
11. Ap Rees T, Beevers H (1960) Pathways of Glucose Dissimilation in Carrot Slices. Plant Physiol 35:830–838.
12. Ap Rees T, Beevers H (1960) Pentose phosphate pathway as a major component of induced respiration of carrot and potato slices. Plant Physiol 35:839–847.
13. Araúzo-Bravo MJ, Shimizu K (2003) An improved method for statistical analysis of metabolic flux analysis using isotopomer mapping matrices with analytical expressions. J Biotechnol 105:117–133.
14. Averill RH, Bailey-Serres J, Kruger NJ (1998) Co-operation between cytosolic and plastidic oxidative pentose phosphate pathways revealed by 6-phosphogluconate dehydrogenase-deficient genotypes of maize. Plant J 14:449–457.
15. Bailey JE (1999) Lessons learned from metabolic engineering for fnctional genomics and drug discovery. Nature Biotechnol 17: 616–918.
16. Bao XM, Focke M, Pollard M, Ohlrogge J (2000) Understanding in vivo carbon precursor supply for fatty acid synthesis in leaf tissue. Plant J 22:39–50.
17. Beevers H., Gibbs M (1954) The direct oxidation pathway in plant respiration. Plant Physiol 29:322–324.
18. Beckles DM, Smith AM, ap Rees T (2001) A cytosolic ADP-glucose pyrophosphorylase is a feature of graminaceous endosperms, but not of other starch-storing organs. Plant Physiol 125:818–827.
19. Blank LM, Kuepfer L, Sauer U (2005) Large-scale ^{13}C-flux analysis reveals mechanistic principles of metabolic network robustness to null mutations in yeast. Genome Biol 6: R49.
20. Boatright J, Negre F, Chen X, Kish CM, Wood B, Peel G, Orlova I, Gang D, Rhodes D, Dudareva N (2004) Understanding in vivo benzenoid metabolism in petunia petal tissue. Plant Physiol 135:1993–2011.
21. Bonarius HPJ, Schmid G, Tramper J (1997) Flux analysis of underdetermined metabolic networks: the quest for the missing constraints. Trends Biotechnol 15:308–314.
22. Calvin M, Benson AA (1948) The path of carbon in photosynthesis. Science 107:476–480.
23. Calvin M (1962) The path of carbon in photosynthesis. Science 135:879–889.
24. Canvin DT, Beevers H (1961) Sucrose synthesis from acetate in the germinating castor bean: kinetics and pathway. J Biol Chem 236: 988–995.
25. Christensen B, Nielsen J. (2002) Reciprocal ^{13}C-labeling: a method for investigating the catabolism of cosubstrates. Biotechnol Prog. 18:163–166.
26. Costenoble R, Muller D, Barl T, van Gulik WM, van Winden WA, Reuss M, Heijnen JJ (2007) ^{13}C-Labeled metabolic flux analysis of a fed-batch culture of elutriated *Saccharomyces cerevisiae*. FEMS Yeast Res 7:511–526.
27. Dasilva PMFR, Eastmond PJ, Hill LM, Smith AM, Rawsthorne S (1997) Starch metabolism in developing embryos of oilseed rape. Planta 203:480–487.
28. Dauner M, Bailey JE, Sauer U. (2001) Metabolic flux analysis with a comprehensive isotopomer model in *Bacillus subtilis*. Biotechnol Bioeng 76:144–156.

29. Dennis DT (1989) Fatty acid biosynthesis in plastids. In: Physiology, Biochemistry, and Genetics of Nongreen Plastids (Boyer, C. D., Shannon, R. C., and Hardison, R. C., eds) pp. 120–129, American Society of Plant Physiologists, Rockville, MD.
30. Dieuaide-Noubhani M, Raffard G, Canioni P, Pradet A, Raymond P (1995) Quantification of compartmented metabolic fluxes in maize root- tips using isotope distribution from C-13-labeled or C-14-labeled glucose. J Biol Chem 270:13147–13159.
31. Dixon RA, Chen F, Guo D, Parvathi K (2001) The biosynthesis of monolignols: a "metabolic grid", or independent pathways to guaiacyl and syringyl units? Phytochemistry 57:1069–1084.
32. Domergue F, Cassagne C, Lessire R (1999) Seed acyl-CoA elongases: the other system of fatty acid synthesis. Ocl-Ol. Corps Gras Lipides 6:101–106.
33. Douce R, Neuberger M (1999) Biochemical dissection of photorespiration. Curr Opin Plant Biol 2:214–222.
34. Edwards S, Nguyen BT, Do B, Roberts JKM (1998) Contribution of malic enzyme, pyruvate kinase, phosphoenolpyruvate carboxylase, and the krebs cycle to respiration and biosynthesis and to intracellular pH regulation during hypoxia in maize root tips observed by nuclear magnetic resonance imaging and gas chromatography-mass spectrometry. Plant Physiol 116:1073–1081.
35. Edwards JS, Covert M, Palsson B (2002) Metabolic modelling of microbes: the flux-balance approach. Environ Microbiol 4:133–140.
36. Emmerling M, Dauner M, Ponti A, Fiaux J, Hochuli M, Szyperski T, Wüthrich K, Bailey JE, Sauer U (2002) Metabolic flux responses to pyruvate kinase knockout in *Escherichia coli*. J Bacteriol 184:152–164.
37. Ettenhuber C, Spielbauer G, Margl L, Hannah LC, Gierl A, Bacher A, Genschel U, Eisenreich W (2005) Changes in flux pattern of the central carbohydrate metabolism during kernel development in maize. Phytochemistry 66:2632–2642.
38. Fait A, Fromm H, Walter D, Galili G, Fernie AR (2008) Highway or byway: the metabolic role of the GABA shunt in plants. Trends Plant Sci 13:14–19.
39. Fatland BL, Nikolau BJ, Wurtele ES (2005) Reverse genetic characterization of cytosolic acetyl-CoA generation by ATP-citrate lyase in *Arabidopsis*. Plant Cell 17:182–203.
40. Fernie AR, Roscher A, Ratcliffe RG, Kruger NJ (2001) Fructose 2,6-bisphosphate activates pyrophosphate: fructose-6-phosphate 1-phosphotransferase and increases triose phosphate to hexose phosphate cycling in heterotrophic cells. Planta 212:250–263.
41. Fischer E, Sauer U. (2003) Metabolic flux profiling of *Escherichia coli* mutants in central carbon metabolism using GC-MS. Eur J Biochem 270:880–891.
42. Fischer E, Zamboni N, Sauer U (2004) High-throughput metabolic flux analysis based on gas chromatography-mass spectrometry derived ^{13}C constraints. Anal Biochem 325: 308–316.
43. Forbes NS, Clark DS, Blanch HW (2001) Using isotopomer path tracing to quantify metabolic fluxes in pathway models containing reversible reactions. Biotechnol Bioeng 74:196–211.
44. Geigenberger P, Stitt M (1991). A futile cycle of sucrose synthesis and degradation is involved in regulating partitioning between sucrose, starch and respiration in cotyledons of germinating *Ricinus communis* L seedlings when phloem transport is inhibited. Planta 185:81–90.
45. Goffman FD, Ruckle M, Ohlrogge JB, Shachar-Hill Y (2004) Carbon dioxide concentrations are very high in developing oil seeds. Plant Physiol Biochem 42:703–708.
46. Graham JW, Williams TC, Morgan M, Fernie AR, Ratcliffe RG, Sweetlove LJ (2007) Glycolytic enzymes associate dynamically with mitochondria in response to respiratory demand and support substrate channeling. Plant Cell 19:3723–3738.
47. Hatzfeld WD, Stitt M (1990) A study of the rate of recycling of triose phosphates in heterotrophic *Chenopodium rubrum* cells, potato tubers, and maize endosperm. Planta 180:198–204.

48. Heinrich R, Schuster S (1996) The Regulation of Cellular Systems. Chapman & Hall, New York.
49. Heinzle E, Matsuda F, Miyagawa H, Wakasa K, Nishioka T (2007) Estimation of metabolic fluxes, expression levels and metabolite dynamics of a secondary metabolic pathway in potato using label pulse-feeding experiments combined with kinetic network modelling and simulation. Plant J 50:176–187.
50. Herrmann KM, Weaver LM (1999) The shikimate pathway. Annu Rev Plant Physiol Plant Mol Biol 50:473–503.
51. Ho CL, Noij M, Saito K (1999) Plastidic pathway of serine biosynthesis. J Biol Chem 274:11007–11012.
52. Hoppe A, Hoffmann S, Holzhütter HG (2007) Including metabolite concentrations into flux balance analysis: thermodynamic realizability as a constraint on flux distributions in metabolic networks. BMC Syst Biol 1:23.
53. Huege J, Sulpice R, Gibon Y, Lisec J, Koehl K, Kopka J (2007) GC-EI-TOF-MS analysis of in vivo carbon-partitioning into soluble metabolite pools of higher plants by monitoring isotope dilution after $^{13}CO2$ labelling. Phytochemistry 68:2258–2272.
54. Iyer VV, Sriram G, Fulton DB, Zhou R, Westgate ME, Shanks JV (2008) Metabolic flux maps comparing the effect of temperature on protein and oil biosynthesis in developing soybean cotyledons. Plant Cell Environ 31:506–517.
55. Junker BH, Lonien J, Heady LE, Rogers A, Schwender J (2007) Parallel determination of enzyme activities and in vivo fluxes in *Brassica napus* embryos grown on organic or inorganic nitrogen source. Phytochemistry 68:2232–2242.
56. Keeling PL, Wood JR, Tyson RH, Bridges IG (1988) Starch biosynthesis in developing wheat grain. Evidence against the direct involvement of triose phosphates in the metabolic pathway. Plant Physiol 87:311–319.
57. Kiefer P, Nicolas C, Letisse F, Portais JC (2007) Determination of carbon labeling distribution of intracellular metabolites from single fragment ions by ion chromatography tandem mass spectrometry. Anal Biochem 360:182–188.
58. Kim HU, Kim TY, Lee SY (2008) Metabolic flux analysis and metabolic engineering of microorganisms. Mol BioSyst 4:113–120.
59. King SP, Badger MR, Furbank RT (1998) CO_2 refixation characteristics of developing canola seeds and silique wall. Aust J Plant Physiol 25:377–386.
60. Kleijn RJ, van Winden WA, van Gulik WM, Heijnen JJ (2005) Revisiting the ^{13}C-label distribution of the non-oxidative branch of the pentose phosphate pathway based upon kinetic and genetic evidence. FEBS J 272:4970–4982.
61. Kleijn RJ, Geertman JM, Nfor BK, Ras C, Schipper D, Pronk JT, Heijnen JJ, van Maris AJ, van Winden WA (2007) Metabolic flux analysis of a glycerol-overproducing Saccharomyces cerevisiae strain based on GC-MS, LC-MS and NMR-derived C-labelling data. FEMS Yeast Res 7:216–231.
62. Krook J, Vreugdenhill D, Dijkema C, van der Plas LHW (1998) Sucrose and starch metabolism in carrot (*Daucus carota* L.) cell suspensions analysed by ^{13}C-labelling: indications for a cytosol and a plastid-localized oxidative pentose phosphate pathway. J Exp Bot 49:1917–1924.
63. Kruger NJ, Huddleston JE, Le Lay P, Brown ND, Ratcliffe RG (2007) Network flux analysis: impact of ^{13}C-substrates on metabolism in *Arabidopsis thaliana* cell suspension cultures. Phytochemistry 68:2176–2188.
64. Kruger NJ, Le Lay P, Ratcliffe RG (2007) Vacuolar compartmentation complicates the steady-state analysis of glucose metabolism and forces reappraisal of sucrose cycling in plants. Phytochemistry 68: 2189–2196.
65. Kümmel A, Panke S, Heinemann M (2006) Putative regulatory sites unraveled by network-embedded thermodynamic analysis of metabolome data. Mol Syst Biol 2:2006.0034.
66. Lichtenthaler HK, Schwender J, Disch A, Rohmer M (1997) Biosynthesis of isoprenoids in higher plant chloroplasts proceeds via a mevalonate independent pathway. FEBS Lett 400:271–274.

67. Libourel IGL, Gehan JP, Shachar-Hill Y (2007) Design of substrate label for steady state flux measurements in plant systems using the metabolic network of Brassica napus embryos. Phytochemistry 68:2211–2221.
68. Ma H, Zeng AP (2003) Reconstruction of metabolic networks from genome data and analysis of their global structure for various organisms. Bioinformatics 19:270–277.
69. Malloy CR, Sherry AD, Jeffrey FMH. (1988) Evaluation of carbon flux and substrate selection through alternate pathways involving the citric acid cycle of the heart by 13C NMR spectroscopy. J Biol Chem 263:6964–6971.
70. Matsuda F, Wakasa K, Miyagawa H (2007) Metabolic flux analysis in plants using dynamic labeling technique: Application to tryptophan biosynthesis in cultured rice cells. Phytochemistry 68:2290–2301.
71. McClung CR, Hsu M, Painter JE, Gagne JM, Karlsberg SD, Salome PA (2000) Integrated temporal regulation of the photorespiratory pathway. Circadian regulation of two *Arabidopsis* genes encoding serine hydroxymethyltransferase. Plant Physiol 123:381–392.
72. McNeil SD, Rhodes D, Russell BL, Nuccio ML, Shachar-Hill Y, Hanson AD (2000) Metabolic modeling identifies key constraints on an engineered glycine betaine synthesis pathway in tobacco. Plant Physiol 124:153–162.
73. Möllney M, Wiechert W, Kownatzki D, de Graaf AA (1999) Bidirectional Reaction Steps in Metabolic Networks: IV. Optimal Design of Isotopomer Labeling Experiments. Biotechnol Bioeng 66:86–103.
74. Morgan JA, Rhodes D (2002) Mathematical modeling of plant metabolism. Metabol. Eng. 4:80–89.
75. Nöh K, Grönke K, Luo B, Takors R, Oldiges M, Wiechert W (2007) Metabolic flux analysis at ultra short time scale: isotopically non-stationary ^{13}C labeling experiments. J Biotechnol 129:249–267.
76. Ohta D, Fujimori K, Mizutani M, Nakayama Y, Kunpaisal-Hashimoto R, Münzer S, Kozaki A (2000) Molecular cloning and characterization of ATP-phosphoribosyl transferase from *Arabidopsis*, a key enzyme in the histidine biosynthetic pathway. Plant Physiol 122: 907–914.
77. Orlova I, Marshall-Colón A, Schnepp J, Wood B, Varbanova M, Fridman E, Blakeslee JJ, Peer WA, Murphy AS, Rhodes D, Pichersky E, Dudareva N (2006) Reduction of benzenoid synthesis in petunia flowers reveals multiple pathways to benzoic acid and enhancement in auxin transport. Plant Cell 18:3458–3475.
78. Ohlrogge JB, Kuhn DN, Stumpf PK (1979) Subcellular localization of acyl carrier protein in leaf protoplasts of Spinacia oleracea. Proc Natl Acad Sci USA 76:1194–1198.
79. Plaxton WC, Podesta FE (2006) The functional organization and control of plant metabolism. Crit Rev Plant Sci 25:159–198.
80. Pleite R, Pike MJ, Garces R, Martinez-Force E, Rawsthorne S (2005) The sources of carbon and reducing power for fatty acid synthesis in the heterotrophic plastids of developing sunflower (*Helianthus annuus* L.) embryos. J Exp Bot 56:1297–1303.
81. Rantanen A, Rousu J, Kokkonen JT, Tarkiainen V, Ketola RA (2002) Computing positional isotopomer distributions from tandem mass spectrometric data. Met Eng 4:285–94.
82. Rantanen A, Rousu J, Jouhten P, Zamboni N, Maaheimo H, Ukkonen E (2008) An analytic and systematic framework for estimating metabolic flux ratios from 13C tracer experiments. BMC Bioinformatics 9:266.
83. Ratcliffe RG, Shachar-Hill, Y (2006) Measuring multiple fluxes through plant metabolic networks. Plant J 45:490–511.
84. Rawsthorne S (2002) Carbon flux and fatty acid synthesis in plants. Progr Lipid Res 41: 182–196.
85. Rios-Estepa R, Lange BM: Experimental and mathematical approaches to modeling plant metabolic networks. Phytochemistry 68:2351–2374.
86. Romisch-Margl W, Schramek N, Radykewicz T, Ettenhuber C, Eylert E, Huber C, Romisch-Margl L, Schwarz C, Dobner M, Demmel N, Winzenhorlein B, Bacher A, Eisenreich W

(2007) $^{13}CO_2$ as a universal metabolic tracer in isotopologue perturbation experiments. Phytochemistry 68:2273–2289.
87. Rontein D, Dieuaide-Noubhani M, Dufourc EJ, Raymond P, Rolin D (2002) The metabolic architecture of plant cells. Stability of central carbon metabolism and flexibility of anabolic pathways during the growth cycle of tomato cells. J Biol Chem 46: 43948–43960.
88. Roscher A, Kruger NJ, Ratcliffe RG (2000) Strategies for metabolic flux analysis in plants using isotope labelling. J Biotechnol 77:81–102.
89. Roughan PG, Ohlrogge JB (1984) On the assay of acetyl-CoA synthethase activity in chloroplasts and leaf extract. Anal Biochem 216:77–82.
90. Rousu J, Rantanen A, Ketola RA, Kokkonen JT (2005) Isotopomer distribution computation from tandem mass spectrometric data with overlapping fragment spectra. Spectroscopy 19:53–67.
91. Ruuska SA, Schwender J, Ohlrogge JB (2004) The capacity of green oilseeds to utilize photosynthesis to drive biosynthetic processes. Plant Physiol 136:2700–2709.
92. Salon C, Raymond P, Pradet A (1988) Quantification of carbon fluxes through the tricarboxylic acid cycle in early germinating lettuce embryos. J Biol Chem 263:12278–12287.
93. Sauer U (2006) Metabolic networks in motion: ^{13}C-based flux analysis. Mol Sys Biol 2: 62.
94. Sauer U, Lasko DR, Fiaux J, Hochuli M, Glaser R, Szyperski T, Wuthrich K, Bailey JE (1999) Metabolic flux ratio analysis of genetic and environmental modulations of Escherichia coli central carbon metabolism. J Bacteriol 181:6679–6688.
95. Schaefer J, Stejskal EO, Beard CF (1975) Carbon-13 nuclear magnetic resonance analysis of metabolism in soybean labeled by $^{13}CO_2$. Plant Physiol 55:1048–1053.
96. Schilling CH, Letscher D, Palsson BO (2000) Theory for the systemic definition of metabolic pathways and their use in interpreting metabolic function from a pathway-oriented perspective. J Theor Biol 203:229–248.
97. Schmidt K, Carlsen M, Nielsen J, Villadsen J. (1997) Modelling isotopomer distribution in biochemical networks using isotopomer mapping matrices. Biotechnol Bioeng 55:831–840.
98. Schmidt K, Marx A, de Graaf AA, Wiechert W, Sahm H, Nielsen J, Villadsen J (1998) ^{13}C tracer experiments and metabolite balancing for metabolic flux analysis: comparing two approaches. Biotechnol Bioeng 58:254–257.
99. Schmidt K, Nørregaard LC, Pedersen B, Meissner A, Duus JO, Nielsen JO, Villadsen J (1999) Quantification of intracellular metabolic fluxes from fractional enrichment and ^{13}C-^{13}C coupling constraints on the isotopomer distribution in labeled biomass components. Metab Eng 1:166–179.
100. Schultz CJ, Coruzzi GM (1995) The aspartate aminotransferase gene family of *Arabidopsis* encodes isoenzymes localized to three distinct subcellular compartments. Plant J 7:61–75.
101. Schuster S, Dandekar T, Fell DA (1999) Detection of elementary flux modes in biochemical networks: a promising tool for pathway analysis and metabolic engineering. Trends Biotechnol 17:53–60.
102. Schwender J (2008) Metabolic flux analysis as a tool in metabolic engineering of plants. Curr Opin Biotechnol 19:131–137.
103. Schwender J, Ohlrogge JB (2002) Probing in vivo metabolism by stable isotope labeling of storage lipids and proteins in developing *Brassica napus* embryos. Plant Physiol 130: 347–361.
104. Schwender J, Ohlrogge JB, Shachar-Hill Y (2003) A flux model of glycolysis and the oxidative pentosephosphate pathway in developing *Brassica napus* embryos. J Biol Chem 278:29442–29453.
105. Schwender J, Goffman F, Ohlrogge JB, Shachar-Hill Y (2004) Rubisco without the Calvin cycle improves the carbon efficiency of developing green seeds. Nature 432:779–782.
106. Schwender J, Ohlrogge J, Shachar-Hill, Y (2004) Understanding flux in plant metabolic networks. Curr Opin Plant Biol 7:309–317.
107. Schwender J, Shachar-Hill Y, Ohlrogge JB (2006) Mitochondrial metabolism in developing embryos of *Brassica napus*. J Biol Chem 281:34040–34047.

108. Shastri AA, Morgan JA (2007) A transient isotopic labeling methodology for ^{13}C metabolic flux analysis of photoautotrophic microorganisms. Phytochemistry 68:2302–2312.
109. Singh BK (1999) in Plant Amino Acids: Biochemistry and Biotechnology (Singh, B. K., ed.) pp. 227–247, Marcel Dekker, New York.
110. Spielbauer G, Margl L, Hannah LC, Romisch W, Ettenhuber C, Bacher A, Gierl A, Eisenreich W, Genschel U (2006) Robustness of central carbohydrate metabolism in developing maize kernels. Phytochemistry 67:1460–1475.
111. Sriram G, Fulton B, Iyer VV, Peterson JM, Zhou R, Westgate ME, Spalding MH, Shanks JV (2004) Quantification of compartmented metabolic fluxes in developing soybean embryos by employing biosynthetically directed fractional ^{13}C labeling, two-dimensional [^{13}C,^{1}H] nuclear magnetic resonance, and comprehensive isotopomer balancing. Plant Physiol 136:3043–3057.
112. Sriram G, Shanks JV (2004) Improvements in metabolic flux analysis using carbon bond labeling experiments: bondomer balancing and Boolean function mapping. Metab Eng 6:116–132.
113. Sriram G, Fulton DB, Shanks JV (2007) Flux quantification in central carbon metabolism of *Catharanthus roseus* hairy roots by ^{13}C labeling and comprehensive bondomer balancing. Phytochemistry 68:2243–2257.
114. Sriram G, Iyera VV, Fulton DB, Shanks JV (2007) Identification of hexose hydrolysis products in metabolic flux analytes: A case study of levulinic acid in plant protein hydrolysate. Metab Eng 9:442–451.
115. Stelling J, Sauer U, Szallasi Z, Doyle FJ 3rd, Doyle J (2004) Robustness of cellular functions. Cell 118:675–685.
116. Stephanopoulos G, Vallino JJ (1991) Network rigidity and metabolic engineering in metabolite overproduction. Science 252:1675–1681.
117. Stephanopoulos GN, Aristidou AA, Nielsen J (1998) Metabolic Engineering: Principles and Methodologies. Academic Press, San Diego.
118. Stitt M, ap Rees T (1980) Estimation of the activity of the oxidative pentose phosphate pathway in pea chloroplasts. Phytochemistry 19:1583–1585.
119. Studart-Guimarães C, Fait A, Nunes-Nesi A, Carrari F, Usadel B, Fernie AR (2007) Reduced expression of succinyl-coenzyme A ligase can be compensated for by up-regulation of the gamma-aminobutyrate shunt in illuminated tomato leaves. Plant Physiol 145: 626–639.
120. Suthers PF, Burgard AP, Dasika MS, Nowroozi F, Van Dien S, Keasling JD, Maranas CD (2007) Metabolic flux elucidation for large-scale models using ^{13}C labeled isotopes. Metab Eng 9:387–405.
121. Szyperski T (1995) Biosynthetically directed fractional ^{13}C-labeling of proteinogenic amino acids. An efficient analytical tool to investigate intermediary metabolism. Eur J Biochem 232:433–448.
122. Szyperski T, Glaser RW, Hochuli M, Fiaux J, Sauer U, Bailey JE, Wuthrich K (1999) Bioreaction network topology and metabolic flux ratio analysis by biosynthetic fractional ^{13}C labeling and two-dimensional NMR spectroscopy. Metab Eng 1:189–197.
123. Troufflard S, Roscher A, Thomasset B, Barbotin JN, Rawsthorne S, Portais JC (2007) In vivo ^{13}C NMR determines metabolic fluxes and steady state in linseed embryos. Phytochemistry 68:2341–2350.
124. Vandewalle I, Olsson R (1983) The gamma-aminobutyric acid shunt in germinating *Sinapsis alba* seeds. Plant Sci Lett 31:269–273.
125. van Winden WA, Heijnen JJ, Verheijen PJ, Grievink J (2001) A priori analysis of metabolic flux identifiability from (13)C-labeling data. Biotechnol Bioeng 74:505–516.
126. van Winden W, Verheijen P, Heijnen S (2001) Possible pitfalls of flux calculations based on (13)C-labeling. Metab Eng 3:151–162.
127. Van Winden WA, Heijnen JJ, Verheijen PJT (2002) Cumulative bondomers: a new concept in flux analysis from 2D [^{13}C, ^{1}H] COSY NMR data. Biotechnol Bioeng 80:731–745.

128. van Winden WA, van Dam JC, Ras C, Kleijn RJ, Vinke JL, van Gulik WM, Heijnen JJ (2005) Metabolic-flux analysis of Saccharomyces cerevisiae CEN.PK113-7D based on mass isotopomer measurements of (13)C-labeled primary metabolites. FEMS Yeast Res 5: 559–568.
129. Varma A, Palsson BO (1994) Metabolic flux balancing: Basic concepts, scientific, and practical use. Nature Biotechnol 12:994–998.
130. Viola R, Davies HV, Chudeck AR (1991) Pathways of starch and sucrose biosynthesis in developing tubers of potato (*Solanum tuberosum* L.) and seeds of faba bean (*Vicia faba* L.). Elucidation by ^{13}C-nuclear-magnetic-resonance spectroscopy. Planta 183: 202–208.
131. Weitzel M, Wiechert W, Nöh K (2007) The topology of metabolic isotope labeling networks. BMC Bioinf 8:315.
132. Whitfield HV, Murphy DJ, Hills MJ (1993) Subcellular-localization of fatty-acid elongase in developing seeds of *Lunaria annua* and *Brassica napus*. Phytochemistry 32:255–258.
133. Wiechert W (2001) ^{13}C-Metabolic flux analysis. Metabol Eng 3:195–206.
134. Wiechert W. (2002) An introduction to ^{13}C metabolic flux analysis. Genet Eng (NY) 24:215–238.
135. Wiechert W (2007) The thermodynamic meaning of metabolic exchange fluxes. Biophys J 93:2255–2264.
136. Wiechert W, De Graaf AA (1997) Bidirectional reaction steps in metabolic networks: I. Modeling and simulation of carbon isotope labeling experiments. Biotechnol Bioeng 55:101–117.
137. Wiechert W, Siefke C, de Graaf A, Marx A (1997) Bidirectional reaction steps in metabolic networks: II. Flux estimation and statistical analysis. Biotechnol Bioeng 55:118–135.
138. Wiechert W, Möllney M, Isermann N, Wurzel M, de Graaf AA (1999) Bidirectional reaction steps in metabolic networks: III. Explicit solution and analysis of isotopomer labeling systems. Biotechnol Bioeng. 66:69–85.
139. Wiechert W, Mollney M, Petersen S, deGraaf AA (2001) A universal framework for ^{13}C-Metabolic flux analysis. Metabol Eng 3:265–283.
140. Wilkie SE, Warren MJ (1998) Recombinant expression, purification, and characterization of three isoenzymes of aspartate aminotransferase from *Arabidopsis thaliana*. Protein Expression Purif 12:381–389.
141. Williams TCR, Miguet L, Masakapalli SK, Kruger NJ, Sweetlove LJ, Ratcliffe RG (2008) Metabolic network fluxes in heterotrophic *Arabidopsis* cells: stability of the flux distribution under different oxygenation conditions. Plant Physiol, doi:10.1104/pp.108.125195.
142. Wittmann C, Heinzle E (2002) Genealogy profiling through strain improvement by using metabolic network analysis: metabolic flux genealogy of several generations of lysine-producing corynebacteria. Appl Environ Microbiol 68:5843–5859.
143. Yang C, Hua Q, Shimizu K (2002) Metabolic flux analysis in Synechocystis using isotope distribution from C-13-labeled glucose. Metab Eng 4:202–216.
144. Yang C, Hua Q, Shimizu K (2002) Quantitative analysis of intracellular metabolic fluxes using GC-MS and two-dimensional NMR spectroscopy. J Bioscience Bioeng 93: 78–87.
145. Yang TH, Frick O, Heinzle E (2008) Hybrid optimization for 13C metabolic flux analysis using systems parametrized by compactification. BMC Syst Biol 2:29.
146. Young JD, Walther JL, Antoniewicz MR, Yoo H, Stephanopoulos G (2008) An elementary metabolite unit (EMU) based method of isotopically nonstationary flux analysis. Biotechnol Bioeng 99:686–699.
147. Zamboni N, Sauer U (2004) Model-independent fluxome profiling from ^{2}H and ^{13}C experiments for metabolic variant discrimination. Genome Biol 5:R99.
148. Zamboni N, Sauer U (2005) Fluxome profiling in microbes. In: Metabolome Analysis: Strategies for Systems Biology (Vaidayanathan, S., Harrigan. G. G., and Goodacre, R., eds) pp. 307–322, Springer, New York.

149. Zamboni N, Fischer E, Sauer U (2005) FiatFlux – a software for metabolic flux analysis from ^{13}C-glucose experiments. BMC Bioinformatics 6:209.
150. Zhao J, Shimizu K (2003) Metabolic flux analysis of Escherichia coli K12 grown on ^{13}C-labeled acetate and glucose using GC-MS and powerful flux calculation method. J Biotechnol 101:101–117.

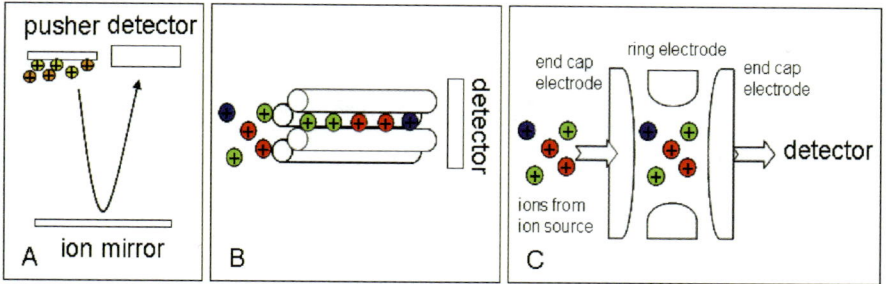

Fig. 3.6 Schematics of (**A**) time-of-flight, (**B**) quadrupole, and (**C**) ion trap mass analyzers

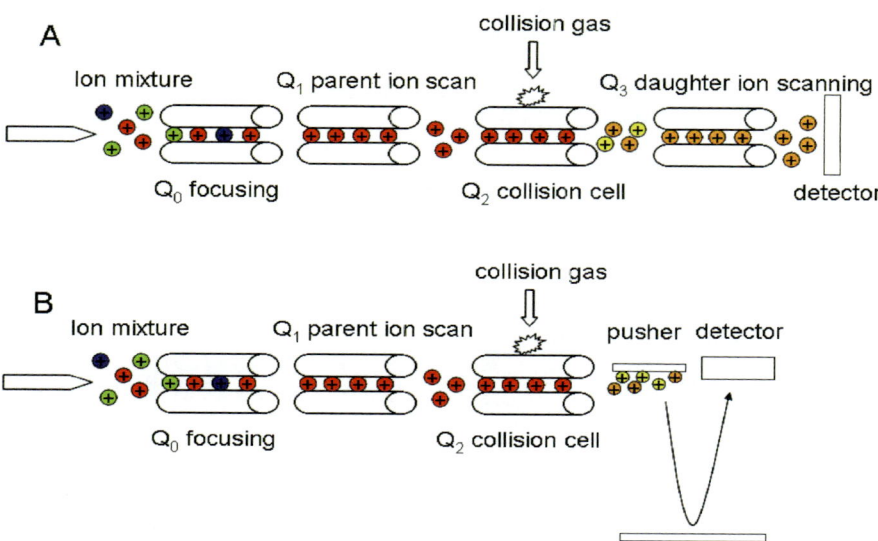

Fig. 3.7 Schemata of tandem MS/MS. (**A**) Triple quadrupole mass spectrometer (QqQ) and (**B**) quadrupole time-of-flight mass spectrometer (QqTOF)

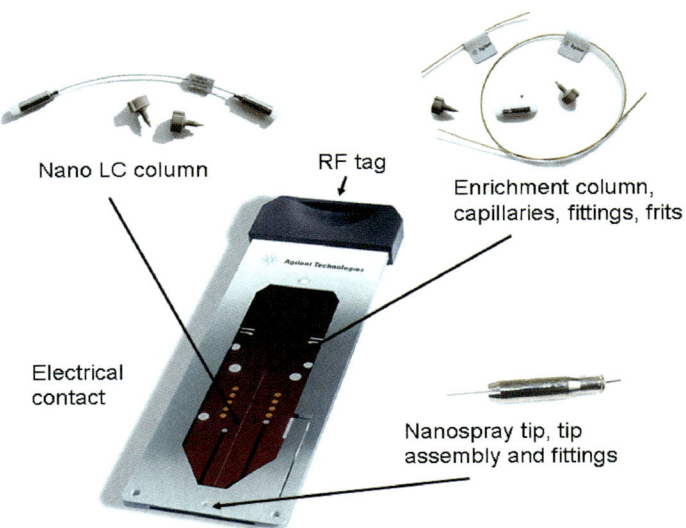

Fig. 3.9 HPLC-Chip and its integrated components related to the components of conventional LC (picture provided in courtesy of Agilent Technologies, Inc.)

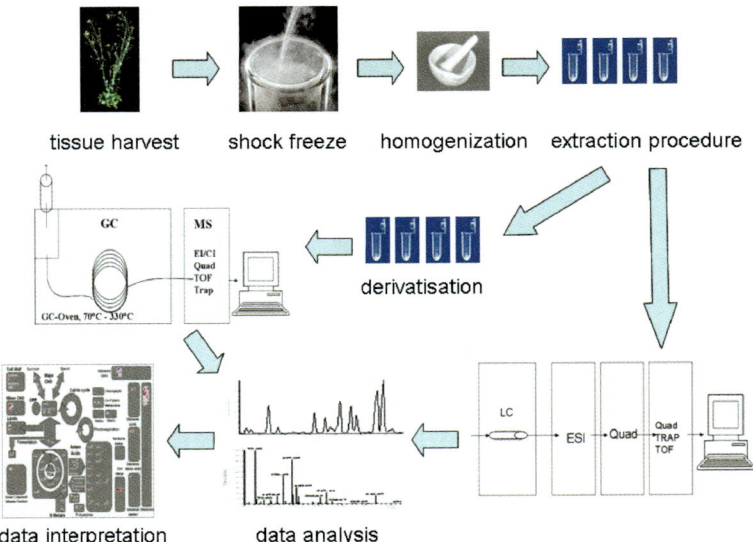

Fig. 3.10 Schematic workflow of a metabolomics approach from tissue harvest to data interpretation using complementary analytical instrumentation for greater comprehensiveness of metabolite detection and quantification

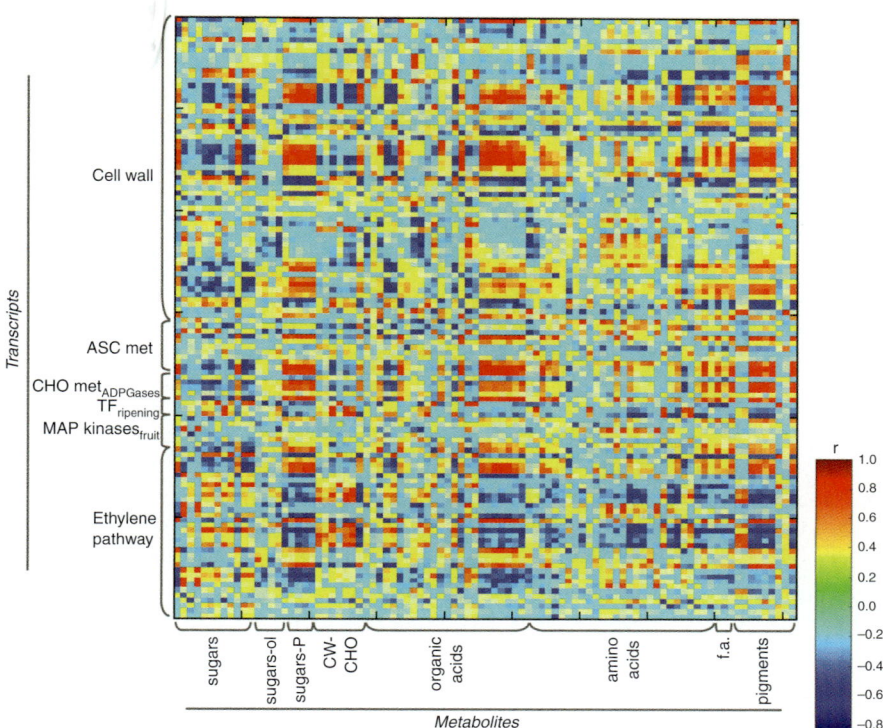

Fig. 6.2 Selected transcript–metabolite correlation visualizations. Heat map surface of selected transcript–metabolite correlations was drawn, and correlation coefficients were calculated. Each dot indicates a given *r* value resulted from a Spearman correlation analysis in a false color scale. Asc met, ascorbate metabolism; CHO met, carbohydrate metabolism; CW, cell wall; sugars-ol, sugar alcohols; sugars-P, sugar phosphates; f.a., fatty acids; TFs, transcription factors [63]

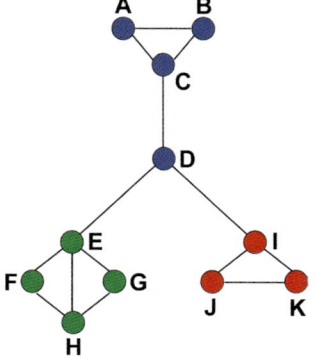

Fig. 7.14 Small hypothetical network (adapted from [92])

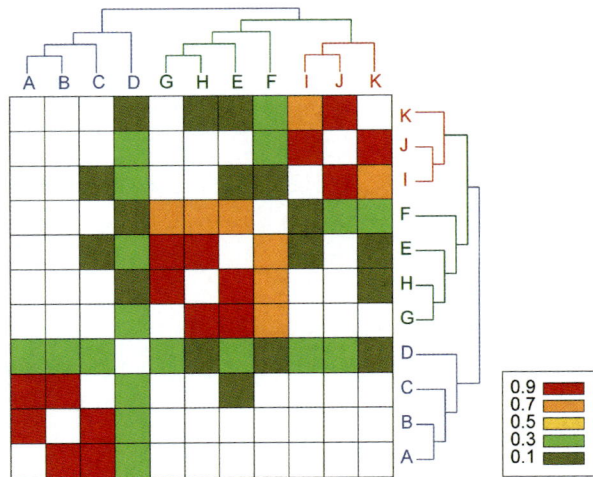

Fig. 7.15 The topological overlap matrix (adapted from [92]) corresponding to the network shown in Fig. 7.13. The topological overlap between each pair of vertices v_i and v_j is defined according to [92] as $O_T(v_i, v_j) = \frac{J_n(v_i,v_j)}{[\min(d(v_i),d(v_j))]}$, where $J_n(v_i, v_j)$ denotes the number of vertices to which both v_i and v_j are linked and $[min(d(v_i), d(v_j))]$ is the smaller of the degrees of v_i and v_j. The degree of topological overlap between the vertices is reflected by the color code. The associated hierarchical tree is obtained by applying an average linkage clustering algorithm to the topological overlap matrix. The given tree reflects three distinct modules contained in the network of Fig. 7.14

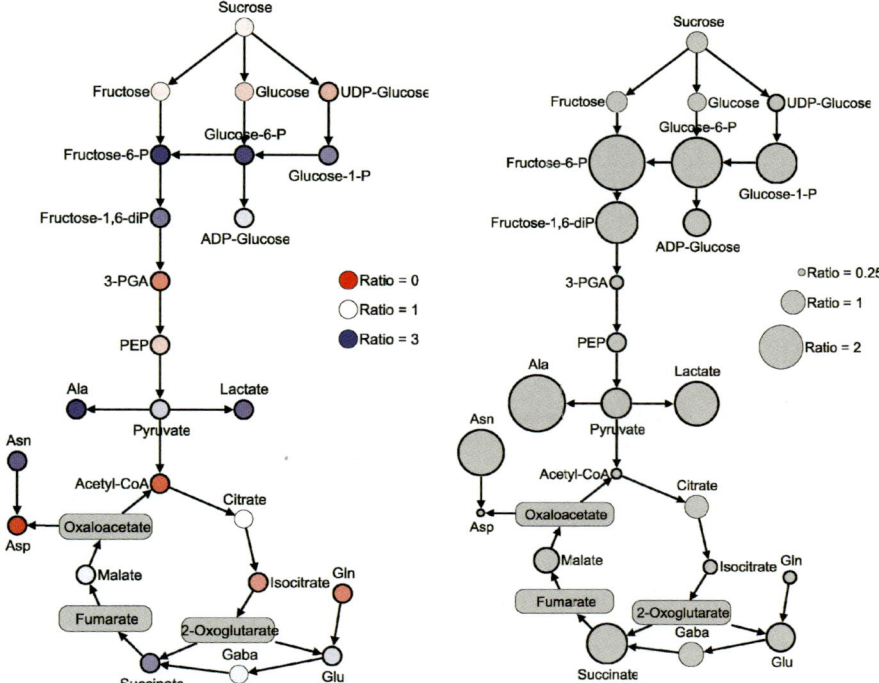

Fig. 7.16 *Left*: Color-coded visualization of the ratio of experimental data from two different environmental conditions (reduced oxygen supply condition data divided by data from normal oxygen supply condition, source of data: [95]). *Right*: Shape-coded visualization of the same data

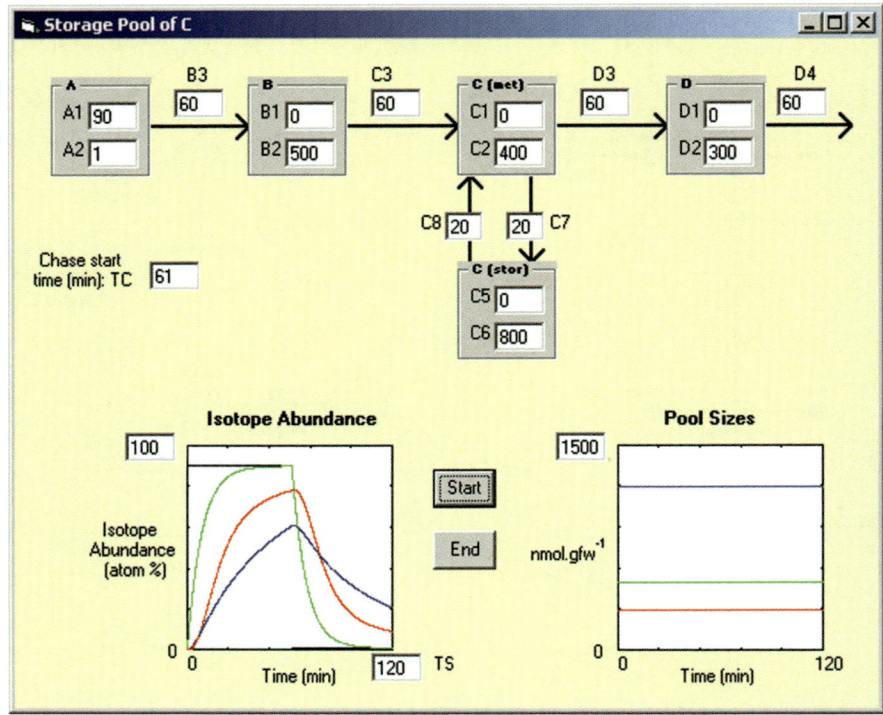

Fig. 10.3 A hypothetical reaction network including both a metabolic (*C* (met)) and storage (*C*(stor)) pool for compound C. The labeling kinetics of the network and pool sizes of intermediates when pulsed with 90% labeled precursor A, then chased with 1% labeled A at 61 minutes, are shown. For pool C, the average labeling of the two pools of $C = ((C1*C2) + (C5*C6))/(C2 + C6)$, and the total pool size of $C = (C2 + C6)$ are shown. Hypothetical network created with Microsoft Visual Basic

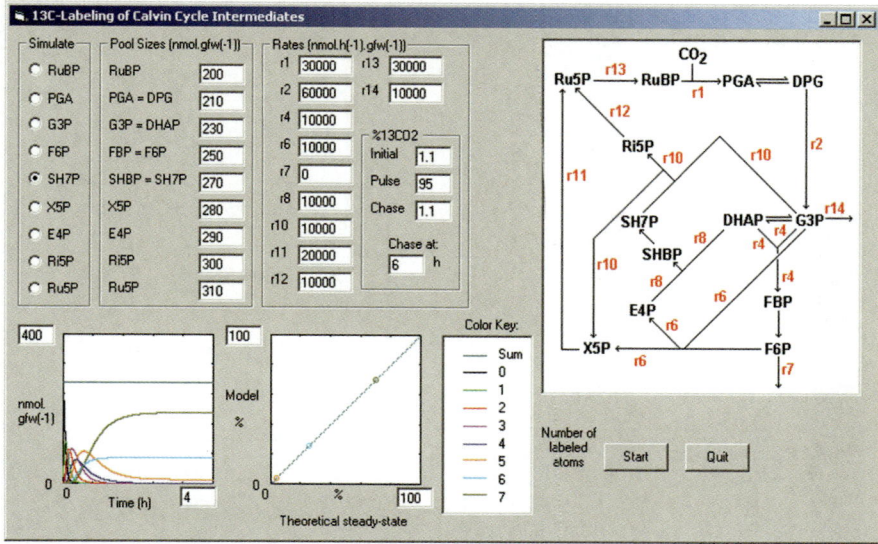

Fig. 10.9 Simulation of the time course of labeling of the Calvin cycle intermediate sedoheptulose-7-phosphate (SH7P) when a photosynthetic tissue is supplied with $^{13}CO_2$ of 95∗ ^{13}C abundance (i.e., $C = 0.95$), using the rates and pool size assumptions, is shown. The simulated time course of absolute pool sizes of SH7P species with 0, 1, 2, 3, 4, 5, 6, or 7 labeled atoms and the sum of all these species are shown in the *lower left panel*. In the *lower right panel*, the model-simulated output at 4 h is compared with the theoretical steady-state percentages of SH7P species with up to 7 labeled atoms calculated assuming uniform labeling as follows:

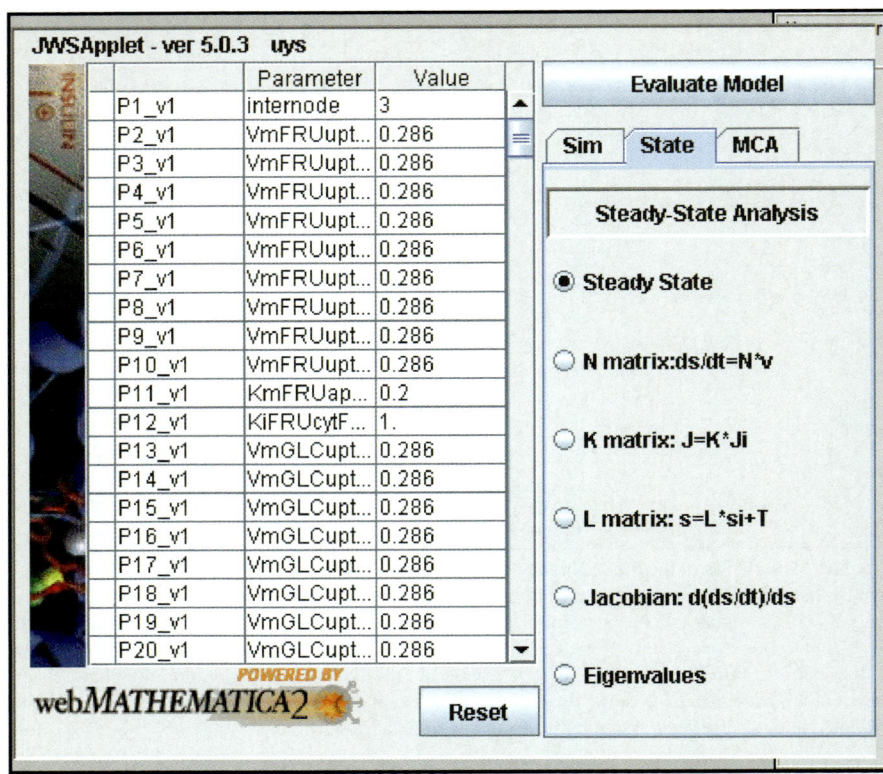

Fig. 11.1 Screenshot of the online sugar cane kinetic model of Uys et al. [44]. In the left window, parameter can be changed. The time-course simulation (Sim) or the steady state of the model can be determined and evaluated (State). Results can be downloaded into text format or Microsoft EXCEL© spreadsheet format. Metabolic control analysis (MCA) is not functional in this particular model. ([20], http://jjj.biochem.sun.ac.za/database/uys/index.html)

Chapter 10
Application of Dynamic Flux Analysis in Plant Metabolic Networks

Amy J.M. Colón, John A. Morgan, Natalia Dudareva, and David Rhodes

10.1 Introduction

Metabolic networks are composed of metabolic pathways that consist of biochemical reaction steps. The functionality of pathways within a network is determined by metabolic fluxes, defined as the amount of converted metabolite per unit of time, per cell (or per unit mass of tissue), through each biochemical step within a metabolic pathway. The flux through a pathway depends on the kinetic properties of enzymes, as well as on their cellular levels and activities, which are regulated by gene expression, posttranscriptional, translational, and/or posttranslational modifications, and enzyme stability. Metabolic flux analysis (MFA) is a tool that has traditionally been used in microbial systems to assess the effects of environmental and targeted genetic changes on in vivo rates of metabolites synthesis. MFA can be used to gauge the degree of success of specific genetic interventions aimed at diverting metabolic flux to desirable products. The techniques of MFA have therefore become important tools in metabolic engineering and systems biology. To fully understand how a cell functions, analysis needs to include not only a description of its molecular parts, which can be obtained from molecular biology methods, but also a description of flux distributions within the complex and dynamic metabolic networks [26]. As will be discussed in the next section, MFA via computer modeling of metabolism can provide information about the proximity of certain compounds to one another, the contribution of a pathway or part of a pathway to end products, the existence of storage pools, and regulation and reversibility of reactions.

The established steady-state metabolic flux analysis methods used in microbial systems are difficult to apply to higher plant metabolic systems due to large, slowly turning-over metabolite pools, a high degree of compartmentation, and duplicate pathways located in different organelles (see Chapter 9). Furthermore, steady-state

---

D. Rhodes (✉)
Department of Horticulture, Purdue University, West Lafayette, IN, USA
e-mail: drhodes@purdue.edu

methods cannot be readily applied to analysis of fluxes in tissues that are undergoing marked diurnal variations in metabolic activities. Flux analysis methods applicable over relatively short time periods (i.e., hours and minutes) are required to obtain snapshots of metabolic processes in these systems. Dynamic flux analysis is a suitable strategy, and essentially this entails supplying a labeled precursor of known isotopic abundance, to the tissue or cells, and monitoring the labeling patterns of intermediates and end product over time. However, a disadvantage of dynamic flux analysis when compared to steady-state methods is that dynamic flux analysis requires multiple samples and measurements of intracellular metabolite concentrations.

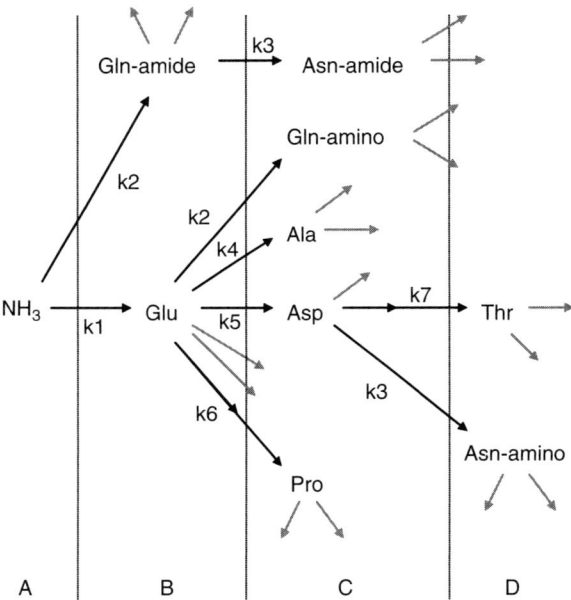

Fig. 10.1 Metabolic pathway of ammonia assimilation in yeast (*Candida utilis*) investigated by Sims and Folkes [28]. NH_3 is assimilated into glutamate (Glu) in the reaction catalyzed by $NADP^+$-glutamate dehydrogenase ($k1$); it is also incorporated into the amide group of glutamine (Gln-amide) in the reaction catalyzed by glutamine synthetase ($k2$). Note that the rate of formation of Gln-amide must be the same as the rate of formation of the amino nitrogen moiety of Gln (Gln-amino) from Glu ($k2$). Glu has numerous metabolic fates, only three of which are shown here: transamination to alanine (Ala) ($k4$), transamination to aspartate (Asp) ($k5$), and conversion to proline (Pro) via negligible pools of glutamyl-5-phosphate, glutamic-5-semialdehyde, and Δ^1-pyrroline-5-carboxylate ($k6$). Asp is further metabolized to threonine (Thr) (via negligible pools of intermediates) ($k7$) and is incorporated into the amino moiety of asparagine (Asn-amino), at the same rate as Gln-amide serves as the amide donor for asparagine in the reaction catalyzed by asparagine synthetase ($k3$). Other metabolic fates of amino acids include incorporation into amino acid residues of protein (not shown). If NH_3 is supplied in labeled form (**A**), then Glu and Gln-amide will behave as primary products (**B**); Gln-amino, Asn-amide, Ala, Asp, and Pro will behave as secondary products (**C**); and Thr and Asn-amino will behave as tertiary products

A classic early example of dynamic flux analysis was its use in yeast to determine fluxes via the amino acid synthesis network in an exponentially growing yeast culture, pulsed with (^{15}NH$_4$)$_2$HPO$_4$ for 30 min, taking samples at 5, 15, and 30 min after transfer. The culture was then chased with unlabeled (NH$_4$)$_2$HPO$_4$ for 30 minutes, again taking samples at 5, 15, and 30 min after the transfer [28]. This labeling study over a time course revealed that glutamic acid and glutamine-amide are the only amino acids that derive their nitrogen directly from ammonia (Fig. 10.1). All of the other amino acids apparently derive their amino N from glutamic acid or glutamine-amide. This was the first study of its kind, and it laid the foundation for dynamic labeling studies. Even though this study was conducted in yeast, we will show in this chapter that the basic principles described by Sims and Folks [28] can be applied to dynamic flux analysis in plant systems.

MFA by dynamic isotopic labeling is a powerful tool for pathway discovery as well as the study of pathway regulation as has been demonstrated in the study of choline synthesis in tobacco [14], the phenylpropanoid pathway in potato tubers [10, 11], and the benzenoid/phenyl-propanoid network in petunia [4, 16]. In this chapter, we will present examples of the use of dynamic flux analysis in plants, discuss different labeling strategies, and finally discuss some of the advantages and disadvantages of this approach compared to a steady-state approach.

10.2 Dynamic Flux Analysis for the Determination of Precursor–Product Relationships in Plants

The essence of dynamic flux analysis is the quantification of the labeling patterns of intermediates and end products of a metabolic pathway in a time course, following the supply (pulse) or removal (chase) of a labeled precursor of known stable isotope abundance or specific radioactivity. Metabolic pool sizes must also be experimentally determined. Ideally, the system should be in metabolic steady state; that is, the rate of synthesis of each compound should be equal to its combined rates of utilization, thus maintaining constant metabolic pool sizes with respect to time. Isotopic labeling of the intermediates and end products can be determined by mass spectrometry (MS) or nuclear magnetic resonance (NMR) spectrometry in the case of stable isotopes, or by scintillation counting of purified intermediates/end products in the case of radioisotopes. Various mathematical and/or computational tools must then be applied to the data to derive metabolic fluxes. We will discuss different approaches to this in subsequent sections.

10.2.1 An Early Example of Dynamic Flux Analysis

As mentioned earlier, Sims and Folkes [28] were the first to use dynamic flux analysis in an elegant study of the time course of labeling of free amino acids and protein amino acids in an exponentially growing culture of yeast, pulsed with ^{15}NH$_4^+$ for 30 min, and then chased with unlabeled NH$_4^+$ for a further 30 min. In this study, they describe differential equations and their solutions for calculating the isotope abundance of intermediates in the amino acid biosynthesis pathways, when the organism is growing exponentially. In this approach, unknown rate values were progressively adjusted until a close fit of simulated to measured data was achieved. This approach is most suitably applied to a fairly uncharacterized pathway where little is known about precursor–product relationships. Their model system consists of branching reaction chains along which materials and isotope pass from one component to the next (Fig. 10.2).

The base equations used by Sims and Folkes assume that the isotopic abundance of any component in the system will depend on its own rate of synthesis as well as the rates of synthesis of each of the components upstream of it in the reaction network. Using a series of differential equations, they derived kinetic equations that describe the labeling patterns of each intermediate as a function of time, pool sizes, and rates of synthesis. A constraint of these equations is that the rate of synthesis must equal the rate of utilization; for example in Fig. 10.2, B3 must equal B4 (the combined utilization fluxes from compound B), and thus the pool sizes of intermediates remain constant with time.

The kinetic equations that describe the labeling patterns of compounds B, C, and D (during the pulse phase) in the hypothetical network shown in Fig. 10.2 are as follows:

$$B' = A' \left(1 - e^{-bt}\right)$$

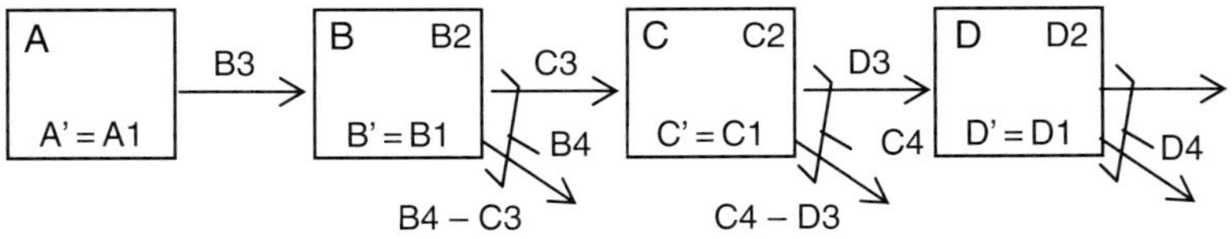

Isotope abundance at t = 0h (atom %)	Pool Sizes (nmol·g FW^{-1})	Rates (nmol·h^{-1}·g FW^{-1})	Turnover con-stants (h^{-1})
A' = A1 = 90%	B2 = 500	B3 = B4 = 5000	b = 10
B' = B1 = 0%	C2 = 500	C3 = C4 = 2000	c = 4
C' = C1 = 0%	D2 = 500	D3 = D4 = 1000	d = 2
D' = D1 = 0%			

Fig. 10.2 A simplified reaction network similar to that described by Sims and Folkes [28], where A is the isotopically labeled precursor ($^{15}NH_4^+$) and B, C, and D are metabolites (amino acids), as indicated in Fig. 10.1. Important parameters for simulating the labeling patterns of intermediates in this metabolic scheme are isotopic abundances at zero time, pool sizes of intermediates, and their rates of synthesis and utilization. If the rate of synthesis of an intermediate is equal to its combined rate of utilization, then the pool of an intermediate will remain constant with time (t), and the labeling of the pool will become a function of the pool's turnover constant (rate of synthesis/pool size) (see text for further details)

$$C' = A' \left(\frac{c\left(1 - e^{-bt}\right) - b\left(1 - e^{-ct}\right)}{c - b} \right)$$

$$D' = A' \left(\frac{cd(c-d)\left(1 - e^{-bt}\right) - bd(b-d)\left(1 - e^{-ct}\right) + bc(b-c)\left(1 - e^{-dt}\right)}{(c-d)(b-d)(b-c)} \right)$$

where

- A' = isotope abundance of precursor (atm%)
- B' = isotope abundance of pool B (atm%)
- C' = isotope abundance of pool C (atm%)
- D' = isotope abundance of pool D (atm%)
- b = turnover constant of pool B (h^{-1}) [rate of synthesis·pool size^{-1}])
- c = turnover constant of pool C (h^{-1}) [rate of synthesis·pool size^{-1}])
- d = turnover constant of pool D (h^{-1}) [rate of synthesis·pool size^{-1}])
- e = exponential function
- t = time (h)

In this series of equations, the isotopic abundance of compound B (B') during the pulse phase is only dependent on the isotopic abundance of its precursor A (A') and its own pool size and rate of synthesis. A convenient expression of both pool size and rate of synthesis (which must equal rate of utilization in a metabolic steady state) is the turnover constant for the pool. Thus, the labeling of compound B (B') is dependent on the isotopic abundance of A (A') and the turnover constant for pool B (b). The isotopic abundance of compound C (C') is dependent on A', b, and c. The isotopic abundance of compound D is dependent on A', b, c, and d. As can be seen, for every compound further along in the reaction chain, the kinetic equation becomes increasingly complex. Different kinetic equations must be used for the chase phase (not shown), but these utilize the same turnover constants [28].

Sims and Folkes systematically adjusted the turnover constants until a satisfactory fit to the experimental data was obtained for all measured compounds. Since the turnover constant of a pool is the rate of synthesis divided by pool size, and since the pool size is experimentally determined, then in essence, this gives rise to the best estimate of synthesis rate (flux). This is similar to the strategy described more recently in Stephanopoulos et al. [29], where iterative adjustments in fluxes are employed until a satisfactory fit between experimentally determined and computed isotopic labeling is achieved.

An alternative approach to simulating the labeling behavior of intermediates in the pathway described by Sims and Folkes [28] is to use a simple iterative computer model described at www.hort.purdue.edu/cfpesp/models/models.htm. This approach is easily applied to fully characterized networks. The output from this approach is virtually identical to that calculated from the kinetic equations of Sims and Folkes:

```
    z = Mx/2000
    For t = 0 To Mx Step z
    B1 = (B1 * B2 + A1 * z * B3)/( B2 + z * B3)
    B2 = B2 + z * B3
    B1 = (B1 * B2 - B1 * z * B4)/( B2 - z * B4)
    B2 = B2 - z * B4
    C1 = (C1 * C2 + B1 * z * C3)/( C2 + z * C3)
    C2 = C2 + z * C3
    C1 = (C1 * C2 - C1 * z * C4)/( C2 - z * C4)
    C2 = C2 - z * C4
    D1 = (D1 * D2 + C1 * z * D3)/( D2 + z * D3)
    D2 = D2 + z * D3
    D1 = (D1 * D2 - D1 * z * D4)/( D2 - z * D4)
    D2 = D2 - z * D4
```

[subroutine to plot $A1$, $B1$, $C1$, and/or $D1$ as a function of t] Next t
where

$A1$ = isotope abundance of precursor (atm%) [supplied]
$B1$ = isotope abundance of pool B (atm%)
$C1$ = isotope abundance of pool C (atm%)
$D1$ = isotope abundance of pool D (atm%)
$B2$ = pool size of pool B (nmol·gfw^{-1})
$C2$ = pool size of pool C (nmol·gfw^{-1})
$D2$ = pool size of pool D (nmol·gfw^{-1})
$B3$ = rate of synthesis of pool B (nmol·h^{-1}·gfw^{-1})
$C3$ = rate of synthesis of pool C (nmol·h^{-1}·gfw^{-1})
$D3$ = rate of synthesis of pool D (nmol·h^{-1}·gfw^{-1})
$B4$ = combined rate of utilization of pool B (nmol·h^{-1}·gfw^{-1})
$C4$ = combined rate of utilization of pool C (nmol·h^{-1}·gfw^{-1})
$D4$ = combined rate of utilization of pool D (nmol·h^{-1}·gfw^{-1})
t = time (h)
z = iteration interval (in this case 1/2000th of total time, Mx = scale of X-axis = 1 h)

Or in more general terms,

$$f_x = \frac{f_x C_x + \sum_{i=1}^{k} f_i^{in} v_i^{in} dt - \sum_{j=1}^{l} f_j^{out} v_j^{out} dt}{C_x + \sum_{i=1}^{k} v_i^{in} dt - \sum_{j=1}^{l} v_j^{out} dt}$$

$$C_x = C_x + \sum_{i=1}^{k} v_i^{in} dt - \sum_{j=1}^{l} v_j^{out} dt$$

where x represents the metabolic pool with fractional enrichment f_x and pool size C_x, and where k represents the input fluxes and l the output fluxes. As an iterative computation, dt would be a small iteration interval.

Through each iteration, or short-time interval (z), material of known isotope abundance is drawn into each pool, mixed, and withdrawn from each pool at specified rates, and new isotope abundance and pool size are calculated and used in the subsequent iteration.

An advantage of the iterative model over the Sims and Folkes kinetic equations is that it can be used in cases where a metabolic steady state is not maintained, i.e., where rates of synthesis and utilization are not equal, and where pools are allowed to expand and deplete with time. The iterative model can be readily applied to consider multiple pools of intermediates with different turnover rates (to be discussed subsequently). Another advantage of the iterative computer modeling approach is that the complex kinetic equations of Sims and Folkes can be replaced with just four short lines of code for each metabolic pool in the reaction network, and chase kinetics can be readily simulated by adding a single line of code.

Often data from a labeling study will reveal a discrepancy between experimental data and what is known about the metabolite sequence in a pathway. For example, labeling data from the Sims and Folkes study revealed some inconsistency in isotopic abundance of aspartic acid, lysine, and histidine and their corresponding protein amino acid residues. Data suggested that only part of the free aspartate amino acid pool is available as an intermediate in protein synthesis because a portion of it is spatially separated in a storage pool, perhaps localized in the vacuole. The storage pool may operate as part of a cellular regulatory mechanism. It was necessary to invoke a storage pool in addition to the metabolically active pool of these amino acids, in order to achieve a good fit for labeling in both free and protein-bound amino acids with a single turnover constant.

The effects of a storage pool on labeling patterns can easily be visualized using the iterative computer model approach (Fig. 10.3). According to the graphical plot of isotopic abundances and pool sizes over a 120-min pulse-chase time course, it appears as though D is more heavily labeled than the bulk pool of C (the sum of C(met) and C(stor)) because of isotope dilution of the metabolic pool of C (C(met)) with its storage pool (C(stor)), which remains mostly unlabeled. From this observation, one would be tempted to assume that D would precede C in the reaction network. Moreover, during the chase phase (Fig. 10.3), label does not flow out of the bulk pool of C as rapidly as the pool of D, again suggesting that it does not precede D in the metabolic pathway. Thus, substantial care should be given to interpreting the labeling time courses in terms of precursor–product relationships. The precise labeling patterns of intermediates are critically dependent on relative sizes of the metabolic and storage pools and the flux rates between them.

The Sims and Folkes approach has been used as a framework to determine isotopic abundance of plant metabolites and quantify flux, as has been successfully performed in a number of plant systems including analysis of ^{15}N-labeling of amino acids in *Lemna minor* [21], tomato [19], and cowpea [13] cell cultures supplied

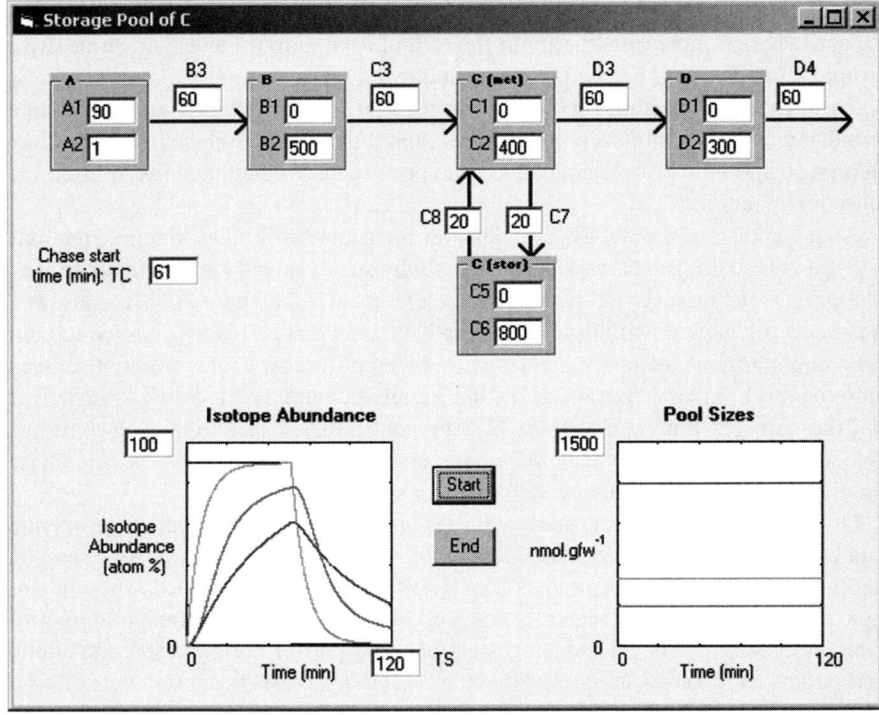

Fig. 10.3 A hypothetical reaction network including both a metabolic (C (met)) and storage (C(stor)) pool for compound C. The labeling kinetics of the network and pool sizes of intermediates when pulsed with 90% labeled precursor A, then chased with 1% labeled A at 61 minutes, are shown. For pool C, the average labeling of the two pools of $C = ((C1*C2) + (C5*C6))/(C2 + C6)$, and the total pool size of $C = (C2 + C6)$ are shown. Hypothetical network created with Microsoft Visual Basic (*see* also Color Insert)

with $^{15}NH_4^+$. It was further extended to consider ^{13}C-labeling of amino acids in *Lemna* (Colón and Rhodes, unpublished data), potato [23], and cultured rice [12] and *Arabidopsis* [3] cells; deuterium labeling of benzenoid intermediates in petunia flowers supplied with deuterium-ring-labeled phenylalanine ($^2H_5$-Phe) [4, 16]; and radioisotope labeling of various metabolic pathways in excised leaf discs of tobacco [14]. We will discuss a few of these examples in subsequent sections.

10.2.2 Dynamic Flux Analysis in Plant Systems: Examples with $^{15}NH_4^+$

In our first plant example, the kinetic equations developed by Sims and Folkes were modified for use in the computer simulation of isotopic abundance of amino acids in the simple plant *Lemna minor* (duckweed) [21]. In this simulation program, the fluxes via each amino acid moiety were estimated using the iterative approach described earlier. The output of the program was in graphical form where curve fitting was performed, and best fit was visually determined. Through the use of the iterative computer simulation, the authors were able to test a series of models to determine the main pathway of ammonia assimilation in *Lemna*. The kinetic experiments in conjunction with ^{15}N-ammonia labeling revealed that at least two intracellular compartments are involved. The chloroplastic compartment contained a glutamine synthetase-glutamate synthase (GS-GOGAT) cycle, and the cytosolic compartment contained a second site of glutamine synthesis. Glutamine-amide N was the first labeled product of ammonia assimilation. Unlike in yeast, glutamate behaved as a product of glutamine-amide rather than as the primary product of ammonia assimilation.

Another example using the concept outlined above was a study comparing the assimilation of ^{15}N into amino acids of water stressed and non-stressed tomato cell cultures [19]. Here, the specific objective was to quantify the amount of N incorporated into each of the amino acids, and to assess the flux changes associated with adaptation to water deficits resulting in a massive accumulation of proline in the tomato cell cultures. Labeling studies with $^{15}NH_4^+$ revealed that this proline accumulation was due to (i) an increase in proline synthesis from its precursor glutamate, (ii) a decreased utilization of proline and other amino acids in protein synthesis, and (iii) an overall decrease in the specific growth rate of the cell culture.

A similar study was conducted on heat-shocked cowpea cell cultures [13]. Here, a large accumulation of γ-aminobutyrate (GABA) was observed in response to heat shock. Labeling with $^{15}NH_4^+$ revealed that the increase in GABA level was due to a marked and rapid increase in biosynthesis rate from glutamate. These results suggested activation of glutamate decarboxylase in response to heat shock. This enzyme was later shown to be activated by Ca^{2+}-calmodulin in plants [8].

The use of N(O,S)-heptaflourobutyryl isobutyl derivatives (HFBI) of amino acids, combined with GC-MS [9, 13, 19], greatly simplified the process and sensitivity of ^{15}N determination of specific nitrogen moieties of amino acids in comparison to earlier methods involving the isolation of each amino acid, conversion of the amino or amide moiety to ammonia, ammonia steam distillation, and conversion to nitrogen gas for ^{15}N analysis by MS [21, 28]. While the amide groups of glutamine and asparagine are lost during HFBI derivatization, the use of *N*-(t-butyldimethylsilyl)-*N*-methyltrifluoroacetamide (MTBSTFA) derivatives [20] preserves these acid labile moieties. Monitoring of the molecular ions of the MTBSTFA amide derivatives therefore permits determination of the relative abundances of unlabeled, single ^{15}N-labeled and double ^{15}N-labeled asparagine and glutamine species. These ratios can be very informative about the occurrence of multiple metabolic pools [20]. Because the amino group of glutamine and asparagine can be specifically determined by HFBI derivatization, the amide ^{15}N abundances can be computed from the relative abundances of unlabeled, single ^{15}N-labeled and double ^{15}N-labeled asparagine and glutamine MTBSTFA species, and amino ^{15}N abundance [20].

10.3 Examples of Dynamic Flux Analysis of Whole Plant Organs

10.3.1 Choosing an Appropriate Labeled Substrate

In the above examples, we have only considered ^{15}N studies in plant cell cultures, but the approach of Sims and Folkes can be applied to ^{13}C metabolism of whole plant organs as well, in spite of some challenges that may arise. For example, labeling experiments with ^{13}C can result in multiply labeled C atoms in a single compound. The number of combinations of positionally labeled carbons is described by the concept of isotopomers (isomers of isotopically labeled compounds). Labeling with ^{13}C can result in up to 2^n isotopomers, where n is the number of C atoms present in a compound (for a more detailed discussion on C isotopomers see Chapter 9 and [17, 33, 35]). Consideration of all isotopomers involves complex mathematical equations. A new approach that dramatically reduces the computational difficulty of the analysis of labeling by single or multiple isotopic tracers is an elementary metabolite units (EMU) based method [2]. The EMU framework is a modeling approach that consists of algorithms that can identify the minimum amount of information needed to simulate isotopic labeling in a metabolic network. An EMU is defined as a distinct subset of a compound's atoms that can exist in a limited variety of mass states, depending on their isotopic compositions [34]. This approach has been shown to decrease the required computational times to determine fluxes by orders of magnitude in bacterial and animal systems.

In spite of some of the challenges that occur with ^{13}C-labeling, it can provide a wealth of new information about the C metabolism of a plant. Several labeling methods have been developed over the years in order to better understand precursor–product relationships in plant metabolic networks [25]. One method uses two chemically different substrates, one labeled and the other unlabeled. This method provides information about the contribution of each substrate to the flux into a certain metabolite of interest. However, this technique can only analyze metabolites that have two distinct routes of synthesis. Another method is the use of positionally labeled substrate(s), which informs about the metabolic fate of the different atomic positions of the substrate molecule. It can also identify fluxes at branch points and those involved in bi-directional reactions. A third method is bond-labeling, which is a technique where a substrate is provided that only a fraction of its molecules are uniformly labeled but the remainder are unlabeled. This technique can give a direct measure of exchange flux. Other labeling possibilities include the combination of these three main methods to achieve a more specific goal. For example, positional and bond-labeling methods can be combined to simultaneously analyze the positional redistribution of label and the isotopomer distribution, which reflects the cleavage of labeled bonds [25].

A study by Roessner-Tunali et al. [23] used the combined methods of ^{13}C isotope labeling and GC-MS as a fast and accurate way to compare kinetic aspects of intracellular metabolism. They used this method to evaluate the exchange of carbon between individual metabolite pools in two transgenic potato lines that have significant variations in primary metabolism in their tubers. Their key findings were that there is starch turnover in tubers and that there is increased carbon partitioning toward several amino acids because of low sucrose content in the transgenics. Using time-dependent labeling experiments and differential equation models, they were able to determine the patterns of the relative unidirectional exchange rates through the major pathways of primary metabolism in both wild type and transgenic potato lines.

The biosynthetic flux of tryptophan in cultured rice cells was determined by dynamic flux analysis, where [1-^{13}C]-serine was supplied over a 24-h time course [12]. The metabolic flux value for Trp synthesis from Ser was estimated by fitting a two-step pathway model describing the labeling dynamics of the pathway to the observed labeling data of wild type rice cell cultures. This method was also applied to determine the Trp biosynthetic flux value in a transgenic line of the rice cells expressing a feedback-insensitive version of the anthranilate synthase α-subunit gene (OASA1D). In these experiments, a constant pool size was assumed for Ser and Trp in the rice cells over the experimental period. The flux value for Trp biosynthesis was estimated to be 6.0 ± 1.1 nmol·g $FW^{-1}·h^{-1}$ in the wild type rice cells. The flux was increased 6-fold in the OASA1D transgenic line and resulted in a 45-fold increase in the cellular level of Trp.

10.3.2 Dynamic Flux Analysis of C Metabolism in Plants: An Example in Lemna minor

As an example of how positionally labeled precursors are used to generate isotopic labeling data over a time course, positionally labeled ^{13}C-Ala (1-^{13}C$_1$, 2-^{13}C$_1$, 3-^{13}C$_1$, and U-^{13}C$_3$-alanine) was supplied to *Lemna minor* at 2 mM concentrations over a 24-h period. Stable isotope labeling and pool sizes of all free amino acids were measured with respect to time by GC-MS using HFBI derivatives.

As can be seen in Fig. 10.4, if the main pathway of synthesis of 2-oxoglutarate, as the carbon skeleton of glutamate (Glu), is via the partial TCA cycle (*bold arrows*), then label from 1-^{13}C$_1$-alanine should not appear in Glu because it is lost in pyruvate decarboxylation to acetyl-CoA (Fig. 10.5). Label from 2-^{13}C$_1$-alanine should appear in the 5-position of Glu. Label for 3-^{13}C$_1$-alanine should appear in the 4-position of Glu, and label from U-^{13}C$_3$-alanine should appear in the 4 and 5 positions of Glu (Fig. 10.5).

The HFBI derivatives of glutamate (Glu) conveniently give fragment ions containing carbons 2–5 (*m/z* 280 and 298), or 2–4 only (*m/z* 252) (Fig. 10.6), and thus comparison of the ^{13}C-labeling patterns of these fragment ions can give important positional labeling information. Actual labeling data for the 252 + 5, 280 + 5, and 298 + 5 fragment ion clusters (expressed as percent Glu species with 0, 1, 2, 3, 4, and 5 labeled atoms) from the time courses of labeling with the four forms of ^{13}C-alanine are shown in Fig. 10.7. Note that the labeling patterns are mostly consistent with expectations of the flux map shown in Fig. 10.4.

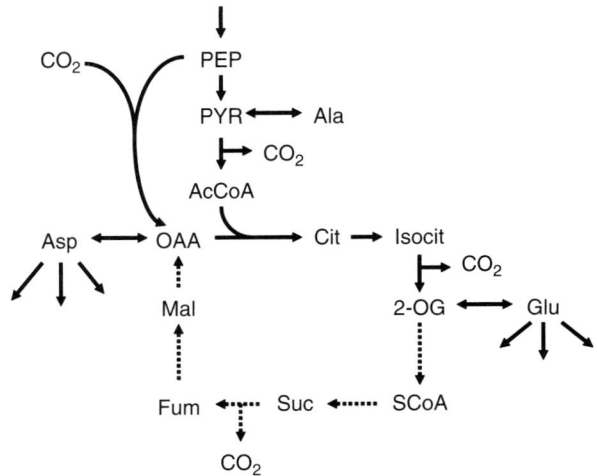

Fig. 10.4 Intermediates of glycolysis and tricarboxylic acid (TCA) cycle relevant to a consideration of labeling patterns of intermediates when supplied with ^{13}C-alanine (Ala) in different positions. Ala can be transaminated to pyruvate (PYR), which is then decarboxylated to acetyl-CoA (AcCoA). The combination of AcCoA and oxaloacetate (OAA) yields citrate (Cit), which is further metabolized to isocitrate (Isocit) and 2-oxoglutarate (2-OG). 2-OG can then serve as carbon skeleton for glutamate (Glu) either by transamination or via the GS-GOGAT cycle. If there is substantial flux to Glu (*bold arrows*), with little recycling of 2-OG to OAA via succinyl-CoA (SCoA), succinate (Suc), fumarate (Fum), and malate (Mal) (*dashed arrows*), then the supply of OAA must be sustained by anaplerotic synthesis of OAA via the catalytic action of phosphoenolpyruvate (PEP) carboxylase

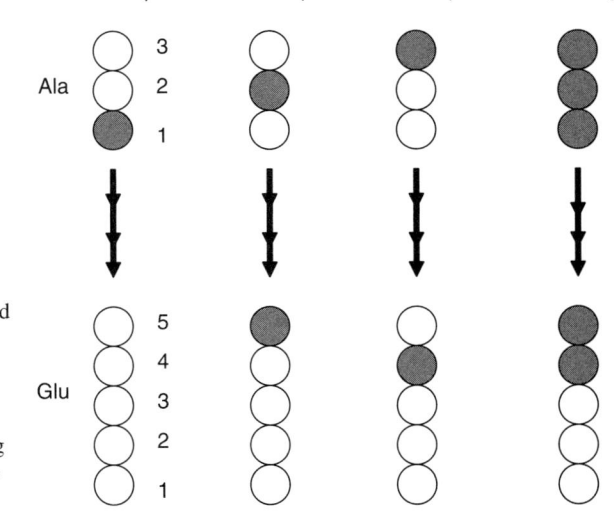

Fig. 10.5 Expected positional labeling of Glu when the metabolic scheme shown in Fig. 10.4 is supplied with Ala labeled in different positions, and when there is no recycling of 2-OG to OAA. Labeling patterns will be more complex if recycling of 2-OG to OAA occurs (see [29])

Fig. 10.6 N-Heptafluorobutyryl isobutyl derivative of glutamate (HFBI-Glu). The molecular weight (M) of the derivative is 455. In electron ionization GC-MS, it fragments as indicated by the arrows to give ions at m/z 354, 298, 280, and 252. Notably, the ions at m/z 298 and 280 include carbons 2–5 of Glu, whereas the ion at m/z 252 includes only carbons 2–4 of Glu

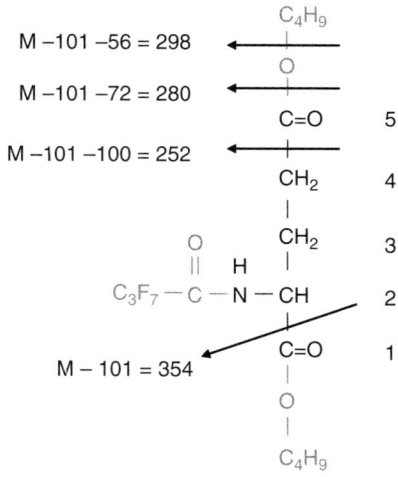

M −101 −56 = 298

M −101 −72 = 280

M −101 −100 = 252

M − 101 = 354

N-HFBI-Glu, M = 455

Moreover, glutamine and GABA were labeled in an almost identical fashion to Glu (not shown), whereas Asp and Asn remained essentially unlabeled regardless of source of ^{13}C-alanine (not shown). The latter is consistent with relatively little flux from 2-oxoglutarate to malate (*dashed line* of Fig. 10.4), and with synthesis of oxaloacetate primarily via PEP carboxylase. This example shows how labeling studies with different positionally labeled substrates can potentially give independent measures of citric acid cycle fluxes, including the anaplerotic flux to oxaloacetate. The combination of the obtained labeling data with the pool size data lays the framework for dynamic flux analysis with the aid of interactive computer simulation models [see 4, 16, 32].

10.3.3 Dynamic Flux Analysis with Radiolabeling

Radiolabeling can have certain advantages over stable isotope labeling in plant metabolic labeling studies. The higher sensitivity of radiolabeling is preferable for naturally small precursor pools to avoid perturbation of the network dynamics that may otherwise occur when the labeling is performed with a high concentration of stable isotope that can cause unnatural swelling of the metabolite pools, forcing increased flux through the reaction network resulting in perturbation of the enzyme kinetics of a network of interest. Matsuda et al. [10] took this into consideration during their feeding experiments with ^{2}H$_{5}$-Phe in potato. This study used the combined methods of stable isotope labeling, radiolabeling, and MFA to determine the effects of β-1,3-glucooligosaccharide elicitor on the phenylpropanoid pathway in potato. Radiolabeling with L-^{14}C(U)tyrosine and 1-^{14}C tyramine revealed that

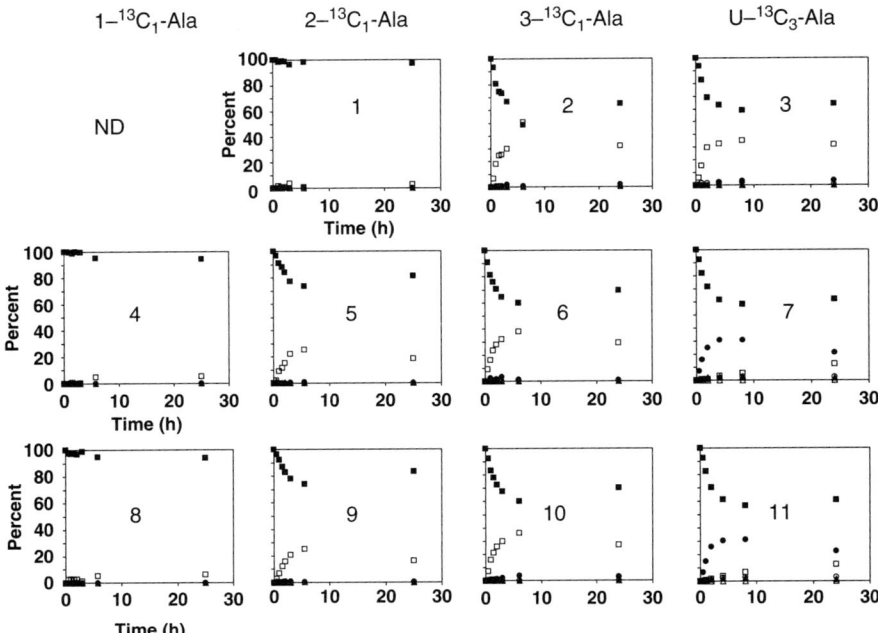

Fig. 10.7 Observed labeling of various fragment ions of HFBI-Glu at different times after supplying positionally labeled ^{13}C-Ala to *Lemna minor*, over a 24-h time course. Shown are the time course of labeling of the *m/z* 252 + 5 fragment ion (**Panels 1–3**), the *m/z* 280 + 5 fragment ion (**Panels 4–7**), and the *m/z* 298 + 5 fragment ion (**Panels 8–12**) of Glu when *Lemna* was supplied with 1-^{13}C$_1$-Ala (panels 4,8), 2-^{13}C$_1$-Ala (panels 1,5,9), 3-^{13}C$_1$-Ala (**Panels 2,6,10**), or U-^{13}C$_3$-Ala (**Panels 3,7,11**). All fragment ion clusters were analyzed for percent unlabeled species (■), and species with 1 (□), 2 (●), 3 (○), 4 (▲) or 5 (△) labeled atoms. The labeling of the *m/z* 252 + 5 fragment ion of Glu when *Lemna* was supplied with 1-^{13}C$_1$-Ala was not determined (ND)

octopamine is synthesized from L-tyrosine via tyramine then incorporated into *N*-p-coumaroyloctopamine (p-CO). Flux analysis also revealed that p-CO has increased synthesis in elicitor treated potatoes compared to control.

McNeil et al. [14] used ^{31}P and ^{14}C radiotracer labeling over a time course, in combination with computer modeling, to analyze choline synthesis in tobacco. The aim of the computer modeling was to identify the major route of biosynthetic flux toward choline (Cho) production. This served as an important foundation for understanding the metabolic constraints on Cho biosynthesis and the challenges associated with metabolic engineering of new metabolic fate of Cho. The major findings were that the primary route of Cho synthesis in tobacco is at the phospho-base level with the first methylation step of phospho-ethanolamine to phospho-monomethylethanolamine, and that free Cho can arise from phosphatidylcholine as well as phospho-choline. These results could not have been attained through graphical analysis of the labeling data alone, but required pool size data and computer-assisted metabolic flux analysis.

10.3.4 Dynamic Flux Analysis of Benzenoid Intermediates in Petunia hybrida

Dynamic flux analysis along with other analysis tools can be used toward the ultimate goal of mathematical modeling of metabolism. The goals of this modeling are to compare metabolic fluxes between species or transgenic and wild type lines, to predict the metabolic effects of environmental or genetic perturbations, and to quantify flux control among different enzymes [15]. The accuracy of metabolic pathway models is limited by the amount of experimental data available.

Mathematical modeling of metabolism has already been applied to plant photosynthesis and central carbon metabolism, but these approaches have great potential to be applied to plant secondary metabolism as well [15].

Dynamic flux analysis has been used in recent years to elucidate the early steps of the benzenoid/phenylpropanoid network in petunia flowers. In a first attempt, the iterative computer modeling of metabolite pool size and isotopic abundance data resulted in generation of a metabolic flux map of the network (Fig. 10.8). Metabolic flux maps contain useful information about the contribution of various pathways to the overall metabolic processes of substrate use and product formation. Comparison of flux maps obtained from genetic perturbations of a species or under different environmental conditions can assess the impact of these changes on a network or can provide information about the importance of particular pathways or reactions in a network [29].

Metabolic fluxes of the benzenoid network in petunia flowers were determined using nearly the same approach as Sims and Folkes in the determination of amino acid synthesis rates. Here, however, the system was supplied with $^2H_5$-Phe, and the flux values (v) were systematically adjusted until a satisfactory fit of computer simulated to the experimentally obtained pool size and labeling data was achieved for all measured compounds in the benzenoid network. Analysis of the petunia benzenoid network fluxes revealed that both the CoA-dependent-β-oxidative and CoA-independent-non-β-oxidative pathways contribute to the formation of benzenoid compounds in petunia flowers, and uncovered that in addition to benzaldehyde, benzylbenzoate is an intermediate between L-Phe and benzoic acid [4].

This approach was further refined and used to compare the change in flux of wild type petunia flowers to a transgenic line, where the gene that encodes the enzyme responsible for the synthesis of benzylbenzoate from benzoyl CoA and benzylalcohol (BPBT) was down-regulated [16]. Data were again obtained by isotopic labeling with $^2H_5$-Phe over a time course, followed by dynamic flux analysis. The comparison of the flux maps revealed that the knockout of BPBT resulted in a redistribution of flux, with increased flux through the non-β-oxidative pathway in the transgenic line compared to the wild type. However, this increase in flux through the non-oxidative pathway was not sufficient to compensate for the loss of flux through the β-oxidative pathway, resulting in an overall decrease in the endogenous benzoic acid pool and a decrease in the emission of the major scent compound methylbenzoate.

In the above applications, the isotopic labeling of phenylacetaldehyde (PhAld) was consistent with synthesis directly from Phe via negligible pools of intermediates [4, 16]. It was previously believed that biosynthesis of PhAld from Phe was a

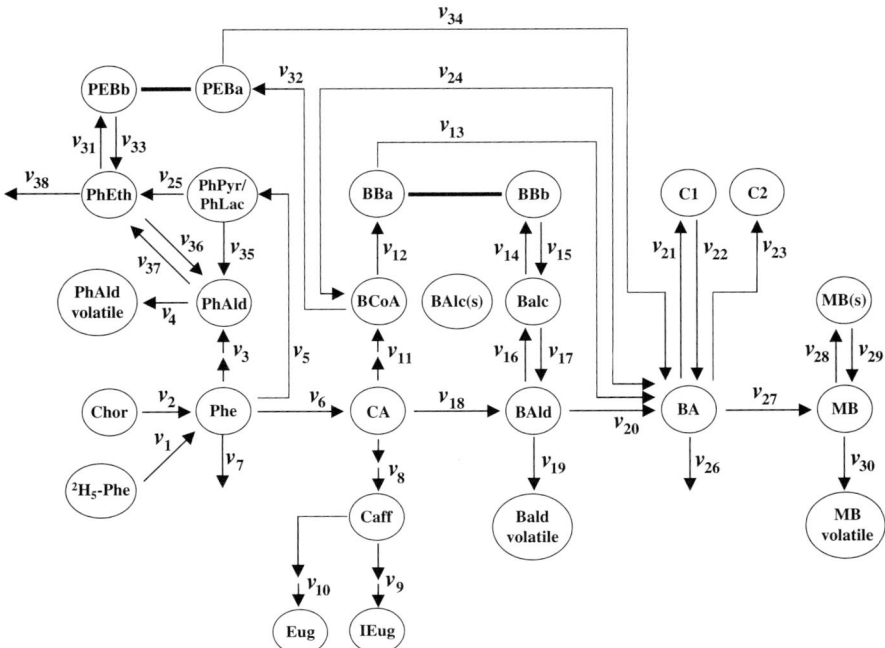

Fig. 10.8 A flux map of the benzenoid and phenylpropanoid network in petunia petal tissue from Boatright et al. [4]. Abbreviations used: ^{2}H$_5$-L-phenylalanine, ^{2}H$_5$-L-Phe; chorismate, Chor; phenylacetaldehyde, PhAld; phenylethanol, PhEth; phenyllactic acid, PhLac; phenylpyruvate, PhPyr; *trans*-cinnamic acid, CA; 4-coumarate plus caffeate, Caff; isoeugenol, IEug; eugenol, Eug; benzoyl-CoA, BCoA; benzylbenzoate, BB [benzoic acid moiety = BBa; benzyl alcohol moiety = BBb]; benzyl alcohol, BAlc; benzaldehyde, Bald; benzoic acid, BA; unidentified non-volatile benzoic acid conjugates, C1 and C2; phenylethylbenzoate, PEB [benzoic acid moiety = PEBa; phenylethanol moiety = PEBb]; and methylbenzoate, MB.

multi-step process via several intermediates, since in yeast its biosynthesis proceeds from Phe to phenylpyruvate via transamination followed by decarboxylation [5, 31]. However, the isolation of PhAld synthase and its biochemical characterization confirmed the modeling conclusions, and it showed that, indeed, PhAld is synthesized in petunia from Phe by a novel single enzyme, catalyzing both decarboxylation and amine oxidation reactions [7].

10.4 Advantages and Disadvantages of Dynamic Flux Analysis

Most of the metabolic flux analysis in microbial systems to date has used steady-state isotopic labeling. Steady-state models reflect structural characteristics of a system because they provide information about time invariant fluxes estimated from steady-state experiments. Isotopic steady state is achieved (1) when the label is allowed to distribute through the metabolic network long enough to reach a stationary distribution in the different metabolite pools and (2) when the rate of incorporation of label into a metabolic intermediate equals its rate of outflow [25]. Here, we must make the distinction between isotopic steady state (defined above) and metabolic steady state. Metabolic steady states are readily achieved when metabolite pool sizes stay constant over the course of the labeling experiment; thus, there is no expansion or contraction of the metabolite pool. Metabolic steady state is readily achieved in yeast and bacterial cell cultures; however, it is quite difficult or may be impossible to achieve in whole plant systems that undergo developmental, diurnal, and seasonal oscillations.

In a classic study by Dieuaide-Noubhani et al. [6], positionally labeled ^{13}C glucose was supplied to 3-mm-excised tips of 3-day-old maize primary roots that were starved of carbohydrates. NMR was used to determine ^{13}C enrichment of various end products, and modeling of the labeling data was used to identify and quantify the major fluxes involved in central carbon metabolism of maize root tips and their response to stress. The significant findings of this study were that the pentose-phosphate pathway (PPP) is active in maize root tips and consumes 32% of the hexose phosphates entering the triose phosphates, with the remaining 68% being consumed in glycolysis. This was the first time that the PPP was identified in maize roots, which provided new insight about the relationship between glycolysis and the citric acid cycle in glucose-metabolizing tissue. However, one method used in this study was to pre-starve the root tips, in order to more rapidly obtain steady state, which may have induced metabolic perturbations.

This example represents the major limitation of the steady-state approach applied to plant systems. It performs best on metabolites that have rapid turnover, but many plant tissues have carbohydrate and protein pools that are slowly and continuously turning over, resulting in isotopic dilution of intermediates [24]. An alternative approach is the use of transient/dynamic isotopic labeling. Steady-state labeling usually only requires a single measurement of abundances, but dynamic labeling requires repeated sampling of the fractional enrichment in metabolites over a time course. Measurements are taken repeatedly during the curvilinear or dynamic phase of the experiment before an isotopic steady state is achieved [15, 27]. A benefit of the dynamic stable isotope MFA is the improved quality of flux estimates; also better estimates of the size of "immeasurable" pools can be made using this approach compared to isotopic steady state.

The fundamentally important case of photo-autotrophic metabolism can only be studied by dynamic labeling methods [27] (Fig. 10.9); otherwise, metabolites become uniformly labeled leading to an uninformative steady state [18]. For example, supplying a plant with labeled ^{13}CO$_2$ will eventually result in uniform labeling of carbons for all metabolites, and the isotopomer compositions achieved will be a simple function of isotope abundance of the supplied ^{13}CO$_2$ (Fig. 10.9). Measurements during the transient phase, before isotopic steady state is achieved, can be extremely informative; the labeling patterns are influenced by uptake rate and isotopic abundance of the labeled substrate, intracellular fluxes, and the pool sizes

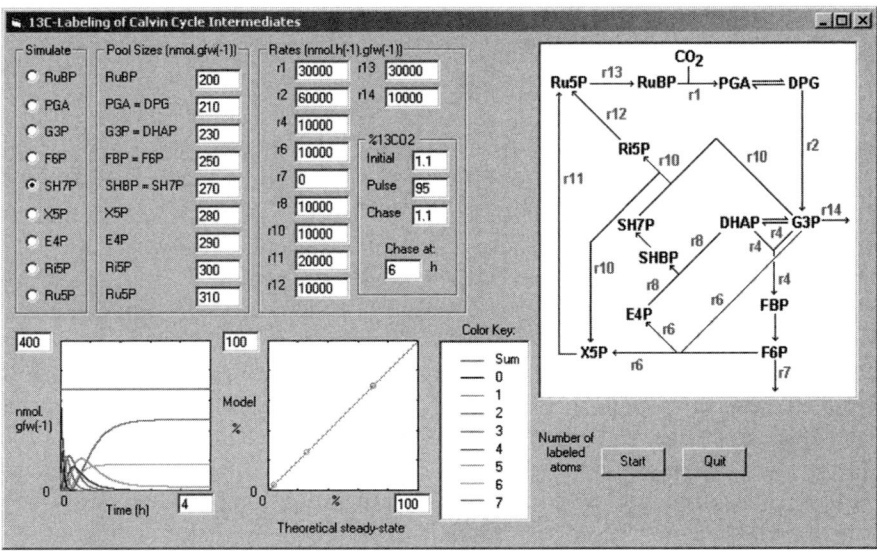

Fig. 10.9 Simulation of the time course of labeling of the Calvin cycle intermediate sedoheptulose-7-phosphate (SH7P) when a photosynthetic tissue is supplied with $^{13}CO_2$ of 95% ^{13}C abundance (i.e., $C = 0.95$), using the rates and pool size assumptions, is shown. The simulated time course of absolute pool sizes of SH7P species with 0, 1, 2, 3, 4, 5, 6, or 7 labeled atoms and the sum of all these species are shown in the *lower left panel*. In the *lower right panel*, the model-simulated output at 4 h is compared with the theoretical steady-state percentages of SH7P species with up to 7 labeled atoms calculated assuming uniform labeling as follows: (*see* also Color Insert)

%0 labeled atoms=$100*(1-C)^7$
%1 labeled atoms=$100*7*C*(1-C)^6$
%2 labeled atoms=$100*21*C^2*(1-C)^5$
%3 labeled atoms=$100*35*C^3*(1-C)^4$
%4 labeled atoms=$100*35*C^4*(1-C)^3$
%5 labeled atoms=$100*21*C^5*(1-C)^2$
%6 labeled atoms=$100*7*C^6*(1-C)$
%7 labeled atoms=$100*C^7$

of the intracellular intermediates (Fig. 10.9). But when compared to steady state, transient labeling experiments require more measurements to obtain precise flux estimates, including multiple measurements of isotopic abundance and metabolite pool sizes over a time course.

One of the major drawbacks of transient MFA when compared to steady state is the difficulty and often inability to obtain constant pool sizes, thus forcing the use of much more complicated equations to quantify flux values. Metabolite pool sizes will change in response to environmental changes such as light intensity, removal of tissue from the plant, floating tissue on liquid media, and especially uptake of a large amount of supplied stable isotope precursor. This variation in metabolite pool size may perturb the system and result in artificial flux estimates. This problem can be overcome with radioisotope labeling with a substrate of high specific activity, which will provide adequate labeling using less substrate (see Section 10.3.3).

A drawback of all MFA studies is that they are valid only based on one concentration of labeled substrate and may not accurately represent in vivo fluxes. Thus, if possible, care should be taken to minimize the perturbation of the intracellular concentration of labeled substrate. Metabolic flux maps generated with this approach provide a snapshot of flux distributions and precursor–product relationships. However, they do not contribute any knowledge about the kinetic properties of enzymes involved in the network. Kinetic modeling of metabolism is required to formulate a predictive model. Since kinetic modeling incorporates kinetic equations for each reaction involved in a network, the model of the system will have some predictive value with respect to alteration of V_{max}, K_m, or regulatory properties of enzymes of the network (for more discussion on kinetic modeling see Chapter 11; [1, 22, 30]).

10.5 Summary and Conclusions

In this chapter, we have illustrated the use of various types of labeling strategies to conduct dynamic flux analysis in plants with several examples, and also have discussed the major advantages and disadvantages of this approach compared to a steady-state approach. The examples presented herein clearly demonstrate that often it is necessary to employ a time course in conjunction with isotopic labeling to obtain information about the transient nature of plant networks, since a steady-state labeling approach does not recognize the dynamic contribution of individual pathways, which can change over plant development, but rather averages their effects. The dynamic flux analysis approach is ideal to determine precursor–product relationships in plant networks that undergo oscillations. In spite of the increased experimental and computational efforts in the transient flux analysis approach, this method will allow us in the future to model complex metabolic networks in plant tissues, where it is often impossible to achieve steady-state status due to the slow turnover of many intermediates and end products.

Acknowledgements This work is supported by the U.S. National Science Foundation (grant numbers: MCB-0615700 for JAM, ND and DR and BES-0348458 for JAM), the U.S. Department of Agriculture (grant numbers 2003-35318-13619 and 2005-35318-16207 for ND and DR), and the Fred Gloeckner Foundation, Inc (ND).

References

1. Alves R, Antunes F, Salvador A (2006) Tools for kinetic modeling of biochemical networks. Nature Biotech 24:667–672.
2. Antoniewicz, MR, Kelleher JK, Stephanopoulos, G (2007) Elementary metabolite units (EMU): a novel framework for modeling isotopic distributions. Metab Eng 9:68–86.
3. Baxter CJ, Liu JL, Fernie AR, Sweetlove LJ (2007) Determination of metabolic fluxes in a non-steady-state system. Phytochemistry 68:2313–2319.

4. Boatright J, Negre F, Chen X, Kish CM, Wood B, Peel G, Orlova I, Gang D, Rhodes D, Dudareva N (2004) Understanding in vivo benzenoid metabolism in petunia petal tissue. Plant Physiol 135:1993–2011.
5. Dickinson JR, Eshantha L, Salgado J, Hewlins MJE (2003) The catabolism of amino acids to long chain and complex alcohols in *Saccharomyces cerevisiae*. J Biol Chem 278:8028–8034.
6. Dieuaide-Noubhani M, Raffard G, Canioni P, Pradet A, Raymond P (1995) Quantification of compartmented metabolic fluxes in maize root tips using isotope distribution from ^{13}C- or ^{14}C-labeled glucose. J Biol Chem 270:13147–13159.
7. Kaminaga Y, Schnepp J, Peel G, Kish CM, Ben-Nissan G, Weiss D, Orlova I, Lavie O, Rhodes D, Wood K, Porterfield DM, Cooper AJL, Schloss JV, Pichersky E, Vainstein A, Dudareva N (2006) Plant phenylacetaldehyde synthase is a bifunctional homotetrameric enzyme that catalyzes phenylalanine decarboxylation and oxidation. J Biol Chem 281:23357–23366.
8. Ling V, Snedden WA, Shelp BJ, Assmann SM (1994) Analysis of a soluble calmodulin binding protein from fava bean roots: identification of glutamate decarboxylase as a calmodulin-activated enzyme. Plant Cell 6:1135–1143.
9. MacKenzie SL, Tenaschuk D (1979) Quantitative formation of N(O,S)-heptafluorobutyryl isobutyl amino acids for gas chromatographic analysis. I. Esterification. J Chromatography 171:195–208.
10. Matsuda F, Morino K, Ano R, Kuzawa M, Wakasa K, Miyagawa H (2005) Metabolic flux analysis of the phenylpropanoid pathway in elicitor-treated potato tuber tissue. Plant Cell Physiol 46:454–466.
11. Matsuda F, Morino K, Miyashita M, Miyagawa H (2003) Metabolic flux analysis of the phenylpropanoid pathway in wound-healing potato tuber tissue using stable isotope-labeled tracer and LC-MS spectroscopy. Plant Cell Physiol 44:510–517.
12. Matsuda F, Wakasa K, Miyagawa H (2007) Metabolic flux analysis in plants using dynamic labeling technique: Application to tryptophan biosynthesis in cultured rice cells. Phytochemistry 68:2290–2301.
13. Mayer RR, Cherry JH, Rhodes D (1990) Effects of heat shock on amino acid metabolism of cowpea cells. Plant Physiol 94:796–810.
14. McNeil SD, Nuccio ML, Rhodes D, Shachar-Hill Y, Hanson AD (2000) Radiotracer and computer modeling evidence that phospho-base methylation is the main route of choline synthesis in tobacco. Plant Physiol 123:371–380.
15. Morgan JA, Rhodes D (2002) Mathematical modeling of plant metabolic pathways. Metab Eng 4:80–89.
16. Orlova I, Marshall-Colon A, Schnepp J, Wood B, Varbanova M, Fridman E, Blakeslee JJ, Peer WA, Murphy AS, Rhodes D, Pichersky E, Dudareva N (2006) Reduction of benzenoid synthesis in petunia flowers reveals multiple pathways to benzoic acid and enhancement in auxin transport. Plant Cell 18:3458–3475.
17. Rantanen A, Rousu J, Kokkonen JT, Tarkiainen V, Ketola RA (2002) Computing positional isotopomer distributions from tandem mass spectrometric data. Metab Eng 4:285–94.
18. Ratcliffe RG, Shachar-Hill Y (2006) Measuring multiple fluxes through plant metabolic networks. Plant J 45:490–511.
19. Rhodes D, Handa S, Bressan RA (1986) Metabolic changes associated with adaptation of plant cells to water stress. Plant Physiol 82:890–903.
20. Rhodes D, Rich PJ, Brunk DG (1989) Amino acid metabolism of *Lemna minor* L. IV. ^{15}N-Labeling of the amide and amino groups of glutamine and asparagine. Plant Physiol 89: 1161–1171.
21. Rhodes D, Sims AP, Folkes BF (1980) Pathway of ammonia assimilation in illuminated *Lemna minor*. Phytochemistry 19:357–365.
22. Rios-Estapa R, Lange BM (2007) Experimental and mathematical approaches to modeling plant metabolic networks. Phytochemistry 68:2351–2374.
23. Roessner-Tunali U, Liu JL, Leisse A, Balbo I, Perez-Melis A, Willmitzer L, Fernie AR (2004) Kinetics of labeling of organic and amino acids in potato tubers by gas chromatography-mass spectrometry following incubation in ^{13}C labeled isotopes. Plant J 39:668–679.

24. Rontein D, Dieuaide-Noubhani M, Dufourc EJ, Raymond P, Rolin D (2002) The metabolic architecture of plant cells. J Biol Chem 277:43948–43960.
25. Roscher A, Kruger NJ, Ratcliffe RG (2000) Strategies for metabolic flux analysis in plants using isotopic labelling. J Biotechnol 77:81–102.
26. Schwender J, Ohlrogge J, Shachar-Hill Y (2004) Understanding flux in plant metabolic networks. Curr Opin Plant Biol 7:309–317.
27. Shastri AA, Morgan JA (2007) A transient isotopic labeling methodology for ^{13}C metabolic flux analysis of photoautotrophic microorganisms. Phytochemistry 68:2302–2312.
28. Sims AP, Folkes BF (1964) A kinetic study of the assimilation of [^{15}N]-ammonia and the synthesis of amino acids in an exponentially growing culture of *Candida utilis*. Proc Roy Soc Lond B Biol Sci 159:479–502.
29. Stephanopoulos GN, Aristidou AA, Nielsen J (1998) Metabolic Engineering Principles and Methodologies. Academic Press, San Diego, CA.
30. Uys L, Botha FC, Hofmeyr JHS, Rohwer JM (2007) Kinetic model of sucrose accumulation in maturing sugarcane culm tissue. Phytochemistry 68:2375–2392.
31. Vuralhan Z, Morais MA, Tai SL, Piper MDW, Pronk JT (2003) Identification and characterization of phenylpyruvate decarboxylase genes in *Saccharomyces cerevisiae*. Appl Environ Microbiol 69:4534–4541.
32. Wiechert W (2002) Modeling and simulation: tools for metabolic engineering. J Biotech 94:37–63.
33. Wahl SA, Dauner M, Wiechert W (2004) New tools for mass isotopomer data evaluation in ^{13}C flux analysis: mass isotope correction, data consistency checking, and precursor relationships. Biotechnol Bioeng 85:259–268.
34. Young JD, Walther JL, Antoniewicz MR, Yoo H, Stephanopoulos G (2008) An elementary metabolite unit (EMU) based method of isotopically nonstationary flux analysis. Biotechnol Bioeng 99:686–699.
35. Zhao J, Shimizu K (2003) Metabolic flux analysis of Escherichia coli K12 grown on ^{13}C-labeled acetate and glucose using GC-MS and powerful flux calculation method. J Biotechnol 101:101–17.

Chapter 11
Kinetic Properties of Metabolic Networks

Jörg Schwender

11.1 Introduction

As seen in Chapters 8–10, models of cell metabolism based on steady-state stoichiometric network simulation have been applied successfully to study plant metabolism. However, such models deliver static views of metabolism and do, therefore, not capture the dynamic behavior of metabolic networks. In addition to reaction stoichiometry, kinetic simulation of metabolic networks considers the concentration of metabolites and enzymes, as well as the kinetic properties of enzymes. Kinetic models can be characterized and interrogated, for example, to predict the effect of changes in enzyme activities, in order to identify possible targets for the re-design of metabolism, which is of central importance for biotechnology.

Overviews of published kinetic models of plant metabolism have been given by Morgan and Rhodes [19] or Rios-Estepa and Lange [30]. Many studies on dynamic simulation of metabolism in plants have been focused on photosynthetic carbon fixation [2, 6, 22, 23, 25, 48]. This certainly owes to the fact that photosynthesis is a process that has been described very well in its dynamics in literature. Under realistic field conditions, physiological factors like light or temperature can change very quickly and the biochemical apparatus has to be able to adjust instantly. From here, the importance of dynamic modeling of the carbon fixation process in photosynthesis becomes clear.

11.2 Example Model

In this chapter, a kinetic model published by Uys et al. [44] will be used to discuss the various basic aspects of kinetic modeling. The model can be found online as part of a model database (Fig. 11.1, http://jjj.biochem.sun.ac.za) [20], where interactive

---

J. Schwender (✉)
Brookhaven National Laboratory, Biology Department, Upton. NY 11973, USA
e-mail: schwend@bnl.gov

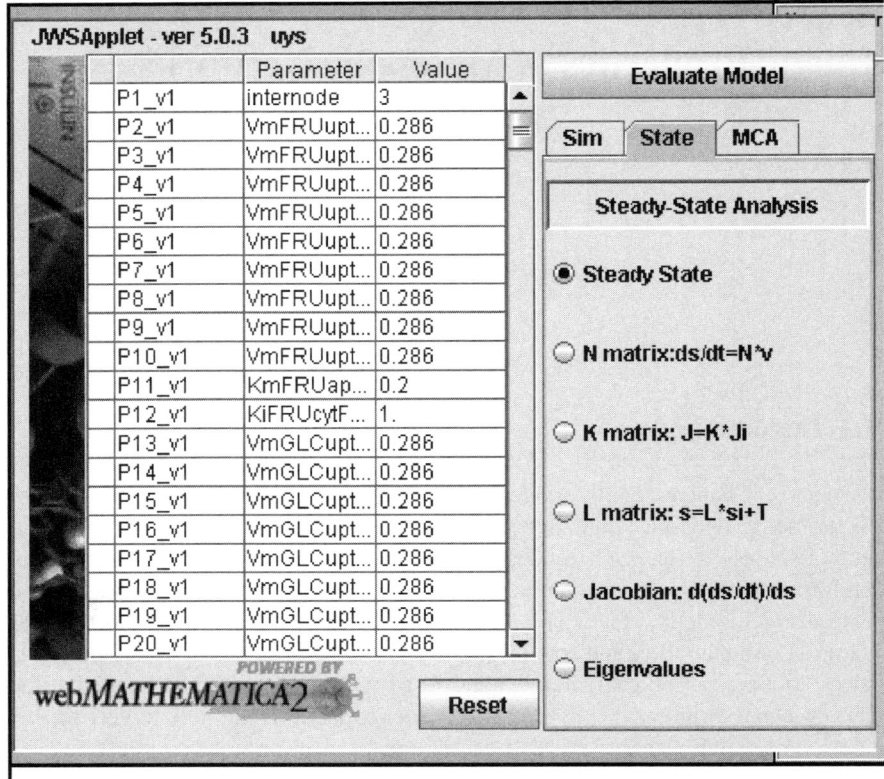

Fig. 11.1 Screenshot of the online sugar cane kinetic model of Uys et al. [44]. In the left window, parameter can be changed. The time-course simulation (Sim) or the steady state of the model can be determined and evaluated (State). Results can be downloaded into text format or Microsoft EXCEL© spreadsheet format. Metabolic control analysis (MCA) is not functional in this particular model. ([20], http://jjj.biochem.sun.ac.za/database/uys/index.html) (*see* also Color Insert)

simulation and analysis over the Internet are possible, while most models can also be accessed in PySCeS or Systems Biology Markup Language (SBML) format.

The authors used this model in order to study the process of carbon partitioning in maturing sucrose-storing internode tissue of sugar cane. In the model, glucose and fructose are taken up and can be transformed into three different products (Fig. 11.2): Sucrose can be exported into the vacuole, triose phosphate is assumed to be consumed by respiration, and UDP-glucuronic acid (UDPGA) is considered to be used in cell wall synthesis. An additional feature that is represented in the model is the cyclic synthesis and degradation of sucrose (sucrose cycling), which has been experimentally observed in this and in other plant tissues. Sucrose cycling can consume a significant fraction of the cellular energy by net hydrolysis of ATP.

The sugar cane model of Uys et al. [44] is based on a previous model by the same group [31]. Several features are added in the recent version. First, isoforms for sucrose synthase and fructokinase are implemented. The isoforms are

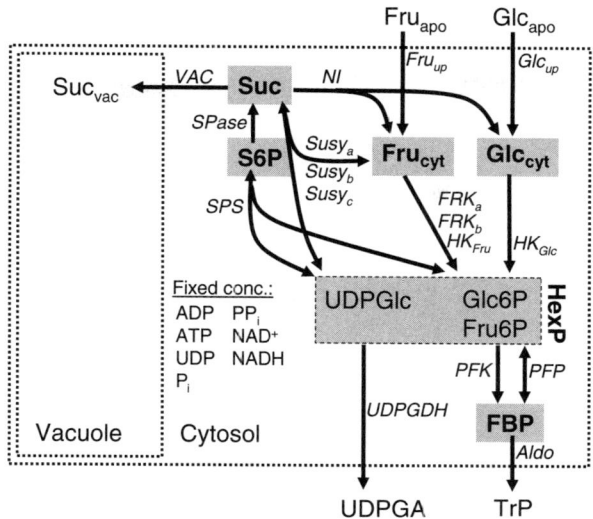

Fig. 11.2 Kinetic model of sucrose metabolism in sugar cane, redrawn from Uys et al. [44]. Dynamically simulated metabolites are bold on grey background. Enzyme names are in *italic*. UDPGlc, Glc6P, and Fru6P are unified as hexose phosphate (HexP) pool (see text). Other metabolites have fixed concentration. Metabolites: Fru, fructose; FBP, fructose 1,6-bisphosphate; Glc, glucose; HexP, hexose phosphates; P_i, phosphate; PP_i, pyrophosphate; S6P, sucrose 6-phosphate; Suc, sucrose; TrP, triose phosphate; UDPGlc, UDP glucose; UDPGA, UDP-glucuronic acid. Enzymes: Aldo, aldolase; FRK, fructokinase; Fru_{up}, fructose uptake; Glc_{up}, glucose uptake; HK_{Fru}, hexokinase acting on fructose; HK_{Glc}, hexokinase acting on glucose; NI, neutral invertase; PFK, phosphofructokinase; PFP, pyrophosphate-dependent PFK; SPase, sucrose 6-phosphate phosphatase; SPS, sucrose phosphate synthase; Susy, sucrose synthase; UDPGDH, UDP glucose dehydrogenase; VAC, transport of sucrose into vacuole. Apo: Apoplast; a,b,c = isoforms

distinguished by differences in kinetic parameters. Second, for the conversion of fructose 6-phosphate to fructose 1,6-bisphosphate, a pyrophosphate-dependent phospho-fructokinase (PFP) is considered in addition to the already present ATP-phospho-fructokinase (PFK, Fig. 11.2). Third, the recent model includes an outflow of hexose phosphates into cell wall synthesis via UDP-glucuronic acid. Furthermore, based on biochemical data, the paper by Uys et al. [44] considers different model variants (distinguished by different maximal enzyme activities, v_{max}), in order to simulate eight stages of internode maturation. In the following, some features of this extended sugar cane model are presented, in order to convey some important issues of kinetic modeling.

11.2.1 Defining the Model

The size of the kinetic model is basically marked by six metabolites that are considered variable, i.e., their concentration is simulated dynamically (Fig. 11.2, see

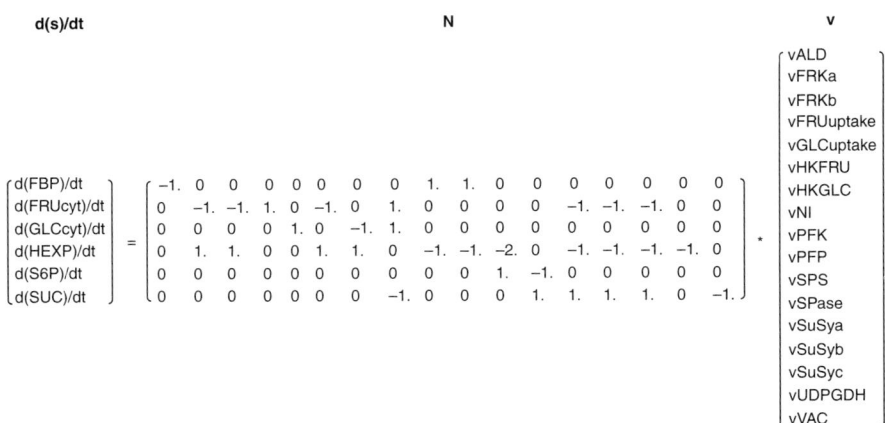

Fig. 11.3 Differential equation system for the sugar cane model [44]. *v*: vector of rate equations; *N*: stoichiometric matrix. Output generated with JWS online simulation tool (http://jjj.biochem.sun.ac.za/database/uys). For abbreviations, see legend of Fig. 11.2

also Fig. 11.3). For seven co-substrates like ATP or NADH, the concentration is set constant according to estimates obtained from literature. In consequence, e.g., physiological effects that change the ATP/ADP ratio are neglected. Five other metabolites are external species and simulated as fixed concentration (Fig. 11.2). The sink metabolites Suc$_{vac}$, UDPGA, and TrP actually do not need to be specified since they are end products of unidirectional reactions. Another simplifying assumption for the model is that the metabolite species glucose 6-phosphate, fructose 6-phosphate, glucose 1-phosphate, and UDP glucose are assumed to be always in chemical equilibrium by fast inter-conversion. Three of the metabolites are shown in Fig. 11.2 unified into what the authors describe as a "hexose phosphate (HexP) equilibrium block." While the concentration of HexP is simulated dynamically, the individual hexose phosphates are coupled to the virtual HexP pool by fixed ratios via equilibrium constants. The assumption of rapid equilibrium between hexose phosphates was justified by experimental data on the respective hexose phosphates in different plant tissues. Without this assumption, three enzymes, namely UDP-glucose pyrophosphorylase, phosphoglucomutase, and phosphoglucoisomerase, would have to be simulated by kinetic rate equations.

11.2.2 Structural Properties of the Stoichiometry

During the construction of a kinetic metabolic model, it is useful to determine the basic functional modes of the system, i.e., to find out if the basic metabolic conversions that are to be simulated are feasible at all. For this purpose, elementary flux mode (EFM) analysis of the sugar cane model was performed on the sugar cane network, using METATOOL [33], as it has been done similarly with the previous sugar cane model [31]. EFM analysis is used to describe all distinct feasible metabolic conversions possible within a certain network (see Chapter 8). EFMs relate to minimal sets of enzymes that are able to perform a certain metabolic conversion at steady state. First, to exclude trivial flux modes, enzyme isoforms were unified to one single reaction (the three sucrose synthase isoforms and two fructokinase isoforms). As a result 14 EFMs were obtained (Table 11.1). All conversions of interest are feasible, always by several alternative modes (i.e., enzyme sets and different substrates involved).

Four modes represent the conversion of extracellular glucose or fructose to triose phosphate. Another four modes result in export of sucrose into the vacuole. Two modes describe the conversion of glucose or fructose to UDP-glucuronic acid, destined to cell wall synthesis. Three modes represent sucrose cycling, i.e., the cyclic synthesis and degradation of sucrose (Table 11.1).

Structural analysis can also reveal so-called conservation relations between metabolites. One typical example is the group of metabolites ATP and ADP. If for each mol of ATP that is used as a substrate within the network, one mol ADP is produced somewhere else, and if none of the two can enter or exit the system, then the sum of concentrations of the two metabolites always remains constant. Such conservation relations restrict concentration range of the respective conserved moieties (metabolites) and therefore limit the dynamic behavior of the system. In the case of the sugar cane model, no conservation relations can be found since ATP, ADP,

Table 11.1 Elementary flux mode analysis of the sugar cane model [44]. Isoenzymes have been removed since they have identical reaction stoichiometry. For abbreviations, see Fig. 11.2

	Net reaction	Function
1:	$ATP + P_i = ADP + PP_i$	Net hydrolysis of ATP by cyclic PFP and PFK
2:	$ATP + UTP = ADP + UDP + PP_i$	Sucrose cycling
3:	$GLC_{apo} + ATP + 2\ NAD = ADP + UDPGA + 2\ NADH$	UDPGA into fiber formation
4:	$ATP = ADP + P_i$	Sucrose cycling
5:	$FRU_{apo} + ATP + 2\ NAD = ADP + UDPGA + 2\ NADH$	UDPGA into fiber formation
6:	$2\ ATP + UTP = 2\ ADP + UDP + P_i + PP_i$	Sucrose cycling
7:	$GLC_{apo} + ATP + PP_i = ADP + P_i + 2\ TrP$	TrP into respiration
8:	$FRU_{apo} + ATP + PP_i = ADP + P_i + 2\ TrP$	TrP into respiration
9:	$GLC_{apo} + 2\ ATP = 2\ ADP + 2\ TrP$	TrP into respiration
10:	$FRU_{apo} + 2\ ATP = 2\ ADP + 2\ TrP$	TrP into respiration
11:	$FRU_{apo} + GLC_{apo} + ATP + UTP = ADP + UDP + PP_i + SUC_{vac}$	Suc into the vacuole
12:	$2\ GLC_{apo} + 2\ ATP + UTP = 2\ ADP + UDP + P_i + PP_i + SUC_{vac}$	Suc into the vacuole
13:	$2\ FRU_{apo} + ATP + UTP = ADP + UDP + PP_i + SUC_{vac}$	Suc into the vacuole
14:	$2\ FRU_{apo} + 2\ ATP + UTP = 2\ ADP + UDP + P_i + PP_i + SUC_{vac}$	Suc into the vacuole

and other cofactors are defined by fixed concentrations, and for the purpose of the network stoichiometry, those are external metabolites.

11.2.3 Kinetic Rate Equations

Figure 11.3 shows the kinetic model in a standard mathematical structure ($d(s)/dt = Nv$). The dynamics of 6 metabolites is considered ($d(s)/dt$), and the 17 columns in v store the kinetic rate equations. The stoichiometric matrix N holds the network structure with each row describing the mass balance for 1 of the 6 metabolites. However, most of the complexity of the model is contained in the rate equations. While in steady-state analysis the vector v would be a variable, it contains here rate 17 Michaelis–Menten-like rate equations that require detailed knowledge on different enzyme reaction mechanisms. As an example, Eq. (11.1) shows the rate equation for neutral invertase (NI) with sucrose as substrate in the sugar cane model. The equation models an irreversible Michaelis–Menten mechanism. The enzyme is inhibited by its products fructose (competitive inhibition) and glucose (uncompetitive inhibition, see Chapter 4). Four kinetic parameters have to be specified, three affinity constants (K_{mSuc}, K_{iGlc}, K_{iFru}) as well as the maximal rate of the enzyme (V_{maxNI}).

$$v_{NI} = \frac{V_{max\,NI}}{1 + [Glc_{cyt}]/K_{iGlc}} \times \frac{[Suc]}{K_{mSuc}\left(1 + [Fru_{cyt}]/K_{iFru}\right) + [Suc]} \qquad (11.1)$$

The four kinetic parameters were derived from in vitro kinetic studies of enzymes purified from sugar cane and carrot (see refs in [31]). Generally, in vitro kinetic studies are performed with purified enzymes, but it should be mentioned that Chapter 4 discusses the validity of "apparent kinetic constants" determined in raw extracts. For the derivation of kinetic rate equations, an introduction into reaction mechanisms and related Michaelis–Menten-like rate equations is given in Chapter 4. Also an interactive web tool is available to derive rate equations (http://www.biokin.com/king-altman/index.html) based on an algorithm by King and Altman [15].

Altogether 38 K_m values, 16 K_i values, and 6 equilibrium constants (K_{eq}) had to be specified in the model. This large number of constants certainly represents one general obstacle for the definition of useful kinetic models, since it can be assumed that the behavior of a kinetic model depends on the values for the kinetic parameters and it may be hard to obtain reliable estimates for all of them. In general, values for kinetic parameters may have limited in vivo relevance [41] since they are usually determined by in vitro assays (see Chapter 4). For example, the parameter values may differ with the exact chemical assay condition (pH, inorganic ions, etc.). Moreover, in vivo complexes of enzymes with other proteins may have kinetic properties that differ from the in vitro properties. In addition, not all important allosteric interactions may be discovered by in vitro studies. In order to specify all parameters needed in a model, the parameter values may be derived from different sources, i.e., literature values that were obtained from different organisms may be used. Hereby it is often assumed that the values of such parameters are similar in closely related organisms. As an exemption, the V_{max} values should be experimentally determined in the particular system that is modeled. The maximum rate (V_{max}) is related to the amount of enzyme present in the tissue/cell type that is under study.

Alternative to the definition of rate equations by the often complex Michaelis–Menten-like kinetics, other definitions of rate equations have been introduced, like the power law formalism in Biochemical Systems Theory [32] or lin-log kinetics [46]. These approaches have both benefits and limitations, which is discussed in detail elsewhere (see [10], [47]). Here, only some general aspects shall be mentioned. In difference to Michaelis–Menten, both approaches have a uniform mathematical formalism which simplifies the construction of kinetic models. As a simplifying assumption, both formalisms imply that substrates, activators, and inhibitors act independently on the reaction rate. In both cases, fewer parameters have to be defined per enzyme. Both approaches are approximations of Michaelis–Menten kinetics that are defined relative to a fixed operating point (steady state). Therefore, it has to be considered that with deviation of enzyme levels (v_{max}) or metabolite levels from the operating point, the results of these approximations become increasingly inaccurate relative to the Michaelis–Menten kinetics [10].

11.2.4 Model Validation

After a kinetic model has been constructed on the basis of enzyme biochemical data, the model has to be validated by independent experimental data. This means that the model has to be checked if it agrees qualitatively and quantitatively with experimental observations. Often it is tested if the model reproduces measured steady-state fluxes and metabolite concentrations. However, first of all, since a biological system should be able to stabilize a metabolic state, the presence of a stable steady state in the model is imperative. Stability means that after a disturbance in the level of a metabolite, the model should return to its steady state. This is demonstrated in Fig. 11.4 where first a steady state was determined, then the fructose concentration (FRU_{cyt}) was raised 10 times at $t = 0$. After 500 min, it can be seen that the system has returned to its original state. While this simulation could demonstrate how a steady state is stabilized, the time scale of the simulation appears to be unrealistically long. In a real experiment, the dynamic changes like those seen in Fig. 11.4 should occur much faster. For example, after a step increase in glucose concentration in yeast cell culture, cellular metabolite levels stabilize within a few minutes [42]. However, although there might be a problem in the time scale of the sugar cane model, it appears that the same dynamic behavior can be simulated at different time domains. For example, scaling all V_{max} values by a factor 10 results in a 10 times faster transition to the same steady-state concentrations, with all steady-state fluxes being exactly 10 times higher (results not shown). This behavior can be explained by the algebraic structure of the differential equation system, in particular by detailed inspection of the 17 rate equations [31, 44]. This reveals that if in the rate equations

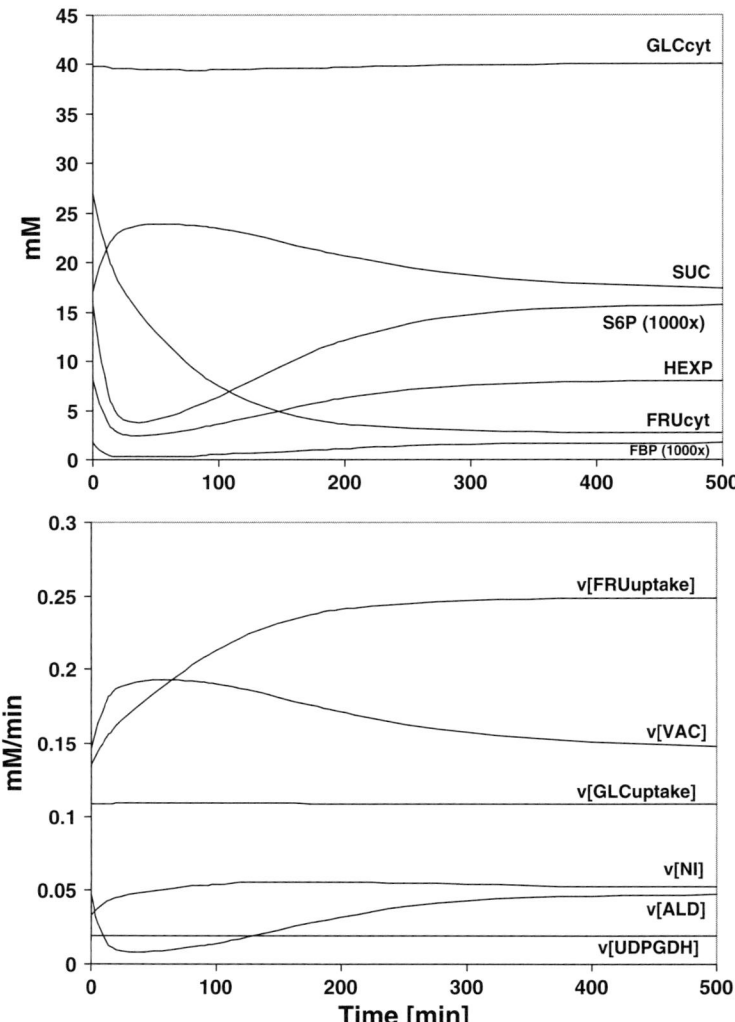

Fig. 11.4 Time-course simulation of the sugar cane model with parameters set for internode 3 [44] with concentrations (*upper panel*) and key fluxes (*lower panel*). At time zero, the metabolite concentrations were at steady state except for Fru$_{cyt}$, which was set to 10 times its steady-state value. For abbreviations, see Fig. 11.2. Values obtained from the JWS online interactive tool (http://jjj.biochem.sun.ac.za/database/uys)

all V_{max} values are multiplied by a scaling factor r, the differential equation system (Fig. 11.3) can be written with r factored out of **V** as: $ds/dt = \mathbf{N}r\mathbf{V}$. This means that r can also be seen as a scaling factor for all steady-state fluxes ($\mathbf{N}r\mathbf{V} = 0$) or for the time differential ($ds/(rdt) = \mathbf{NV}$). It follows that the dynamics of the model can be studied disregarding the actual time units. In consequence, apart from an unrealistic time scale, a kinetic model can be useful if the often problematic conversion of experimental data from a "per fresh weight" basis to real concentration units is omitted.

The online simulation tool also allows testing the stability of a steady state by inspection of the eigenvalues of the Jacobian matrix. The Jacobian matrix is a derivative of the differential equation system relative to an operating point (steady state) and describes the local dynamic behavior of a kinetic model [12]. The eigenvalues indicate if after a small perturbation (deviation in one of the variables) the system returns to its current state or if the perturbation amplifies itself and the system does not return. In the case of the sugar cane model of Uys et al., all eigenvalues of the Jacobian matrix have negative real parts, which is a condition for the existence of a stable steady state [12]. Further details on the interpretation of the Jacobian matrix and eigenvalues can be taken from [12].

In addition, the steady state of the model is similar to ratios of measured metabolic fluxes (relative fluxes into vacuolar sucrose, glycolysis and cell wall synthesis). The model also predicts that close to 22% of synthesized sucrose is degraded by invertase (sucrose cycling), as experimentally observed [31].

Besides existence of steady state and reproduction of fluxes, the validity of a dynamic model furthermore depends on reproduction of measurements of metabolite concentrations. For the sugar cane model, data on the concentration of several metabolites in internodes were available and in part reproduced by the model [31]. However, metabolite measurements are a particular problem in plants (as well as other eucaryotes). While by conventional extraction of plant tissue all subcellular and extracellular compartments are sampled at the same time, many metabolites are present in different subcellular compartments at different concentrations. For the analysis of the metabolite levels in different cell compartments in plant tissue, a non-aqueous fractionation technique [3, 8] is available, but its use has not yet been reported very often. Relative large amounts of tissue are necessary, and the method cannot resolve all subcellular compartments.

For the sugar cane tissue model, metabolite levels were first determined by tissue extraction and expressed on a fresh weight basis. Then it was assumed that 1 g FW has 0.7 ml volume, from which 10% is cytoplasmic and 90% vacuolar volume. Accordingly, the metabolites assumed to be cytosolic (e.g., hexose phosphates) could be converted to cytosolic concentrations [31].

11.2.5 Model Predictions

If a kinetic model has been validated, the response to changing environmental conditions can be studied or different modifications of the model can be simulated. To predict the effect of transgenic alteration, the level of an enzyme can be altered, and the changes in fluxes and metabolite levels can be predicted as shown in Fig. 11.5 for the invertase reaction. With increase in V_{maxNI}, flux through the enzyme increases in a non-linear manner. Please note that considering the enzyme isolated from the

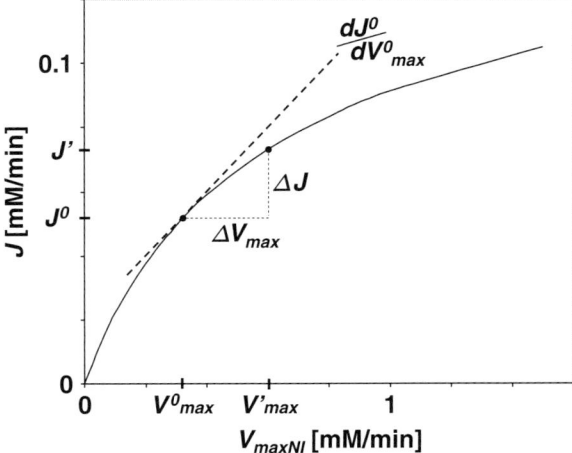

Fig. 11.5 Parameter scan of the sugar cane model. Steady-state fluxes were generated for varying amounts of the enzyme amount of invertase (V_{maxNI}). The steady-state flux through neutral invertase (NI) is shown. As reference point, the model parameters for internode 3 are shown (V^0_{max}, J^0) with its slope (dJ^0/dV^0_{max}). Also, a large-scale increase in V_{max} is indicated (ΔV_{max}, ΔJ). Data generated with JWS online simulation tool (http://jjj.biochem.sun.ac.za/database/uys)

network at constant substrate concentration, the reaction rate increases linearly with V_{max} (Eq. 11.1). In the network an increase in V_{maxNI} causes a change in flux as well as in steady state metabolite levels. This demonstrates the intrinsic non-linear and non-intuitive behavior of metabolic networks.

The model behavior can also be systematically optimized toward a goal such as maximal partitioning of carbon into vacuolar sucrose. Optimizing algorithms have been reviewed, e.g., by Mendes and Kell [16].

11.3 Metabolic Control Analysis

Metabolic control analysis (MCA) [11, 14] is a mathematical framework that allows to compare the influence of different enzymes on fluxes and metabolite concentrations in a pathway. The framework was first described for linear pathways and later generalized to networks. Basic quantities derived in MCA are (1) *control coefficients*, which describe the sensitivity of pathway flux or the concentration of a metabolite in the pathway to a small change in the level of an enzyme and (2) *elasticity coefficients*, which describe the sensitivity of the local reaction rate of an enzyme to a small change in a metabolite. For the comprehensive description of the mathematical framework of MCA, the reader may refer to several textbooks [4, 12, 40]. One important insight resulting from MCA is that control over flux tends to be distributed among all enzymes of a pathway [4, 17]. This means that the biochemical architecture of pathways rarely allows achieving large changes in pathway flux by alteration of only one enzyme. Nevertheless, MCA can help to identify the steps that have most control over flux as the most promising targets for genetic alteration.

The MCA approach strictly relates to infinitesimal small changes (perturbations) of one parameter at a fixed reference state (steady state). For the example in Fig. 11.5, a (infinitesimal) small increase in the activity of invertase (V^0_{max}) in the network will cause a certain increase in flux through the enzyme (J) at steady state, which is quantified as a *flux control coefficient* (FCC):

$$C^J_{V^0_{max}} = \frac{\frac{dJ^0}{J^0}}{\frac{dV^0_{max}}{V^0_{max}}} = \frac{dJ^0}{dV^0_{max}} \frac{V^0_{max}}{J^0} \qquad (11.2)$$

The equation relates the slope in the reference state (dJ^0/dV^0_{max}) to the reference state itself (J^0/V^0_{max}, Fig. 11.5). A consequence of this scaled definition is that FCCs assume dimensionless values between 0 and 1, which corresponds to the enzyme having no control and full flux control, respectively[1]. Also, the FCCs of a pathway can be related to each other by the summation theorem, which states that the sum of the FCCs of all enzymes in a pathway equals one [11, 14].

Kinetic models of plant systems have been used frequently to determine control and elasticity coefficients [6, 23, 24, 25, 44, 48]. For the sugar cane model, FCCs for invertase flux (C^{NI}) and sucrose export (C^{VAC}) (Fig. 11.2) were determined for all 17 enzymes relative to the steady states of all internodes. Since newly formed sucrose may be either exported into the vacuole or degraded (sucrose cycling), the authors describe the difference C^{NI}–C^{VAC} as "futile cycling control coefficient" (FCCC) [31]. This analysis showed that the enzymes HK_{Glc} and NI have the largest positive control over sucrose cycling, while Glc_{up}, Fru_{up}, and VAC have most negative control. According to these predictions, the maximal activity of Fru_{up}, for example, was increased up to 5-fold with the effect that the model partitioned less carbon into sucrose cycling and more into the vacuole. However, as characteristic for large-scale increase of enzyme activity, the 5-fold increase in Fru_{up} resulted in less then 2-fold increase in sucrose flux into the vacuole (VAC).

11.3.1 Experimental MCA

Besides determination of control coefficients with mathematical models, there is also a related experimental MCA approach. For an overview of experimental determination of control coefficients and elasticities in plants, the review by ap Rees and Hill is recommended [29].

The control of photosynthetic carbon fixation has been extensively studied using transgenic plants [26]. Another example for experimental MCA in plants can be found for the study of starch synthesis in potato tubers. Summarizing multiple

---

[1] As opposed to linear pathways, in networks, FCCs can also be negative.

studies, flux control coefficients were estimated based on transgenic changes in 13 enzymes and 2 plastidial transporters involved in starch synthesis in potato tubers [7]. Hereby, flux was inferred via final starch content in mature tubers or via short-time incorporation of ^{14}C-glucose in tuber tissue. As a result, dominant control over starch synthesis was assigned to the plastidial ATP/ADP translocator and the ADP-glucose pyrophosphorylase [7]. Accordingly, over-expression of these proteins represents some of the successful approaches reported in literature to increase starch content in potato tubers [36, 43]. In addition to experimental MCA based on transgenic modulation of enzyme activity, specific inhibitors have also been used as well to reduce apparent enzyme activity and to determine control coefficients [27, 28].

11.3.2 Adaptation of MCA for Large Deviations

A general problem for the experimental MCA approach is that control coefficients are mathematically derived considering infinitesimal changes [11, 14], while experimental determination of FCCs must rather consider large-scale changes, e.g., changes in enzyme level in transgenics. Small and Kacser [34, 35] introduced an approach that allows to estimate control coefficients based on large-scale perturbations. The *deviation index* was defined by Small and Kacser [34] in analogy to the FCC. Referring to Fig. 11.5 where a large-scale increase in V_{max} is considered, the deviation index is written as follows:

$$D_{V'_{max}}^{J'} = \frac{\frac{\Delta J}{J'}}{\frac{\Delta V_{max}}{V'_{max}}} = \frac{\Delta J}{\Delta V_{max}} \frac{V'_{max}}{J'} \tag{11.3}$$

Note that D is referenced to the second state (V'_{max}, J') in Fig. 11.5, while C in Eq. (11.2) is related to the reference state (V^0_{max}, J^0). If some simplifying assumptions on the enzyme kinetics are applied, then FCC and deviation index are mathematical identical. In short, the simplifying assumptions relate to a linearization of the rate equations and to the enzyme being largely unsaturated by substrate [34]. The authors show by experimental case studies that the experimentally determined deviation index often closely approximates the FCC. Accordingly, in case of the numerical example in Fig. 11.5, D and C can be calculated from the simulations, and one can see that both values come close to each other:

$$C_{V^0_{max}}^{J^0} = 0.61 \qquad D_{V'_{max}}^{J'} = 0.63$$

In consequence, the sugar cane model suggests that with a roughly 2-fold increase in enzyme activity (ΔV_{max}), the experimentally determined deviation index should well approximate the real FCC.

11.3.3 The Top-Down Approach of MCA

From a practical point of view, experimental MCA appears to be tedious since the control coefficients relate to single enzymes. This means that single enzymes have to be manipulated one at a time, e.g., by a transgenic approach. To mention here is the concept of top-down MCA [1] which allows to determine control coefficients for groups of reactions via the elasticities of metabolites that connect the groups. For example, in the study by Ramli et al. [27], the synthesis of complex lipids in cell cultures was divided into two blocks: the plastidial fatty acid synthesis and the incorporation of fatty acids into complex lipids. The two blocks are connected by the metabolite pool of acyl-CoAs. Via experimental determination of elasticity coefficients, the group flux control coefficients of the two groups could be determined [27].

11.3.4 Network Rigidity

Metabolic networks regularly resist attempts to alter flux due to regulatory properties apparently in place to support and stabilize the "usual" metabolic performance of a cell [38]. This network rigidity can often be attributed to rigidity of particular nodes in the network. For more flux to be diverted toward a desired product, the ratio of fluxes emanating from such a principal node has to change [38, 39]. Mechanisms of allosteric regulation may cause a rigid node to tightly control the emanating fluxes relative to each other. Strongly rigid nodes involve activation of an enzyme in a branch by a product of the competing branch [38]. Depending on the biochemical architecture, e.g., the glucose 6-phosphate node may behave rigid in *Bacillus subtilis* [5] while in *Corynebacterium glutamicum* the same branch point is flexible [45]. In the latter case, the in vivo activity of glucose 6-phosphate dehydrogenase appears to be mainly controlled by the concentration ratio of NADP and NADPH, indicating negative feedback inhibition of the oxidative pentose phosphate pathway by its product NADPH [18]. Such feedback inhibition in one branch allows flexible adaptation of the flux to the demands of the product.

Nodal flexibility can be studied by applying, in separate experiments, perturbations to different branches of a node. Hereby, different kinds of flux perturbations can be combined, e.g., change of substrates in medium, genetic mutants, and specific inhibitors [9]. Flux control coefficients can then be determined based on flux measurements made under the different conditions [9]. For a node with three branches, nine resulting group flux control coefficients characterize the sensitivity of each flux to changes in an enzyme in each of the branches [37, 9].

11.4 Software for Dynamic Simulation

Kinetic models of biochemical networks can be simulated and analyzed by using mathematical programming languages like MatLab© (http://mathworks.com), which has been used, e.g., by Zhu et al. [48]. In addition, several more dedicated software packages are available that allow dynamic analysis of biochemical networks and include features like stability analysis or control analysis. For example, PySCeS is a console-based application (http://pysces.sourceforge.net) [21] to be used under the programming language Python (http://www.python.org). COPASI (http://www.copasi.org) [13] can easily be used under a graphical user interface. Both applications are available under open source license and available for different operating systems.

References

1. Brown GC, Hafner RP, Brand, MD (1990) A "top-down" approach to the determination of control coefficients by metabolic control theory. Eur J Biochem 188:321–325.
2. Farquhar, GD, von Caemmerer, C, Berry, JA (1980) A biochemical model of photosynthetic CO_2 assimilation in leaves of C3 species. Planta 149:78–90.
3. Farre EM, Tiessen A, Roessner U, Geigenberger P, Trethewey RN, Willmitzer L (2001) Analysis of the compartmentation of glycolytic intermediates, nucleotides, sugars, organic acids, amino acids, and sugar alcohols in potato tubers using a nonaqueous fractionation method. Plant Physiol 127:685–700.
4. Fell DA (1997) Understanding the Control of Metabolism. Portland Press, London.
5. Fischer E, Sauer U (2005) Large-scale *in vivo* flux analysis shows rigidity and suboptimal performance of *Bacillus subtilis* metabolism. Nat Genet 37:636–640.
6. Fridlyand LE, Scheibe R (2000) Regulation in metabolic systems under homeostatic flux control. Arch Biochem Biophys 374:198–206.
7. Geigenberger P, Stitt M, Fernie AR (2004) Metabolic control analysis and regulation of the conversion of sucrose to starch in growing potato tubers. Plant Cell Environ 27: 655–673.
8. Gerhardt R, Stitt M, Heldt HW (1987) Subcellular metabolite levels in spinach leaves. Plant Physiol 83:399–407.
9. Heijnen JJ, van Gulik WM, Shimizu H, Stephanopoulos G (2004) Metabolic flux control analysis of branch points: an improved approach to obtain flux control coefficients from large perturbation data. Metabol Eng 6:391–400.
10. Heijnen JJ (2005) Approximative kinetic formats used in metabolic network modeling. Biotechnol Bioeng 91:534–545.
11. Heinrich R, Rapoport RA (1974) A linear steady-state treatment of enzymatic chains. general properties, control and effector strength. Eur. J. Biochem. 42:89–95.
12. Heinrich R, Schuster S (1996) The Regulation of Cellular Systems. Chapman & Hall, New York.
13. Hoops S, Sahle S, Gauges R, Lee C, Pahle J, Simus N, Singhal M, Xu L, Mendes P, Kummer U (2006) COPASI – a COmplex PAthway SImulator. Bioinformatics 22:3067–3074.
14. Kacser H, Burns JA (1973) The control of flux. Symp Soc Exp Biol 27:65–104.
15. King EL, Altman C (1956) A schematic method of deriving the rate laws for enzyme-catalyzed reactions. J Phys Chem 60:1375–1378.
16. Mendes P, Kell D (1998) Non-linear optimization of biochemical pathways: applications to metabolic engineering and parameter estimation. Bioinformatics 14:869–883.
17. Morandini P, Salamini F (2003) Plant biotechnology and breeding: allied for years to come. Trends Plant Sci 8:70–75.
18. Moritz B, Striegel K, de Graaf AA, Sahm H (2000) Kinetic properties of the glucose-6-phosphate and 6-phosphogluconate dehydrogenases from *Corynebacterium glutamicum* and their application for predicting pentose phosphate pathway flux *in vivo*. Eur J Biochem 267:3442–3452.

19. Morgan JA, Rhodes D (2002) Mathematical modeling of plant metabolic pathways. Metab Eng 4:80–89.
20. Olivier BG, Snoep JL (2004) Web-based kinetic modelling using JWS Online. Bioinformatics 20:2143–2144.
21. Olivier BG, Rohwer JM, Hofmeyr JHS (2005) Modelling cellular systems with PySCeS. Bioinformatics 21:560–561.
22. Pettersson G, Ryde-Pettersson U (1988) A mathematical model of the Calvin photosynthesis cycle. Eur J Biochem 175:661–672.
23. Pettersson G, Ryde-Pettersson U (1989) On the regulatory significance of inhibitors acting on non-equilibrium enzymes in the Calvin photosynthesis cycle. Eur J Biochem 182:373–377.
24. Poolman MG, Fell DA, Thomas S (2000) Modelling photosynthesis and its control. J Exp Bot 51:319–328.
25. Poolman MG, Olçer H, Lloyd JC, Raines CA, Fell DA (2001) Computer modelling and experimental evidence for two steady states in the photosynthetic Calvin cycle. Eur J Biochem 268:2810–2816.
26. Raines CA (2003) The Calvin cycle revisited. Photosynth Res 75: 1–10.
27. Ramli US, Salas JJ, Quant PA, Harwood JL (2002) Control analysis of lipid biosynthesis in tissue cultures from oil crops shows that flux control is shared between fatty acid synthesis and lipid assembly. Biochem J 364:393–401.
28. Ramli US, Salas JJ, Quant PA, Harwood JL (2005) Metabolic control analysis reveals an important role for diacylglycerol acyltransferase in olive but not in oil palm lipid accumulation. FEBS J 272:5764–5770.
29. ap Rees T, Hill SA (1994) Metabolic control analysis of plant metabolism. Plant Cell Environ 17:587–599.
30. Rios-Estepa R, Lange BM (2007) Experimental and mathematical approaches to modeling plant metabolic networks. Phytochemistry 68:2351–2374.
31. Rohwer JM, Botha FC (2001) Analysis of sucrose accumulation in the sugar cane culm on the basis of in vitro kinetic data. Biochem J 358:437–445.
32. Savageau MA. 1976. Biochemical system analysis. Addison-Wesley Publishing Company, Reading, MA.
33. Schuster S, Dandekar T, Fell DA (1999) Detection of elementary flux modes in biochemical networks: a promising tool for pathway analysis and metabolic engineering. Trends Biotechnol 17: 53–60.
34. Small JR, Kacser H (1993) Responses of metabolic systems to large changes in enzyme activities and effectors. 1. The linear treatment of unbranched chains. Eur J Biochem 213: 613–624.
35. Small JR, Kacser H (1993) Responses of metabolic systems to large changes in enzyme activities and effectors. 2. The linear treatment of branched pathways and metabolite concentrations. Assessment of the general non-linear case. Eur J Biochem 213:625–640.
36. Stark DM, Timmerman KP, Barry GF, Preiss J, Kishore GM (1992) Regulation of the amount of starch in plant-tissues by ADP glucose pyrophosphorylase. Science 258:287–292.
37. Simpson TW, Shimizu H, Stephanopoulos G (1998) Experimental determination of group flux control coefficients in metabolic networks. Biotechnol Bioeng 58:149–153.
38. Stephanopoulos G, Vallino JJ (1991) Network rigidity and metabolic engineering in metabolite overproduction. Science 252:1675–1681.
39. Stephanopoulos G, Simpson TW (1997) Flux amplification in complex metabolic networks. Chem Eng Sci 52:2607–2627.
40. Stephanopoulos GN, Aristidou AA, Nielsen J (1998) Metabolic Engineering: Principles and Methodologies. Academic Press, San Diego, CA 1998.
41. Teusink B, Passarge J, Reijenga CA, Esgalhado E, van der Weijden CC, Schepper M, Walsh MC, Bakker BM, van Dam K, Westerhoff HV, Snoep JL (2000) Can yeast glycolysis be understood in terms of in vitro kinetics of the constituent enzymes? Testing biochemistry. Eur J Biochem 267:5313–5329.

42. Theobald U, Mailinger W, Baltes M, Rizzi M, Reuss M (1997) In vivo analysis of metabolic dynamics in Saccharomyces cerevisiae : I. Experimental observations. Biotechnol Bioeng 55:305–16.
43. Tjaden J, Möhlmann T, Kampfenkel K, Henrichs G, Neuhaus E (1998) Altered plastidic ATP/ADP-transporter activity influences potato (*Solanum tuberosum* L.) tuber morphology, yield and composition of tuber starch. Plant J 16:531–540.
44. Uys L, Botha FC, Hofmeyr JH, Rohwer JM (2007) Kinetic model of sucrose accumulation in maturing sugarcane culm tissue. Phytochemistry 68:2375–2392.
45. Varela C, Agosin E, Baez M, Klapa M, Stephanopoulos G (2003) Metabolic flux redistribution in *Corynebacterium glutamicum* in response to osmotic stress. Appl Microbiol Biotechnol 60:547–555.
46. Visser D, Heijnen JJ (2003) Dynamic simulation and metabolic redesign of a branched pathway using lin-log kinetics. Metabol Eng 5:164–176.
47. Voit EO (2000) Computational analysis of biochemical systems. A Practical Guide for Biochemists and Molecular Biologists, Cambridge University Press, Cambridge.
48. Zhu XG, de Sturler E, Long SP (2007) Optimizing the distribution of resources between enzymes of carbon metabolism can dramatically increase photosynthetic rate: a numerical simulation using an evolutionary algorithm. Plant Physiol 145: 513–526.

Index

A
Abundance-based correlation analysis, 158
Allen, D. K., 105–142
Anabaena cylindrica, nitrogen fixation study in, 11
Analysis of variance (ANOVA), 55–56
  See also Metabolomics
Arabidopsis spp., 26, 59
  batch-learning self-organizing map algorithm (BL-SOM algorithm), 164
  ecotypes Col0∗C24 crossing, 160
  genomic data
    and kaPPA-View, 58
  RIL populations, 158
Arginine-TBDMS GCMS fragmentation, 134
Arita, M., 180, 183
Arthospira platensis cofactor limitation during growth exopolysaccharide production, 216

B
Beard, D. A., 228, 229
Beckles, D. M., 39–64
Benson, A. A., 12, 251
Berger, S., 124
BioCarta software systems, 196
Biological homogeneity index (BHI) and biological stability index (BSI), 19
Biomass formation equation and FBA, 220–221
  See also Flux balance analysis (FBA)
Biomolecule analysis
  amino and organic acids
    FAME and McLafferty fragments, 132–133
    TBDMS derivatives, 134
  storage lipids, 132
  sugars and storage carbohydrates, 135–136

Bipartite and hyper-graphs, 174
Bond connectivity information, 107
Bondomer concept, 270
  See also Flux estimation, labeling data in
Bow-tie structure of metabolic networks, 179
Boyle, N. R., 211–239
BRENDA database, 88
Briggs, G., 75
Broadhurst, D., 56

C
Calvin, M., 12, 251
Calvin cycle, 10
Canonical correlation analysis (CCA), 158
Capillary electrophoresis and mass spectrometry (CE-MS) technique, in metabolites measurement, 50–51
Carbon-concentrating mechanisms (CCM), 16
16-and 18-Carbon fatty acids, separation of derivatives, 126
Card, S. K., 200
Carrari, F., 164
CellNetAnalyzer, 195
CentiBiN tool in software system, 183–184
Centrality analysis, 181
13CFLUX software, 275
Chlorella pyrenoidosa, stoichiometric model, 216
Chloroplasts, endosymbiotic origin, 23
^{13}C-Labeling experiments, 105
  ^{13}C-constrained flux analysis (NETTO), 275
  ^{13}C NMR, 106
    NMRView software, 118
    spectrum, 111–112
    usage of, 11
  COSY and TOCSY experiments, 116
  ^{1}H NMR method, 111–112, 115
  mass isotopomer, 107

323

^{13}C-Labeling experiments (*cont.*)
  sample preparation, 107–108
  and steady-state assumption, 255–256
Clustering techniques, in metabolic profiling, 160–162
^{13}C-Metabolic flux analysis (^{13}C-MFA), 252
  isotopomer concept, 253
  software tools for, 275–276
  steady-state isotopic labeling experiment, 253–255
    and dynamic labeling approaches, 253
C metabolism in plants dynamic flux analysis
  benzenoid intermediates in *Petunia hybrida*, 299
  flux map of, 300
  intermediates of glycolysis and TCA cycle, 296
  in *Lemna minor*, 295
  N-heptafluorobutyryl isobutyl derivative of glutamate (HFBI-Glu), 297
  positional labeling of Glu in metabolic scheme, 296
  with radiolabeling, 297–298
CO_2 autotrophy flux analysis, 251–252
Colón, A. J. M., 285–303
Color coding techniques, 201
Cook, D., 63
Correlation analysis, in metabolic profiling, 155
  abundance-based correlation analysis, 158
  association between pair of variables, 157
  canonical correlation analysis (CCA), 158
  comparative analysis of tissue types in species, 156
  quantitative trait locus (QTL) profiles, 159
  Spearman/Kendall test statistics, 157
Corynebacterium glutamicum
  ^{15}N NMR analysis of time course, 109
C_4 photosynthesis, biochemical CO_2 pump in bundle sheath cells
  CO_2 fixation by Rubisco in, 17
  malate/pyruvate exchanger in, 19
  plastidic electron transfer chain in, 17
  dual-cell C_4 plants, 16–17
  and intercellular transport, 18
  Kranz anatomy and, 16
  Laisk–Edwards model, 18
  $NADP^+$-malic enzyme type C_4 carbon-concentrating mechanism, 17
  OAA transporter (OAT) and 2-oxoglutarate/malate translocator DiT1, 19

PEP carboxylase (PEPC), 16
PEP/phosphate translocator (PPT), 18–19
phosphate/triosephosphate translocator (TPT) and fluxes, 18
pyruvate transport, 19
Cyanobacteria and plastid metabolism, 10
Cytoscape software platform, 189

D

Data
  integration in metabolic profiling
    clustering techniques, 160–162
    correlation analysis, 155–159
    correlation-based datasets integration, 162–166
    data normalization, 152–153
    limitations of, 166
    outliers detection and missing values, 154–155
    principle and independent component analyses, 159–160
  mining methods, in metabolomics applications, 57–58
  normalization, in metabolic profiling, 152–153
Datta, S., 161
Dauner, M., 131, 133
DBE-Gravisto system, 202
Dead end metabolites, 236–237
Dieuaide-Noubhani, M., 247, 267, 268, 301
Dogrusoz, U., 200
Dudareva, N., 285–303
Dynamic flux analysis, in plant metabolic networks, 285–287
  advantages and disadvantages, 300
  metabolic flux maps, 303
  photo-autotrophic metabolism, 301
  simulation of time course of labeling of Calvin cycle intermediate sedoheptulose-7-phosphate (SH7P), 302
  steady-state approach, 301
  precursor-product relationship, determination of
    culture of yeast with $^{15}NH_4^+$, study of, 287–288, 292–293
    hypothetical reaction network, 292
    reaction network, 288
    Sims and Folkes kinetic equations, 288–291
    storage pool on labeling patterns, 291
  in whole plant organs
    C metabolism, 295–297

labeled substrate, 294–295
Petunia hybrida, benzenoid intermediates in, 299–300
radiolabeling, 297–298
Dynamic simulation softwares, 319–320
Dynamic visualization systems, 197–198
See also Metabolic network

E

E. coli cells, TOCSY spectrum of biomass hydrolysate, 117
Eccentricity centrality, 181
Ederer, M., 194
Edwards, G. E., 18, 19
Electrons
  electrospray ionization, 43
  impact ionization, 42
Elementary flux-modes analysis of seed metabolism from glucose to fatty acids, 236
Endosymbiont metabolism, 26–28
Enzyme kinetics
  continuous and discontinuous assays, 91
  coupling reactions, 92–93
  cycling assays, 92
  functions of, 74–75
  logistics
    automation, 97
    extracts preparation, 96
    reagents, 95–96
    sample handling, 96
    stability of, 96–97
    temperature and time, 97
    types of, 95
  measurement of, 87–89
  Michaelis–Menten equation, 75–78
    graphical determination of, 79–80
    multisubstrate reactions, 80–82
    parameters of, 78–79
  network, 175
  quantification techniques
    Beer–Lambert law, 90–91
    electrochemical detection, 90
    mass spectrometry methods, 89–90
    radioactivity, 89
  in raw extracts, determination of, 93–95, 98–100
  reaction rates and reaction order, 72–74
  regulation
    allosteric enzymes, 85–86
    competitive inhibition, 83–84
    cooperativity/homoallostery, 86–87
    enzyme inhibition, 82
    heteroallostery, 87
    noncompetitive inhibition, 84
    substrate and product inhibition, 85
    uncompetitive inhibition, 84–85
  sensitivity, 91
Ettenhuber, C., 114
ExPASY software systems, 196

F

Fait, A., 151–167
FANMOD tool and motif-detection algorithm, 190
Fast atom bombardment (FAB), 42, 128
Fatty acid methyl ester (FAME) derivatives and McLafferty fragments, 132
  comparison of, 133
Fell, 179, 182
Fernie, A. R., 151–167
FiatFlux open source software package, 275
First-order irreversible and reversible reaction, 72–74
Floyd algorithm, 194
  See also Network clustering
Flügge, U. I., 19
FluxAnalyzer software for EM computation, 231
Flux balance analysis (FBA), 5, 216
  applications of
    metabolic engineering of organisms, 223–224
    mixotrophic growth yields and fluxes, 223
    Synechocystis spp. PCC 6803 metabolism, 222
  dynamic
    batch and fed batch growth of *E. coli* on glucose, 230
    pseudo-steady-state analysis, 229
  formulation of linear optimization constraints in
    biomass formation equation, 220–221
    maintenance energy, 219–220
    objective function, 221
    oxidative phosphorylation and photophosphorylation, 218–219
    typical growth and non-growth-associated maintenance values, 220
  linear optimization, 216
    graphical representation of, 217
    stoichiometric models, 218
  metabolic engineering, application in, 223–224

Flux balance analysis (FBA) (*cont.*)
    regulatory constraints
        minimization of metabolic adjustment (MOMA) and regulatory on/off minimization (ROOM), 224
        optimization principle, 225
        transcriptional regulation, 226
    thermodynamic constraints, 226
        calculation of *in vivo* free energy change in, 227
        E. coli iJR904 model, 229
        genome-scale stoichiometric model of *Helicobacter pylori*, 228–229
        structural properties of network, 227
        theory of oriented matroids, 228
        thermodynamics-based metabolic flux analysis (TMFA), 227–228
Flux estimation, labeling data in
    ^{13}C-constrained flux analysis (NETTO), 275
    13CFLUX software package, 275
    ^{13}C-MFA, software tools for, 266, 275–276
    data interpretation of
        carbon and isotopomer balancing of biochemical reactions, 267
        fluxome profiling approach, 267
        metabolic flux ratio (METAFoR) analysis, 267
        model-independent comparison of labeling signatures, 265
    flux ratio analysis (RATIO), 275
    flux values, 271
        and label measurements, 272
        reliability of, 272–273
    network scale label balancing
        alternative descriptions of labeling states used in network flux analysis, 269–271
        cumomer simulation, computational speedup, 269
        isotopomer network simulation, 268–269
        network simulation by positional ^{13}C-enrichment, 267–268
    NMR2Flux software, 276
    statistical analysis, 273
        linearized statistics, 274
        Monte Carlo stochastic simulation, 274
        statistical errors, measurement of, 274
        true confidence intervals, 275
Flux ratio analysis (RATIO), 275
Folks, B. F., 286–299

Fourier transform ion cyclotron resonance mass spectrometry (FT-ICR-MS), 41, 43–44
Free energy diagram for reversible exothermic reaction, 75
Frommer, W. B., 1, 253, 260, 269, 274, 275

G
Gagneur, J., 194
Gas chromatography and mass spectrometry (GC-MS) technique, in metabolites measurement, 46–48, 106
GCMS fragmentation of arginine-TBDMS derivatives, 134
    See also Biomolecule analysis
Genomics
    bi-directional best-hit (BBH), pairwise genome comparisons, 29
    definition of, 28
    genetically modified plants (GMOs), 61
    genome scale metabolic networks
        application of, 238–239
        model reconstruction, 235–238
        network reconstruction, 237
Gibon, Y., 53, 71–100
Glucitol hexa-acetate derivative of glucose spectrum, 130
Gluconeogenesis, 11
^{13}C-Glucose tracer analysis, 21
Glutamine synthetase/glutamate synthase pathway, 12
Glycerol-3-phosphate cycling, 92
    See also Enzyme kinetics
Glyoxylate cycle, 11
Gnanadesikan, R., 162
Gong, Q., 63
Grafahrend-Belau, E., 173–203

H
Haldane, J., 75
Hatch, M. D., 19
Hatzfeld, W. D., 246
Heldt, H. W., 18
Helicobacter pylori, genome-scale stoichiometric model, 228
Henry, C. S., 227
Hexose phosphate (HexP) pool, 11, 310
HI programs, 195
Hirai, M. Y., 164
Holme, P., 195
Hordeum vulgare, GC-MS technology, 59
2 D^{13}C,^{1}H-COSY (HSQC) method for isotopomer analysis
2D INADEQUATE pulse sequence, 119

E. coli analysis, 117
non-overlapping and overlapping multiplets, 118
protein hydrolysates from plant cell extracts, 118
spectrum of hydrolysate from *Catharanthus roseus,* 119
tool for analyzing fluxes, pathways of central metabolism in microbes, 118
Huber, S. C., 18, 19
Hyper-graphs, 173–175

I

Independent component analysis (ICA), 159–160
Ionization methods
chemical ionization (CI), 129
desorption, 128
evaporative ionization, 128–129
Ion monitoring
and ion trap mass analyzers, 43
selective ion monitoring (SIM), 129
total ion monitoring (TIM), 129–130
Isotope label quantification
fractionation during gas chromatography, 125
isotope labeling networks (ILN), 270
isotopic tracers techniques, 246–247
mass spectrometry, 120–123
biomolecules analysis, 132–136
chromatographic parameters, 123–126
natural abundance corrections, 136–141
spectrometric detection, 126–132
nuclear magnetic resonance (NMR) spectroscopy
NMR-detectable isotopes, 109–110
one-dimensional NMR, 110–116
two-dimensional NMR, 116–120
in vivo NMR, 120
sample preparation, 105–108
Isotopomer
analysis by GCMS, 123–124
mass isotopomers
abundance, 131
mass enrichment, 107
two fragments from alanine, 122
network simulation
and prediction of labeling measurements, 269
steady-state abundance, 270

J

Jeong, H., 178, 179
Johnson, S. C., 162

Junker, B. H., 173–203

K

Kacser, H., 318
Kaplan, A., 163
Kaplan, F., 63
Karp, P. D., 198
Keeling, P. L., 246
KEGG automatic annotation server (KAAS), 29
KEGG-LIGAND database, 183
Kell, D., 316
Kell, D. B., 56
Keurentjes, J. J. B., 62
Klukas, C., 173–203
KNApSAcK database, 60
Koschützki, D., 173–203
Kummer, U., 199

L

Laisk, A., 18
Laisk–Edwards model, 18
See also C_4 photosynthesis, biochemical CO_2 pump in
Lange, B. M., 307
Laser-induced fluorescence detection (LIF detection), 50
Lee, W. N. P., 140
Leegood, R. C., 18
Levenstien, M. A., 162
Lineweaver–Burke plot of uninhibited enzyme, 81
Linum usitatissimum, in vivo NMR, 120
Liquid chromatography and mass spectrometry (LC-MS) technique, in metabolites measurement, 49–50
Lotus japonicus, GC-MS technology, 59
Lowry, C. V., 92

M

Ma, H. W., 179, 183, 193
Manchester, J., 93
MapMan software systems, 196
Mass spectrometry (MS), 120
atmospheric pressure chemical ionization (APCI), 42
atmospheric pressure photon ionization (APPI), 42
biomolecules analysis
amino and organic acids, 132–135
storage lipids, 132
sugars and storage carbohydrates, 135–136
chromatographic parameters, 123

Mass spectrometry (MS) (*cont.*)
    column stationary phase, 124–125
    derivative groups, 125–126
    isotopomer measurements for
        carbon fragments, 121
        theoretical fragmentation and mass distribution of three-carbon compound, 122
    mass spectral library (MSRI), 59
    matrix-assisted laser desorption (MALDI), 128
    natural abundance corrections
        assumptions for, 140–141
        elemental correction matrices, 138
        elements for, 138
        ion abundances contributions, 139–140
        isotopomers, relative contributions, 137
        mass isotopomer distribution vector (MDV), 136
        serine-TBDMS with fragmentation, 139
        size of matrix and, 137–138
    secondary ion mass spectrometry (SIMS), 42
    spectrometric detection
        ionization methods, 128–129
        ion monitoring, 129–132
        mass spectrometers types, 126–128
    tool in flux analysis and network studies, 121
Matsuda, F., 297
MAVisto algorithm and network motifs visualization, 190
Mavrovouniotis, M. L., 227
McNeil, S. D., 298
Mendes, P., 316
Menton, M., 75
Metabolic control analysis (MCA), 316
    adaptation for large deviations
        deviation index, 318
        experimental, 317–318
    network rigidity
        nodal flexibility and flux control coefficients, 319
        top-down approach, 319
Metabolic network
    centralities in
        concepts of, 180–182
        metabolite ranking, 182–183
        tools, 183–184
    clustering of, 190–192
        network, 192–195
        tools, 195
    fingerprinting approach, 60
    kinetic properties of, 307–309
        kinetic rate equations, 312–313
        model definition, 309–310
        model validation, 313–316
        stoichiometry, structural properties of, 310–312
    and linear optimization formulation, 218–221
    metabolic flux analysis (MFA), 214, 245
    metabolic flux ratio (METAFoR), 267
    Metabolic Network Exchange (MetNet), 58
    motifs in, 187–189
        concepts of, 184–186
        statistics of, 186–187
        tools, 189–190
    online resources for reconstruction, 164
    properties of, 177–178
        global properties, 178–180
        local properties, 178
    reaction reversibility in, 261
    solute transport, 12
        CO_2 pump of C_4 photosynthesis, 16–19
        photorespiratory C_2 oxidation cycle, 13–15
    visualization of, 195–196
        interaction times and techniques for, 197–198
        network-integrated data visualization, 200–203
        pathway layout, 198–200
        static and dynamic visualizations, 196
    See also Plant metabolic network
Metabolites
    data analysis
        analytical instrumentation and statistical methods, 54–57
        mining, classification and visualization, 57–58
    metabolomics approaches, 58–60
    network, 175
    in plant science, 60–63
    role of, 39
    technologies for, 40
        capillary electrophoresis and mass spectrometry, 50–51
        gas chromatography and mass spectrometry, 46–48
        GC-MS analysis of metabolites using TMS and TBS derivatization, workflow for, 48
        GC-MS setup in metabolomics approaches, 41

Index 329

HPLC-chip and integrated components, 50
IMS technique, 52–53
LC-MS technology, 49–50
LC-tandem MS setups in metabolomics approaches, 41
mass spectrometry, 41–46
NMR spectroscopy, 51–52
transcript–metabolite correlation visualizations, 165
transporters, 13
Metabolomics, 39
data
Student's t test and analysis of variance (ANOVA), 55–56
metabolons channeling
peroxisomes photorespiratory metabolism, metabolic channeling in, 21–22
substrate channeling and membrane transport, 20–21
workflow of approach, tissue harvest to data interpretation, 59
METATOOL software for EM computation, 231
Mfinder tool and network motif detection, 189
Michaelis, L., 75
Michaelis-Menten equation in enzyme kinetics, 75–78
graphical determination of, 79–80
key parameters of, 78–79
Michaelis–Menten-like kinetics, 313
multisubstrate reactions, 80–82
Michal, G., 199
Microchannel Plate Detectors, 45–46
Minimization of metabolic adjustment (MOMA), 5, 224
Missing value algorithms on artificial data with non-linear structure, 155
Monte Carlo stochastic simulation, 274
See also Flux estimation, labeling data in
Morgan, J. A., 211–239, 285–303, 307

N
Network clustering
average-linkage clustering algorithm, 193
biological systems, 193
degree-based method, 194–195
dendrogram of distance matrix, 192
Floyd algorithm and Ward's clustering, 194
hypothetical, 193
modularity and design principles, 192
Mycoplasma pneumoniae, study for, 194

ordinary differential equation (ODE) model, 194
reaction/enzyme network representation, 193
tools for, 195
topological overlap matrix, 194–195
Nitrogen assimilation pathways in organisms, 215
NMR2Flux software, 276
Nuclear magnetic resonance spectroscopy (NMR spectroscopy), 41, 51–52
in isotope labeling
NMR-detectable isotopes, 109–110
one-dimensional NMR methods, 110–116
two-dimensional NMR methods, 116–120
in vivo NMR, 120
See also Metabolites

O
Oliver, S. G., 39
OptKnock and OptStrain softwares, in metabolic engineering, 223–224
See also Flux balance analysis (FBA)
Orbitrap mass analyzer, 44
Outlaw, W. H., 93
Oxidative pentose phosphate pathway (OPPP), 247, 249

P
Pajek software system, 184
Palsson, B. O., 230
Pathway Analysis Tool for Integration and Knowledge Acquisition (PATIKA), 200
Pathway Tools Omics Viewer (PTOV), 57
Peroxisomes photorespiratory metabolism, 21–22
See also Metabolomics
Petunia hybrida, dynamic flux analysis in, 299–300
2-Phosphoglycolate (2-PG) and 3-phosphoglycerate (3-PGA) in photorespiratory carbon cycle, 13
Photophosphorylationis, 219
Photorespiration, 13
Photorespiratory C_2 oxidation cycle, 13–15
See also Plant metabolic network
Photosynthetic carbon oxidation cycle, 13
Plant(s)
chemical compounds, analysis of, 40
metabolic network, 1–3
metabolic pathways in, 12–13

Plant(s) (cont.)
  analysis, 230–235
  biochemical CO_2 pump of C_4 photosynthesis, 16–19
  photorespiratory C_2 oxidation cycle, 13–15
  metabolism
    complexity of, 3–4
    selective partitioning of organic metabolites, 12
    unicellular algae usage in, 10–11
  organic metabolites partitioning, 12
  plastids and
    chloroplast–cytoplasm metabolic connection, 24–26
    endosymbiont metabolism, integration of, 26–28
    endosymbiotic origin of chloroplasts, 23
    plastidial phosphate translocators in, 25
  seed metabolism, elementary flux-modes analysis of, 236
  transport processes in, 12–13
Plant metabolism(ic) network, 1–3, 173
  centralities in
    concepts of, 180–182
    metabolite ranking, 182–183
    tools, 183–184
  clustering of, 190–192
    network, 192–195
    tools, 195
  experimental data and visualization of, 195–196
    interaction times and techniques for, 197–198
    network-integrated data visualization, 200–203
    pathway layout, 198–200
    static and dynamic visualizations, 196
  graph notation and types, 173–175
  motifs in
    concepts of, 184–186
    metabolic networks, 187–189
    statistics of, 186–187
    tools, 189–190
  network properties, definition of, 175–177
  properties of, 177–178
    global, 178–180
    local, 178
Plasmodium falciparum, apicoplast-containing malaria parasite, 26
Poolman, M. G., 233
Price, N. D., 228

Principle component analysis (PCA), 154, 159–160
Protein hydrolysate
  gas chromatogram of TBDMS-derivatized amino acids, 124
  HSQC method in, 118
Proteins
  extraction of, 108
  protein–protein interaction (PPI), 185

Q

Quadrupole time-of-flight mass spectrometer (QqTOF), 45
  quadrupole mass filters, 126, 128

R

Ramli, U. S., 319
Random-ordered mechanism, 81
  See also Enzyme kinetics
Ratcliffe, R. G., 105–142
Ravasz, E., 193
Rhodes, D., 285–303, 307
Ribulose bisphosphate carboxylase oxygenase (RubisCO) and ribulose 1,5-bisphosphate (RuBP), in carbon cycle, 13
Riesz, P., 124
Rios-Estepa, R., 307
Roessner, U., 39–64
Roessner-Tunali, U., 61, 294
Rogers, A., 71–100
Rohlf, F. J., 157

S

Sauer, U., 131, 133
Scenedesmus spp.
  carbon in photosynthesis, path of, 10
  isoprenoids, biosynthesis of, 11
Schauer, N., 62
Scholz, M., 153, 154
Schreiber, F., 173–203
Schuster, S., 194
Schweissgut, O., 1, 253, 260, 269, 274, 275
Schwender, J., 1–6, 245–276, 307–320
Schwöbbermeyer, H., 173–203
Second-order reaction, 74
SEPARATOR programs, 195
Serine synthesis, carbon position and isotopomer balances, 268
Shastri, A. A., 211–239
Sims, A. P., 286–299
Small, J. R., 318
Sokal, R. R., 157

Index

Solute transport and metabolic networks, 12–13
Stadler, P. F., 183
Steady-state isotopic labeling experiment and dynamic labeling approaches, 253
and flux parameter fitting, 254
Stitt, M., 18, 246
Stoichiometric modeling
　applications of, 215–216
　fundamentals of
　　coefficient, 212–213
　　cofactor pairs and transhydrogenases, 215
　　formulation of, 213
　　fully determined and underdetermined system, 214
　　matrix, 214
　　nitrogen assimilation pathways, 215
　　over-determined system, 215
　MFA role in, 214
Student's t test, 55–56
　See also Metabolomics
Sugar cane
　and differential equation system, 310
　kinetic model
　　elementary flux modes analysis of, 311
　　parameter scan, 316
　　screenshot of, 308
　　sucrose metabolism in, 310
　　time-course simulation of, 314
Synechocystis sp. PCC 6803, model systems for study metabolic fluxes, 12

T

Takanaga, H., 1, 253, 260, 269, 274, 275
TargetP software, for SP detection, 237
Taylor, J., 160
Tohge, T., 60
Transient metabolic flux analysis, 12
Triosephosphate/phosphate translocator, 24–26
Triple quadrupole mass spectrometer (QqQ), 45
Tuikkala, J., 154

U

Unicellular algae, in plant metabolism study, 10–12
Uys, L., 307–309

V

Varma, A., 230
Visone tool in software system, 183–184
Visualization and Analysis of Networks with related Experimental Data (VANTED), 58
Voltage-gated anion channel (VDAC), 21
von Caemmerer, S., 18

W

Wagner, A., 179, 182
Warburg, O., 92
Ward's clustering, 194
　See also Network clustering
Weber, A. P. M., 9–29
Wegner, K., 199
Wiechert, W., 1, 253, 260, 269, 274, 275
Wilzbach, K. E., 124
Wuchty, S., 183

Y

Yang, C., 216
Yeast, metabolic pathway of ammonia assimilation, 286

Z

Zeng, A. P., 179, 183
Zhang, H., 4
Zhao, X. M., 193
Z-score transformation, 153

Printed in the United States of America

DATE DUE